Safety and Risk in Society Series

CONSUMER PRODUCT SAFETY ISSUES

SAFETY AND RISK IN SOCIETY SERIES

Chemical Safety Board
Robert W. Talford (Editor)
2009. ISBN 978-1-60692-586-7

Consumer Product Safety Issues
Peter A. Varga and Mate D. Pinter
2009. ISMN 978-1-60456-826-4

Safety and Risk in Society Series

Consumer Product Safety Issues

Péter A. Varga
and
Máté D. Pintér
Editors

Nova Science Publishers, Inc.
New York

For permission to use material from this book please contact us:
Telephone 631-231-7269; Fax 631-231-8175
Web Site: http://www.novapublishers.com

Library of Congress Cataloging-in-Publication Data

Consumer product safety issues / [edited by] Péter A. Varga and Máté D. Pintér.
p. cm.
Includes index.
ISBN 978-1-60456-826-4 (hardcover)
1. Product safety. 2. Consumer protection. 3. Product safety--Law and legislation. 4. Consumer protection--Law and legislation. I. Varga, Péter A. II. Pintér, Máté D.
TS175.C67 2009
363.19--dc22

2009007876

Published by Nova Science Publishers, Inc. ✢ New York

CONTENTS

PREFACE

Consumer safety has taken, over the last decades, an ever-growing and important position in our society, in particular in developed countries.This new book focuses on some of the key issues in this field which touches virtually everyone including child seats, firearms, food safety, and plastics. In addition, laws and federal agencies and their activities are reviewed and discussed.

Chapter 1 - Motor vehicle crash (MVC) related injury is a major preventable global public health problem as described in a recent report by the World Health Organization (WHO). Despite considerable advances in road safety, road trauma is still the leading cause of death and disability for children in highly motorized countries such as the US, Canada, UK, Japan, New Zealand and Australia. Scientific evidence indicates that the single most significant risk factor for injury in a MVC is the non-use or misuse of a restraint. International research and experience has demonstrated that the use of a child restraint seat (CRS) significantly reduces the risk and severity of injury and the number of deaths resulting from MVC. Estimates of effectiveness with respect to reductions in the number of deaths in MVC with CRS use are 71% for infants and 54% for children 1 to 4 years of age. A recent study found that for children up to 4 years of age, a correctly installed CRS reduces injury hospitalization by 69%. A systematic review on CRS-related research provides ample evidence in support of the notion that the proper use of a CRS significantly reduces the extent of child morbidity and mortality on the highway.

Due to high rates of non-use and misuse, the available literature shows the benefit of using a CRS has not been understood. Research in this area has shown that there is no universal use of CRS and that these restraints are frequently used inappropriately. This inappropriate use decreases the effectiveness of the CRS in preventing injuries or death. Contributing to the improper use of a CRS is the incorrect installation of the restraint within the vehicle. Installation of a CRS is complicated, and differs depending upon the model, the vehicle, the vehicle safety belt system, and diverse vehicle seating arrangements. The high frequency of improper CRS use suggests that there is still much work to be conducted in order to facilitate ease and correct use by parents and caregivers.

The goals of this chapter are to quantify the frequency of improper CRS use and to identify the most common mistakes, so that priorities for design innovations and other promising strategies to reduce misuse can be identified and achieved. A general increase in the correct use of CRS can be achieved through the application of multiple strategies, including community awareness and education, media campaigns, public policies, enforcement of laws and further CRS design improvements.

Chapter 2 - Despite the ubiquity of firearms in American homes, and the abundance of evidence linking access to firearms at home with the risk of suicide by firearm, information on consumer product safety issues and policies, such as placement of triggers locks, that may protect household members against risk of suicide by firearm is conspicuously sparse. Here we report results from the first study to estimate the protective effect of storing firearms locked and/or unloaded (i.e., safer storage practices). We analyzed data from the latest National Mortality Followback Survey (1993 NMFS), a nationally representative survey of next-of-kin of 22,957 decedents representing 2.2 million individuals. In our analysis we partially control for respondents' intention to die and focus on suicides that are likely to have been impulsive. Relative to decedents who did not practice safer storage of firearms, risk of suicide by firearms was less 67% among those who store their firearms locked and/or unloaded. Our findings strongly support the utility of consumer product safety devices and regulations intended to reduce the likelihood of unauthorized or impulsive use of firearms.

Chapter 3 - Anti-technology lobbyists continue to promulgate and perpetuate various misapprehensions about a wide range of products and technologies, from pesticides and other chemicals to fat substitutes. Recombinant DNA technology (also known as "gene splicing" or "genetic modification" (GM)) applied to agriculture to create plants with new or selectively enhanced traits has been among the most abused -- accused repeatedly of being untested, unproven, unsafe, unwanted and unregulated. However, anti-biotechnology activists ignore our vast experience and the scientific consensus that gene-splicing is an extension, or refinement, of less precise, less predictable techniques; that the acreage increases each year because farmers in two dozen countries are so pleased with the results; and that Americans have consumed more than a *trillion* servings of foods that contain gene-spliced ingredients, with not a single untoward reaction.

What makes false alarms about biotech – or any new technology – hard to expose is the virtual impossibility of demonstrating the absolute safety of any activity or product: There is always the possibility that we haven't yet gotten to the *n*th hypothetical risk or to the *n*th dose or the *n*th year of exposure, when the risk will finally be demonstrated. It is logically impossible to prove a negative, and all activities pose some nonzero risk of adverse effects.

Anti-technology, anti-business activists – whether they are found in NGOs, government or the media – claim to fear a world in which exploitative, rapacious, multinational corporations conspire to strip away individual choice from the world's farmers and consumers. Yet it is *they* who are guilty of the mendacity and manipulation they imagine they see in others; *they* who are guilty of stripping away the freedom of researchers to research, doctors to doctor, and consumers to consume vaccines and drugs that can be life-saving. Were anti-technology, anti-business activists to have their way, unwise, regressive public policy could reduce significantly the pursuit of knowledge and production of wealth worldwide.

Chapter 4 - As the consumption of fresh-cut produce increases, there has been an increasing number of serious food poisoning outbreaks linked to these products. The primary pathogens of concern are *Escherichia coli* O157:H7, *Salmonella* and *Listeria monocytogenes*; there are also important emerging threats from viral and protozoan pathogens. With no pasteurisation or kill step during processing, this food system relies on good agricultural and manufacturing practices, an incomplete decontamination process (dipping in chlorinated water or its equivalent), and refrigerated storage, as the only hurdles to prevent contamination and growth of human pathogens. These hurdles are ineffective, as demonstrated by the large number of foodborne outbreaks associated these products over the past 10 years.

This chapter describes the processing and packaging technologies involved in the production of fresh-cut produce. The effects of unit operations, such as antimicrobial washing and decontamination treatments, on the survival of pathogens are discussed. The implications of gas atmospheres and storage conditions on the growth of associated pathogens are reviewed in detail. While much progress has been made in the past decade in our understanding of the microbiological safety of these complex food systems, the need for further research and additional control measures are highlighted in order to assure a safer food supply and to restore confidence in the fresh-cut produce sector.

Chapter 5 - In the field of consumer safety our competences are especially involved in contaminants residues in foods and, in particular, our researches are focused to investigate the presence of polycyclic aromatic hydrocarbons (PAHs), polychlorinated biphenyls and organochlorine pesticides. In this research paper the toxicological risk caused by PAH levels in several foods, specifically in fish and fishery products are highlighted. The attention will be paid to fresh matter in which the PAH content is a consequence of natural pollution, but also to processed products such as smoked, grilled or hydrogenated foods. PAHs, of which benzo(*a*)pyrene (BaP) is the most commonly studied and investigated, are formed by the incomplete combustion of organic matter. Human exposure occurs mainly by inhalation of airborne particulates containing PAHs and ingestion of food products contaminated by environmental pollution or by food processing (drying, smoking) or food cooking (roasting, grilling and frying). The major dietary sources are cereals, vegetables, oil, fats, fish and meat. The main concern for consumer safety is linked to PAHs carcinogenic potential as reported in the "Opinion of the Scientific Committee on Food on the risks to human health of Polycyclic Aromatic Hydrocarbons in food". Several agencies have classified a limited number of PAHs as probable or possible carcinogens. The U.S. Environmental Protection Agency proposed, for the risk assessment, to separate them in two subclasses consisting of the carcinogenic and the noncarcinogenic PAHs and to apply a cancer slope factor to the subclass of carcinogenic PAHs. The development and the establishment of toxic equivalency factors (TEFs) for PAHs similar to the TEF concept used in the assessment of mixtures containing polychlorinated biphenyls and dioxins could help to characterize more precisely the carcinogenic properties of PAH mixtures. Our contribution will focus on the significance and availability of TEF application for a detailed evaluation of consumer safety. The European Regulation established a maximum limit only for BaP as a marker for the occurrence and effect of carcinogenic PAHs in food, but this approach is limited because of the difficulty of defining safe levels of such a complex mixture.

Chapter 6 - Numerous federal, state, and local agencies share responsibilities for regulating the safety of the U.S. food supply, which many experts say is among the safest in the world. Nevertheless, critics view this system as lacking the organization and resources to adequately combat foodborne illness — as evidenced by a series of widely publicized food safety problems, including concerns about adulterated food and food ingredient imports, and illnesses linked to various types of fresh produce and to meat and poultry products. Numerous bills addressing various aspects of food safety have been introduced into the first session the 1 10th Congress, including proposals to reorganize oversight authorities, increase funding, tighten regulation of imported foods, and establish new authorities for enforcement, recall, and notification regarding adulterated foods, among others. Portions of several proposals have become law; more sweeping changes in the federal food safety system are still under consideration in 2008.

Chapter 7 – Consumer safety has taken, over the last decades, an important position in our society, in particular in developed countries. To ensure consumer safety, the European Union as well other countries have issued laws to ensure that the whole food chain is under control. The final objective will be the full implementation of the HACCP-concept and principles at the primary sector of production, in our case the dairy production units. As we already know, the European consumers are rapidly sensibilized by problems in areas including food safety and public health, as well as animal welfare and animal health. To illustrate this, we only need to refer to the Salmonella survey in several species in the last years. European consumers have a high impact on animal production methods, and since the 1990s we observe a growing and permanent pressure on the production sector that drives the producers to start the implementation of on-farm control programmes. (Cullor, 1997; Noordhuizen, 2004a; Radostits, 2004).European consumer concern with food goes far beyond the sole nutritional aspects or product quality issues. The possibly transmittable diseases are of major concern nowadays. Examples are salmonella as already mentioned, E. coli strains, Listeriosis, among others. So the different sectors of production of food of animal origin are very active in developing and undertaking control activities which should increase the confidence of consumers. (Vieira et al., 2007).

Chapter 8 - Public alarm about the spate of recent product recalls throughout 2007, particularly of toys and other products used by children, has focused attention on the Consumer Product Safety Commission (the CPSC or the Commission). This scrutiny led to consideration of major amendments to the Consumer Product Safety Act (CPSA), which established and authorized the CPSC in 1972 in response to growing concerns about protecting the public from unsafe, defective consumer products. Jurisdiction over the administration and enforcement of several existing consumer safety statutes was transferred from other agencies to and consolidated under the CPSC. However, in the years since its establishment, the staff and resources of the CPSC have been considerably reduced, leading many observers to doubt its ability to fulfill its mission effectively.

Consequently, Congress considered major reform legislation to address organizational and systemic deficiencies. Legislative proposals in the 110th Congress included provisions targeting specific consumer product defects and hazards. On July 29, 2008, H.Rept. 110-787, the Conference Report for H.R. 4040, the Consumer Product Safety Improvement Act of 2008 (CPSIA), was released after several months of negotiations in the conference committee to reconcile differences between the House and Senate versions of the bill. The bill passed the House of Representatives and the Senate on July 30, 2008 (424-1) and July 31, 2008 (89-3), respectively. On August 14, 2008, President Bush signed the bill into law as P.L. 110-314. CPSC Chairman Nord and Commissioner Moore each expressed approval of the final legislation, with Chairman Nord expressing a desire for Congress to appropriate further funding to carry out the new mandates of the legislation.

This chapter provides an overview of the prior authority of the CPSC to establish consumer product safety standards and to inspect and recall unsafe consumer products, and discusses P.L. 110-314, the Consumer Product Safety Improvement Act of 2008, reforming the CPSC and strengthening enforcement of consumer product safety standards. For an overview and context of the current issues facing the Commission, see CRS Report RS2282

1, *Consumer Product Safety Commission: Current Issues*, by Bruce K. Mulock. For an overview of issues regarding safety of consumer products imported from China, see CRS Report RS22713, *Health and Safety Concerns Over U.S. Imports of Chinese Products: An Overview*, by Wayne M. Morrison. For an overview of the issue of phthalates in children's products, see CRS Report RL34572, *Phthalates in Plastics and Possible Human Health Effects*, by Linda-Jo Schierow and Margaret Mikyung Lee.

Chapter 9 - Public alarm about the spate of recent product recalls, particularly of toys and other products used by children, has focused attention on the Consumer Product Safety Commission (the CPSC or the Commission). The Consumer Product Safety Act (CPSA) established and authorized the CPSC in 1972 in response to growing concerns about protecting the public from unsafe, defective consumer products. Jurisdiction over the administration and enforcement of several existing consumer safety statutes was transferred from other agencies to and consolidated under the CPSC. However, in the years since its establishment, the staff and resources of the CPSC have been considerably reduced, leading many observers to doubt its ability to fulfill its mission effectively.

Consequently, Congress is considering major reform legislation to address organizational and systemic deficiencies. Legislative proposals include provisions targeting specific consumer product defects and hazards. On December 19, 2007, the House of Representatives passed H.R. 4040, the Consumer Product Safety Modernization Act, which was subsequently placed on the Senate calendar. On December 5, 2007, the Senate Committee on Commerce, Science, and Transportation reported S. 2045, the CPSC Reform Act of 2007, which was placed on the Senate calendar. On February 15, 2008, Senate leaders announced a bipartisan agreement for compromise legislation that was introduced and placed on the calendar as S. 2663. On March 6, 2008, the Senate passed H.R. 4040, after amending it to substitute the text of S. 2663 as amended on the floor.

Conferees from each chamber were designated to reconcile differences between the House and Senate versions of the bill. On July 28, 2008, the conferees announced that a final agreement had been reached, and the conference report was released on July 29, 2008. As of the date of this report, it remains uncertain whether the bill will be brought up for a floor vote in both chambers before the August recess, although it appears likely. This report summarizes information drawn from secondary sources and will be updated to reflect actual conference language.

This chapter provides an overview of the current authority of the CPSC to establish consumer product safety standards and to inspect and recall unsafe consumer products, and discusses the summary of the conference agreement as well as provisions of the Senate-passed and House-passed versions of H.R. 4040 reforming the CPSC and strengthening enforcement of consumer product safety standards. For an overview and context of the current issues facing the Commission, see CRS Report RS22821, *Consumer Product Safety Commission: Current Issues*, by Bruce K. Mulock. For an overview of issues regarding safety of consumer products imported from China, see CRS Report RS22713, *Health and Safety Concerns Over U.S. Imports of Chinese Products: An Overview*, by Wayne M. Morrison. For an overview of the issue of phthalates in children's products, see CRS Report RL34572, *Phthalates in Plastics and Possible Human Health Effects*, by Linda-Jo Schierow and Margaret Mikyung Lee.

Chapter 10 - Roughly a dozen chemicals known as phthalates are used to make the plastics found in thousands of consumer products, ranging from medical tubing to automotive dashboards to bath toys. These phthalates are not tightly held by the plastics and are released into the environment over time. Congress is concerned about possible human health effects from exposure to six of these chemicals: di-(2-ethylhexyl) phthalate (DEHP), dibutyl phthalate (DBP), benzyl butyl phthalate (BBP), diisononyl phthalate (DINP), diisodecyl phthalate (DIDP), and di-n-octyl phthalate (DnOP). DEHP, DBP, BBP, and (to less extent) DINP are known to be toxic to the reproductive systems of rodents. Recent experiments demonstrate that pre-natal exposure at a sufficient level to these same phthalates disrupts the normal action of hormones and can cause malformations of the reproductive organs of offspring (especially males).

Disruption of hormonal functions in humans is known to result in abnormal reproductive development. Many scientists believe that the phthalates toxic to rodents might be able to cause similar malformations in humans. However, human health effects of phthalate exposure have not been conclusively demonstrated. Very few studies have looked at possible effects in humans, but their results have been consistent with the results of rodent experiments. More research would be needed to test this hypothesis. Recent surveys have found almost universal exposure to phthalates. Individuals may be exposed to high enough levels of phthalates to cause reproductive abnormalities. Scientists at the National Toxicology Program have expressed "serious concern" about human male infants undergoing intensive medical procedures, and "concern" about development of human males less than a year old who are exposed to DEHP. In light of these concerns, the National Academy of Sciences is evaluating the risk of aggregate human exposure to multiple phthalates.

Federal agencies have taken several actions, some as early as the mid 1980s, to evaluate and regulate phthalates, but no product to date has been banned outright. The agency responsible for regulating toys and most other child-care products is the Consumer Product Safety Commission (CPSC). In March 2008, the Senate approved an amendment to H.R. 4040, the Consumer Product Safety Commission Reform Act, that would restrict the use of six phthalates in toys and child-care products. The House version had no phthalate amendment. On July 29, 2008, the conferees announced approval of an amended version of the Senate provision.

The scientific basis for concerns about human health risks appears to be strong in the case of some phthalates (such as DEHP), adequate with respect to others (perhaps DINP), and weak for the remaining chemicals (for example, DIDP and DnOP). The strongest evidence with respect to developmental effects has been produced since about the year 2000. The Senate amendment would codify the voluntary agreements reached by CPSC with product manufacturers and reduce exposure to one particular phthalate. New formulations for toys and child-care products may pose greater or fewer risks than current formulations.

Chapter 11 - Bisphenol A (BPA) is used to produce certain types of plastic. Containers made of these plastics may expose people to small amounts of BPA in food and water. Some animal experiments have found that fetal and infant development may be harmed by small amounts of BPA, but scientists disagree about the value of the animal studies for predicting harmful effects in people. At least one regulatory decision in the face of the scientific disagreement has led to a congressional inquiry into the extent to which the decision was based on good science. Legislation proposed in April 2008, S. 2928, would prohibit use of BPA in some products intended for use by children. Legislation proposed in June 2008, H.R. 6228, would prohibit the use of BPA in food and beverage containers regulated by the Food and Drug Administration (FDA).

Chapter 12 - The Consumer Product Safety Commission (CPSC), one of the agency's two current Commissioners stated recently, is at a crossroads. According to CPSC Commissioner Thomas Moore, either Congress will reverse the agency's steady decline or the CPSC will become so weakened as to be irrelevant.

Created in the early 1970s, the CPSC is a small agency by Washington standards (slightly more than 400 employees and a $62.5 million FY2007 appropriation) charged with a seemingly overwhelming responsibility: to protect American consumers against death or injury from unsafe products, including imported products. A series of high- profile product recalls — most notably of children's toys imported from China — has focused attention on the CPSC's resources, including its legal authority, for the task at hand. Legislation (H.R. 4040 and S. 2663) to strengthen the Commission is being considered.

The CPSC is but one part of growing congressional concern about the protection being afforded American consumers from unsafe imports. Food (pet food as well as that intended for human consumption), drugs, and products fall under the jurisdictions of other federal regulatory agencies. This report will be updated as events warrant.

Chapter 13 - China is a major source of U.S. imports of consumer products (such as toys) and an increasingly important supplier of various food products. Reports of unsafe seafood, pet food, toys, tires, and other products imported from China over the past year or so have raised concern in the United States over the health, safety, and quality of imported Chinese products. This report provides an overview of this issue and implications for U.S.-China trade relations and will be updated as events warrant.

In 2007, China overtook Canada to became the largest source of U.S. imports (at $322 billion); about 17% of all U.S. imports now come from China. Over the past year or so, numerous recalls and warnings have been issued by U.S. firms over various products imported from China, due to health and safety concerns. This has led many U.S. policymakers to question the adequacy of China's regulatory environment in ensuring that its exports to the United States meet U.S. standards for health, safety, and quality; as well as the ability of U.S. government regulators, importers, and retailers to identify and take action against unsafe imports (from all countries) before they enter the U.S. market.

Chapter 14 -The Food and Drug Administration (FDA) is responsible for overseeing the safety and effectiveness of human drugs that are marketed in the United States, whether they are manufactured in foreign or domestic establishments. FDA inspects foreign establishments to ensure that they meet the same standards required of domestic establishments. Ongoing concerns regarding FDA's foreign drug inspection program recently were heightened when FDA learned that contaminated doses of a common blood thinner had been manufactured at a Chinese establishment that the agency had never inspected. FDA has announced initiatives to improve its foreign drug inspection program.

In November 2007, GAO testified on weaknesses in FDA's foreign drug inspection program (GAO-08-224T). This statement presents preliminary findings on how FDA's initiatives address the weaknesses GAO identified. GAO interviewed FDA officials and analyzed FDA's initiatives. GAO examined reports and proposals prepared by the agency, as well as its plans to improve databases it uses to manage its foreign drug inspection program.

In: Consumer Product Safety Issues
Editors: P. A. Varga, M. D. Pintér
ISBN: 978-1-60456-826-4

Chapter 1

CHILD RESTRAINT SEATS: USE, MISUSE AND STRATEGIES FOR IMPROVED ACTION

Ediriweera Desapriya[1, 2], Giulia Scime[1], Shelina Babul[1, 2], Peter Cripton[3], Takeo Fujiwara[2], Sayed Subzwari[1] and Ian Pike[1,2]

[1]British Columbia Injury Research and Prevention Unit, Centre for Community Child Health Research, Child and Family Research Institute,Vancouver, BC, Canada
[2]Department of Pediatrics, University of British Columbia, Vancouver, BC, Canada
[3]Department of Mechanical Engineering, University of British Columbia, Vancouver, BC, Canada

ABSTRACT

Motor vehicle crash (MVC) related injury is a major preventable global public health problem as described in a recent report by the World Health Organization (WHO). Despite considerable advances in road safety, road trauma is still the leading cause of death and disability for children in highly motorized countries such as the US, Canada, UK, Japan, New Zealand and Australia. Scientific evidence indicates that the single most significant risk factor for injury in a MVC is the non-use or misuse of a restraint. International research and experience has demonstrated that the use of a child restraint seat (CRS) significantly reduces the risk and severity of injury and the number of deaths resulting from MVC. Estimates of effectiveness with respect to reductions in the number of deaths in MVC with CRS use are 71% for infants and 54% for children 1 to 4 years of age. A recent study found that for children up to 4 years of age, a correctly installed CRS reduces injury hospitalization by 69%. A systematic review on CRS-related research provides ample evidence in support of the notion that the proper use of a CRS significantly reduces the extent of child morbidity and mortality on the highway.

Due to high rates of non-use and misuse, the available literature shows the benefit of using a CRS has not been understood. Research in this area has shown that there is no universal use of CRS and that these restraints are frequently used inappropriately. This inappropriate use decreases the effectiveness of the CRS in preventing injuries or death. Contributing to the improper use of a CRS is the incorrect installation of the restraint within the vehicle. Installation of a CRS is complicated, and differs depending upon the model, the vehicle, the vehicle safety belt system, and diverse vehicle seating arrangements. The high frequency of improper CRS use suggests that there is still much work to be conducted in order to facilitate ease and correct use by parents and caregivers.

The goals of this chapter are to quantify the frequency of improper CRS use and to identify the most common mistakes, so that priorities for design innovations and other promising strategies to reduce misuse can be identified and achieved. A general increase in the correct use of CRS can be achieved through the application of multiple strategies, including community awareness and education, media campaigns, public policies, enforcement of laws and further CRS design improvements.

INTRODUCTION

Motor vehicle crash (MVC) related injuries and fatalities comprise a major global public health problem as described in a recent report by the World Health Organization (WHO) (Peden et al., 2004). Worldwide, MVCs are responsible for many child deaths and serious injuries (Turner et al., 2005; WHO, 2002). Despite considerable advances in road safety, road trauma is still the leading cause of death and disability for children in highly motorized countries such as the United States (US), Canada, Japan, New Zealand and Australia (UNICEF, 2001). The widely cited United Nations Children's Fund's (UNICEF) report (2001) highlighted that injury is the leading cause of mortality in children in every industrialized country, claiming more than 20,000 lives annually. According to this report, the most prevalent cause of death is traffic crashes (41%), followed by other unintentional injuries (16%) and drowning (15%) (Sibbald, 2001).

Increased attention to evidence-based decision making in public health as well as a global commitment to the millennium development goals has generated renewed interest in strengthening child-health epidemiology as a foundation for improved efforts to reduce mortality in children younger than 5 years of age (UNICEF, 2001).

However, developing countries are becoming rapidly motorized even though data related to child injuries are woefully lacking. With rapid motorization, MVC have become increasingly responsible for many child deaths and serious injuries in a majority of countries worldwide. Therefore, prevention of pediatric vehicle occupant injuries is of great importance to both individuals and society, but unfortunately, has largely been either excluded from receiving attention or treated in an inappropriate manner (Sehgal et al., 2004).

This is evident in the topic of child safety seats overall. A recent survey conducted by the United Nations Economic Commission for Europe (UNECE) (2005) revealed that in Burundi, Cameroon, Central African Republic, Mauritius Morocco, Azerbaijan, Bulgaria, Georgia, Kazakhstan, Kyrgyzstan, Serbia and Montenegro (all are rapidly motorizing countries in Europe, Africa and Asia), there are no CRS laws, enforcement or education around child safety seats and no public information campaigns to increase awareness of the safety value of CRS (UNECE, 2005).

In contrast, injury prevention experts in highly motorized countries were polled about the priorities for discussion of injury prevention issues at pediatric preventive care visits. All of those polled, agreed that pediatric occupant protection should receive the highest priority on the traffic safety injury prevention agenda (Cohen et al., 1997).

TRAFFIC SAFETY BENEFITS OF CRS

Motor vehicle occupant injury is the leading cause of death in US children, adolescents, and young adults aged 3 to 33 years and of unintentional injury-related deaths for persons of all ages (US Preventative Services Task Force, 2007). Proper use of motor vehicle (MV) occupant restraints (CRS, booster seats, and lap-and-shoulder belts) is associated with a 45% to 70% reduction of fatality risk. However, improper use reduces the efficacy of restraints substantially (US Preventative Services Task Force, 2007).

National Highway Traffic Safety Administration (NHTSA, 1996) had estimated the effectiveness of CRS with respect to a reduction in the number of deaths in MVC with CRS use are 71% for infants and 54% for children aged 1 to 4 (NHTSA, 1996). For children up to 4 years of age, correctly installed CRS reduce injury hospitalization by 69% (NHTSA, 2002). Belt-positioning booster seats, suitable for children aged 4–8 years (approximately) provide 59% greater protection to the child than do adult seatbelts (Durbin et al., 2003).

Scientific evidence indicates that the single strongest risk factor for injury in a MVC is the nonuse or misuse of a restraint. International research and experience has shown that the use of a CRS significantly reduces the risk and severity of injury and the number of deaths that result from MVC (Turner et al., 2005).This is particularly true as restraint systems are effective in preventing occupant ejection during a crash, which is a major cause of fatal and incapacitating injuries. Biomechanical research during the 1940s proved that properly harnessed occupants could better survive crashes than those who were unrestrained (Stapp, 1957).

The effectiveness of safety restraints in protecting children from MV-related injuries has been well documented prompting many countries to require that child passengers ride in appropriate restraints. Although the prevalence of child restraint devices has increased following the passage of child restraint laws, many drivers tend to use these restraints incorrectly (Zaza et al., 2001).

A systematic review on CRS-related research provides ample evidence in support of the proposition that the appropriate use of a CRS reduces the extent of child morbidity and mortality on the highway (Johnston et al., 1994; US Preventative Services Task Force, 2007; Zaza et al., 2001). Studies have proven that any restraint use is better than no restraint use; however the best protection for child passengers can be achieved by using CRS that are appropriate for the child's age and body size (Decina and Knoebel, 1997; Desapriya et al., 2004; Johnston et al., 1994; US Preventative Services Task Force, 2007), and which are used and installed correctly.

The injury prevention benefits of CRS and booster seats can only occur if the seat is used appropriately. That is, the seats should be age- weight- and height-appropriate and should be installed and placed into the vehicle correctly in the rear seat (Decina and Knoebel, 1997; Durbin et al., 2003; Lane et al., 2000; US Preventative Task Force, 2007; Zaza et al., 2001). Table 1 provides a detailed description of the age, type of CRS and the guidelines associated with each one. As shown, specific guidelines are in place for each type of CRS.

Table 1. Type of car seat and user guidelines

Age	Type of seat	Guideline
Infants	Infant-only rear facing convertible seat	All infants should always ride rear-facing until they are 1 year of age and weigh at least 20-22 pounds. There are some infant seats used for special needs purposes which are designed to allow the infant to lie flat.
Toddlers/Preschoolers	Convertible, combination-forward facing seat	Children 1 year of age and at least 20 pounds can ride forward-facing. It is best to ride rear-facing as long as possible. Models with higher weight limits are ideal for children over 40 lbs. who are under ages 3-4 years or not ready to sit properly in a booster seat with lap/shoulder belt.
School-aged children	Booster seat	Booster seats are for older children who have outgrown their forward-facing car safety seats. Children should stay in a booster seat until the adult seat belt fits correctly (usually when a child reaches about 4' 9" in height and is between 8 and 12 years of age and up to 80 lbs. or upper weight limit listed on seat).
Older children	Seat belts	Children who have outgrown their booster seats should ride in a lap and shoulder belt. They should ride in the back seat until they are at least 12 years of age.

(Source: NHTSA child safety seat information: Available from: http://www.alavoices.org/Files/AlaVoices/documents/CPS-NHTSA%202004%20Child%20Safety%20Seat%20Guidelines.pdf)

Child Restraint Misuse

To correctly define 'child restraint misuse', we must first define 'correct use'. A recent definition provided by Williams and colleagues (2007) states that appropriate use is defined by age, weight, height and location (mid rear seating) as recommended by international traffic safety research evidence (Williams et al., 2007). Appropriate use of CRS is the most effective way to protect young children involved in a MVC and is a crucial step in reducing the risk of childhood injuries as a result of MVC. This effort involves encouraging parents and caregivers to become familiar with and to adhere to the guidelines and best practices for the use of CRS.

To gain the highest levels of protection from each type of child restraint seats, restricting use to the recommended size range of the child to that which each type of restraint designed, is required. Child restraint misuse occurs when parents and caregivers do not adhere to the guidelines (Desapriya et al., 2004; Williams et al., 2007). A growing body of literature supports the premise that inappropriate CRS use results in reduced protection and increases the chances of severe injury. The effect of inappropriate CRS use on the risk of injuries and fatalities has been consistently highlighted in the scientific literature, and these studies were supported by crash-related injury data (Desapriya et al., 2006; Durbin et al., 2003). Research on CRS use has shown that there is far from universal use of CRS and that they are frequently used inappropriately (Bull et al., 1988; Margolis et al., 1992). A contributing factor is that the proper installation and use of CRS is complicated, and differs for different models of CRS, different models of cars, different vehicle safety belt systems, and different seating arrangements (Margolis et al., 1992).

Table 2. Common errors in child restraint seat use

1. CRS not tightly secured in the vehicle
2. CRS not securely fastened in the seat (loose seat belt)
3. CRS is not reclined to 45 degree angle (rear facing seats)
4. CRS tethers not used correctly, or used at all
5. Incorrect placement of the vehicle shoulder harness in the CRS
6. Loose vehicle shoulder harness
7. CRS harness straps not snuggly fitted to the child
8. Locking clip not used correctly
9. Twisted harness straps
10. Safety belt not in locked mode
11. Safety belt not routed correctly
12. CRS placed in front of active air bag
13. CRS placed in the front seat
14. CRS is not in upright position (rear-facing seats)
15. Child is in the incorrect type of CRS for his/her weight, height and age

The aim of this book chapter is to quantify the frequency of improper CRS use in several countries, to identify common and most critical misuses (in traffic safety literature, critical misuse is defined as age, weight and height mismatched use) in CRS. Our objective is to highlight the issues so that priorities for anticipatory guidance regarding common misuse and critical misuse can be identified. The high frequency of universal common misuse and critical misuse suggests the need to counsel parents and caregivers about the proper use of CRS as an urgent public health issue. Experts in the field, including authors of this chapter, have expressed concern about the limited quality and lack of recent research (outside the US) in this body of evidence (Grossman and Garcia, 1999; Williams, 1999), particularly given the magnitude of this public health burden (US Preventative Services Task Force, 2007).

Therefore, we direct our chapter to identify greater scope regarding the issue of critical misuse and common misuses. This will include the prevalence of booster seat use and misuse, the effectiveness of booster seats, and the worldwide emerging trend towards booster seat legislative efforts to increase appropriate use, which we explore in detail. It is known that legislative efforts are key to the success of appropriate CRS usage. However, we have found that even current CRS laws lack accompanying efforts to provide comprehensive information on appropriate use. We will discuss these shortcomings, utilizing information on current CRS laws. As evident in the child traffic safety literature, US researchers are world leaders in child occupant safety research and therefore, much of the evidence referred to and exemplified in this chapter is from US-based research.

Prevalence of Misuse: The Issues

For CRS to be optimally effective, they must be installed correctly, yet we know from numerous studies that improper restraint use is alarmingly prevalent, ranging from 79% to 94% (Decina and Knoebel, 1997; Lane et al., 2000). The underlying reasons for such a high rate of misuse for this important child safety device are not well understood. Possible contributors include:

1) Engineering/design problems;
2) Physical difficulty with installation;
3) Poor comprehension of installation instructions.

There have been improvements in engineering and design over the years in an attempt to improve ease of use. Legislation has also been enacted to improve uniformity and ease of installation, including the mandated development of upper tether straps and lower anchoring systems. As of September 2002, all new cars made in the US were required to feature these safety enhancements. However, these advances will not yield immediate benefits as older cars and older CRS will remain in circulation, particularly among lower-income families. Therefore, the comprehensibility of current installation instructions is likely to influence parent compliance with this important safety recommendation in the foreseeable future.

Prevalence of Misuse and Impact on Child Occupant Safety

Nearly 30% of children under the age of 4 years do not ride in a proper CRS and are at twice the risk for fatal and nonfatal injuries. Of those children riding in CRS approximately 85% are improperly restrained (Zaza et al., 2001). Improper use of CRS may decrease the effectiveness of these seats in preventing injuries or death (Kohn et al., 2000). Child safety seat "checkups" or "clinics" in which an expert in CRS use and installation examines whether a particular seat is appropriate for a specific child, how that seat is attached in the vehicle, and how the child is secured in the seat, may be one way to promote proper use.

Prevalence of Use/Misuse in Different Countries

Observational studies, fatality, and serious injury-related crash data have established high levels of incorrect or no use of CRS in various countries. In 2004, we reported that in many countries misuse and nonuse of CRS is highly prevalent (Desapriya et al., 2004). Based on available evidence, we provide a brief summary of the prevalence of use/misuse in several countries.

Japan

A Japanese national observational survey jointly conducted by the National Police Agency (NPA) and the Japanese Automobile Federation (JAF) found that 7 out of 10 CRS were loosely fitted (Desapriya et al., 2005). A survey conducted by JAF in 1998 revealed that a mere 8.5% of children under the age of 14 years used any type of safety seat at all (Rowling, 1999).

Canada

Rothestein and colleagues (2004) demonstrated that road traffic injury is the leading cause of death among Canadian children. In 1998, Transport Canada conducted a roadside study and found that 25% of CRS were misused (Transport Canada, 1998). A recent Canadian national study showed that the rate of CRS misuse is as high as 41.6% (Snowden, 2006). Misuse and nonuse of CRS remains a significant issue and many child vehicle occupant deaths can be prevented by the proper use of CRS.

Mexico

A study by Mock et al., (2002) showed that even for the upper and middle income levels, CRS use was below 50% in Mexico. This study further demonstrated that CRS were available in stores in the study area, but priced beyond what most parents can afford. There is a need to consider the availability and cost of CRS especially for those with limited economic resources, and several authors have suggested a need to consider efforts to collaborate with manufacturers in an initiative to make CRS more affordable to lower income communities (Brixey et al., 2008).

UK

Large studies conducted in the UK found that only 25-40% of CRS were fitted correctly (Britax/GMTV, Pampers). In 2002, a campaign conducted in Scotland (Scottish In-Car Safety Initiative) found that 40% of CRS were fitted incorrectly.

Italy

Valent and Barbone (2003) found that even though 98% of children were restrained when traveling in a car, less than 90% were using the CRS appropriately for the height and weight of the child occupant. This study found that half of the respondents had never received any information on the appropriate use of CRS.

France

There is very little published literature on the misuse of CRS in France. However, in a small field study, of 307 vehicles, 73% of children were seated in an incorrect, or incorrectly installed CRS (UNECE, 2005).

Singapore

A review that appeared in the Singapore Medical Journal has shown that CRS use is currently very low in Singapore and authors have cited that it is attributed to a lack of comprehensive CRS law and lack of law enforcement (Ong et al., 2003).

Michigan – US

The Michigan statewide survey was designed to determine the frequency of CRS use for children under the age of 4 years, and to determine types of misuse (Eby and Kostyniuk, 1999). The study results indicate that 74.5% of children under 4 years of age in Michigan are in safety seats, with use highest in vehicles driven by belted drivers, females, and drivers under 60 years of age. Some degree of seat misuse was found in 88.5% of the inspections. The most common types of misuse were related to snugness of fit (both in installing the seat into the vehicle and in securing the child), as well as use of the safety belt locking clip, and the harness positioning clip. The study reported that drivers who had a high occurrence of misuse, as compared with drivers with lower misuse, had lower educational levels, removed the seat frequently from their vehicle, were not the parent of the child, and had children who were both younger and smaller (Eby and Kostyniuk, 1999). Data from the US child restraint checkpoints suggested that as many as 30% of parents install rear-facing restraints at the incorrect recline angle (Taft et al., 1999). Table 3 identifies results from studies that examined the correlates of CRS misuse (Eby and Kostyniuk, 1999;Taft et al., 1999; Winston et al., 2006).

Table 3. Common factors that correlate with CRS misuse

Unrestrained driver
Installing the seat without instructions
Narrow or deeply contoured vehicle seats
Lower education and awareness of parents and caregivers
Lower socioeconomic status (SES) of parents and caregivers
Race and ethnicity
Lack of fluency in English

(Eby and Kostyniuk, 1999;Taft et al., 1999; Winston et al., 2006)

Child safety seat errors included incorrect seat orientation, incorrect use of vehicle seatbelt, incorrect use of locking clip, harness not buckled, incorrect routing of harness, and incorrect use or fit of the chest clip. It is reported that in the US, more than 95% of the CRS inspected have one or more errors (NHTSA, 2000).

Kohn and colleagues (2000) documented the frequent misuse of forward- and rear-facing child safety seats. Overall, 94% of the CRS were used incorrectly and 37% were misused in a way that had the potential to extensively compromise the seat's ability to prevent injury. Further Kohn et al., (2000) called urgent attention to the persistence of this problem despite its recognition. This article also documented that the 3 most frequent errors in use included the seat not belted into the vehicle tightly, the seat harness straps not snug, and the harness retainer clip not at armpit level. Nonuse remains a significant issue in the US; in 2000, 47% of passenger occupant fatalities among children under 5 years of age involved unrestrained children.

More recently, Winston et al., (2006) highlighted several parent driver demographics and socioeconomic characteristics that are associated with the use of suboptimal restraints for child passengers less than 9 years of age in the US. The authors defined suboptimal restraint as the use of forward-facing child safety seats for infants under 1 year of age or weighing fewer than 20 lbs, and seatbelt use for children less than 9 years old. Importantly, Winston et al., (2006) reported that suboptimal restraint use was more common among children between the ages of 4 and 8 years (65%) compared to those less than 1 year (18%) and those aged 1–3 years (5%). Their multivariate analysis revealed several independent risk factors for suboptimal restraint among children aged 4–8 years as follows:

1) Parent drivers aged 35 years or above (compared with drivers aged 24 or below, RR = 1.33, 95% CI: 1.05, 1.79);
2) Less educated parent drivers (compared with college degree or above, RR for vocational training/some college = 1.15, 95% CI: 1.04, 1.29; RR for high school degree = 1.20, 95% CI: 1.08, 1.36; RR for less than high school degree = 1.26, 95% CI: 1.03, 1.49);
3) Non-Hispanic black parent drivers (compared with non-Hispanic white drivers as reference, RR = 1.19 95% CI: 1.02, 1.34); and
4) Parent drivers with income below $20,000 (compared with $50,000 or above as reference, RR = 1.26, 95% CI: 1.10, 1.42).

Winston et al., (2006) concluded that, the importance of poverty in predicting suboptimal restraint use underscores the importance of child restraint and booster seat disbursement and education programs, particularly aimed at low income parent drivers. Conversely, a recent study conducted in the US showed that 67% of children aged 4 to 7 years were inappropriately restrained, even after they had been provided with a free CRS, through a give away program (Brixey et al., 2008). The barriers that parents and caregivers cited included the bulky size of the CRS, the time it takes to properly fit the seat, and limited knowledge of how to appropriately use the CRS. Brixey and colleagues (2008) concluded that appropriate use of CRS for inner-city children is suboptimal and below state and national rates, despite the provision of a free CRS and education. These communities tend to face multiple barriers in using CRS, even when cost is eliminated. Therefore, further social and behavioral research based on outreach methods to increase appropriate CRS use is required in lower income communities.

Technical Reasons for Child Safety Seat Misuse

According to Blue Ribbon Panel on Child Restraint and Vehicle Compatibility (BRPCRVC) one important contributing factor to CRS misuse is the complexity of properly attaching and using different CRS properly in different models and types of vehicles (BRPCRVC, 1995; NHTSA, 1995). Models of CRS attach differently, and different types and models of vehicles require differing attachment mechanisms. Recognition of this problem has led to the adoption of federal legislation in the US (2002) requiring that all models of vehicles and all models of CRS to have a universal and simplified mechanism for attachment (BRPCRVC, 1995; NHTSA, 1995).

Emerging Technology to Reduce CRS Misuse: LATCH (Low Anchors and Tethers for Children)

LATCH is an attachment system that eliminates the need to use seatbelts to secure the CRS. As an engineering and design innovation, LATCH may prove to be the most important factor in decreasing misuse. Uniform standards such as upper tether straps and lower anchorage systems (which allow seats virtually to be snapped in place) have the potential to make CRS installation a much simpler process, requiring very little instruction.

According to NHTSA, the LATCH system is expected to prevent up to 50 deaths and 3,000 injuries to children each year. Traffic safety experts believe that a loose CRS in a vehicle constitutes serious misuse and that the LATCH system can simplify CRS installation and reduce misuse of this type (Stewart et al., 2005).

However, many vehicles and CRS with earlier manufacture dates will still be used for many years to come. Thus, it appears that CRS misuse is likely to continue as a problem that will not be resolved immediately with changes in technology (BRPCRVC, 1995). Further, the LATCH system does not address numerous other errors that can occur when positioning a child in a CRS. Placing the retainer clip too low, routinely using harness straps incorrectly, securing the harness straps too closely, or facing an infant in the forward direction (Eby and Kostyniuk, 1999) are all errors that are likely to continue.

Effectiveness of the LATCH System

One Australian study (Charlton et al., 2004) examined the performance of 3 rear-facing and 2 forward-facing CRS with 3 anchorage systems: standard seatbelt, LATCH (flexible) and ISOFIX (rigid). Frontal (64 km/h) and side impact (15 km/h) HYGE sled tests were conducted using a sedan Buick vehicle. Preliminary findings suggested superior performance of the rigid ISOFIX system over seatbelt and flexible anchorages, particularly in side impact crashes. The authors have suggested that these findings have important implications for the proposed introduction of changes to Australian Standards for new anchorage systems.

Emerging Issues with the LATCH System

One recent study has explored the effectiveness of the LATCH system on CRS-related crashes in the US (IIHS, 2003). Results of this study showed that CRS misuse is likely to be a problem that will not be resolved immediately with changes in LATCH technology (Decina and Lococo, 2007). However, this study has shown that the misuse rate has declined somewhat following the introduction of this engineering design modification.

An additional study by Arbogast and Jermakian (2007) provides the first real-world experience of this new restraint system. Cases were selected from the Partners for Child Passenger Safety (PCPS) child-specific crash surveillance system based in the US. An in-depth review of these data found that misuse remained a frequent occurrence even in LATCH assisted CRS. These cases demonstrate that even after introducing the LATCH technology, misuse-related injury may still occur.

CRS misuse observational studies conducted by child occupant injury prevention experts have found that LATCH misuse was associated with tether and lower anchor connections, and importantly, there were several cases of parents or caregivers using the vehicle safety belt and the lower anchor attachments together (Decina and Lococo, 2004; 2005).

Reduction and Prevention of CRS Misuse: Promising Strategies

Clinical counseling in conjunction with community-based interventions has been effective in increasing the proper use of CRS. Over the past decade, legislation and enforcement have contributed substantially to the increasing trends in CRS and seatbelt use. A comprehensive strategy that includes legislation, enforcement, community-based interventions and counseling in primary care settings, are critical to the improvement of proper CRS use and a decrease in the incidence of MV occupant injury (US Preventative Task Force, 2007).

There is good evidence to suggest that community and public health interventions, including legislation, law enforcement campaigns, car seat distribution campaigns, media campaigns, and other types of community-based interventions, are effective in improving the proper use of child safety seats, booster seats, and seatbelts (US Preventative Task Force, 2007).

Unrestrained Child Occupant Fatalities: Statistics for Thought

US

Four hundred and seventy-one children under the age of 5 died as occupants in light vehicle crashes in 2003. Of those 471 fatalities, an estimated 166 (35%) were totally unrestrained. Research shows that CRS, when used correctly, can reduce fatalities among children less than 1 year of age by 71% in passenger vehicles (NHTSA, 2004).

Japan

Five years after enactment of the compulsory CRS law in 2005, 143 out of 198 (72%) child MV occupant fatal cases in the 0-5 age group were not restrained in CRS. During the period between 2000 and 2005, lack of CRS use contributed to 43% of all front seat child passenger fatalities and 54% of rear-seated passenger fatalities among children aged 0-5 years. In addition, more than 500 child MV occupants under the age of 5 years who were seriously injured were not using any type of child restraint (Desapriya et al., 2008).

Canada

More Canadian children die of road traffic injuries than from any other cause, and nonuse or misuse of CRS is a major contributing factor. The National Occupant Restraint Program (NORP) in Canada reported that 38% of children ages 0-4 years involved in fatal crashes were not restrained in any type of device (NORP, 2004). Optimizing protection of children against injuries and in particular ejection injuries requires better understanding of how restraints can prevent injuries and the universal use of restraints.

CRS Critical Misuse and Older Child Occupant Fatalities

Critical Misuses

Many studies have demonstrated that children do not achieve an adequate seated height for good adult seatbelt fit until they reach a standing height of 145 cm (Weber 1993; Winston et al., 2005). The use of adult seatbelts by children under this height is likely to result in suboptimal levels of protection to child MV occupants.

Previous literature shows that safety seat misuses are not equally dangerous. Margolis et al., (1992) categorized each instance of misuse according to the degree to which it compromised the efficacy of the seat to prevent injuries. Based on a review of the literature, Kohn et al., (2000) further developed the child safety seat misuse index first developed by Margolis et al., in 1992. This index has 3 levels of seat misuse: minor misuse or correct use, partial misuse, and extensive misuse or critical misuse (Kohn et al., 2000).

Evidence of Critical Misuse Rates

Australia

Charlton et al., 2006 found that only 24% of children 4-11 years of age were traveling in a booster seat and 93% of booster seats were appropriately restrained. However, the study also showed that those who used booster seats were significantly more likely to be sitting in the front seat. In Australia in 2005, 72 children aged 16 years and younger died and 1,850 were seriously injured as MV occupants (Charlton et al., 2006).

One recent Australian study (Koppel et al., 2008) showed that of the 195 children who met the height–weight criteria for a booster seat, 44% were correctly seated in a booster seat, while 56% had been prematurely shifted into a seatbelt. This is consistent with other international findings that have reported high proportions of seatbelt use rates (i.e., up to

75%) in the booster seat age group (Cameron et al., 2006; Decina and Knoebel, 1997; Ramsey et al., 2000). Not surprisingly, children traveling in booster seats were significantly younger, lighter in weight and shorter in height than children traveling in seatbelts. This is consistent with the findings reported by Durbin et al., 2003, which demonstrated that child restraint use is generally higher for younger children.

US

Much of the research has shown that child occupants aged 4-8 years frequently sit unrestrained or are placed prematurely in adult seatbelts. Evidence-based practices in pediatric occupant safety recommend that children in this age group be restrained properly in booster seats (CDC-P, 2000). Decina and Knoebel (1997) found that nearly 2 thirds of fatally injured children were unrestrained at the time of the crash. Winston et al., (1999) found that 4-6% of children aged 4-8 years used booster seats when riding in MV. In addition, a report by the CDC-P (2000) shows that in 1998, restraint use among fatally injured children aged 4-8 was 38.1%.Winston and Durbin (1999) found that 62% of children 4 to 8 years of age were inappropriately restrained in adult seatbelts. Similarly, NHTSA's national sample obtained through the MV Occupant Safety Survey in 2003 revealed that only 21% of children aged 4 to 8 years were restrained in a booster seat.

Knowledge regarding appropriate restraint use is particularly important for the majority of countries as the current legislation regarding child restraint use is considerably less stringent than in few other countries where booster seat legislation is effective. For example, in 2006 the UK implemented comprehensive CRS legislation which stated that children under age 9 years and 146 cm in height should be placed in booster seats while traveling. In the UK, the booster seat legislation is not comprehensive, as rear seating is not required by this law. This is a major set back in new US, UK (European directives for booster seats legislation) and Canadian booster seat legislation. Shortcomings of the current booster seat legislation are discussed taking several examples from the US booster seat legislation.

Many of the country, state and provincial child restraint laws are currently under review and consequently there is an important opportunity for extensive evidence presented in this chapter on the effectiveness and shortcomings of booster seat laws. Our goal is to help guide and inform future comprehensive global child occupant policy and legislation.

A Focus on Booster Seats

The Prevalence of Booster Seat Use/Misuse

Despite the obvious benefit of booster seats, children and their families are not commonly using them; only 10–28% of children observed or recorded at crash sites have been properly restrained in a booster seat (Durbin et al., 2001; Kunkel et al., 2001; Ramsey et al., 2000). Another study estimated that 6 to 19% of 4 to 8 year old children currently ride in booster seats, while the majority uses adult seatbelts (CDC-P, 2002; Cody et al., 2002; Ebel et al., 2003). Unpublished data obtained in 2000 at the Cincinnati Children's Hospital Medical Center (CCHMC) showed that only 8% of children were observed to be in belt-positioning booster seats after 2,300 car seat checks. A recent study that examined child passenger restraint use and emergency department (ED)-reported injuries, through the National

Electronic Injury Surveillance System in the US, found that 77% of inappropriate restraint use occurred among children 4-8 years of age (those who should be in a booster seat), as they were using seatbelts (Lee et al., 2008). This study also showed that unrestrained children were 3 times more likely to require hospitalization as compared to restrained children, and they were 1.7 times as likely to suffer multiple injuries (Lee et al., 2008). Rivara et al., (2001) reported that parents tended to be confused about the appropriate weight and age of children who should be placed in a booster seat (Rivara et al., 2001). Ramsey and colleagues provided another explanation for dismal CRS use as they showed that parents believed children fit appropriately in a lap/shoulder belt (Ramsey et al., 2000).

Necessity of Booster Seats

Data from mechanical sled tests, (Klinich et al., 1994) computer crash simulations, and epidemiological data (Durbin et al., 2001; Winston et al., 2000) all suggest that booster seats are protective for children who do not yet fit into seatbelts. Therefore, safety experts recommend that children who have outgrown child harness seats and do not yet fit into adult seatbelts should use a booster seat (AAP, 2002; NHTSA, 2007).

Cost-Effectiveness of Booster Seats

One study conducted in the US has estimated that booster seats cost \$30, plus \$167 for maintenance and time spent on installation and use. This study shows that the investment saves \$1,854 per seat, a return on investment of 9.4 to 1. The study concluded that belt-positioning booster seats offer a sound return on investment. The study proposed that booster seat use laws should be passed, publicized, and enforced universally (Miller et al., 2006).

Legislation and Booster Seat Use

The Powerful Story of Anton Skeen

The first booster seat legislation in US law was named in memory of Anton Skeen, a 4 year old boy who was killed in a rollover crash in Washington State in 1996. Anton, who was sitting in the right front seat of his family's vehicle, and was wearing a lap/shoulder belt, was ejected from the vehicle, and died. The belt remained buckled even after Anton had been ejected. After losing her son, Autumn Alexander Skeen, a local journalist, did extensive research on booster seats and prompted the Washington State legislature to pass the country's first mandatory booster seat provision, also dubbed "Anton's Law." She was also instrumental in advocating for the enactment of Federal legislation (NHTSA, 2006; US Library of Congress, S.980 "Anton's Law", passed December, 2003).

Optimal performance of restraint systems depends on an adequate fit between the restraint system and the occupant at the time of the crash. Increased risk for injury in young children restrained in seatbelts as a result of improper loading of deceleration forces has been documented for over 50 years (Agran et al., 1987; Garrett and Braunstein, 1962; Kulowski and Rost, 1956). CRS offer improved fit of restraints for children who are too small for the adult-sized seatbelt, thereby affording a mechanical protective advantage over seatbelts.

Importantly, CRS including booster seats are proven to be a considerable safety advantage over seatbelts for biomechanical reasons. CRS are designed to reduce risk of ejection during a crash, better distribute the load of the crash through structurally stronger bones rather than soft tissues, limit the crash forces experienced by the occupant by prolonging the time of deceleration, and potentially limit the contact of the occupant with intruding vehicle structures (Howard, 2002).

Belt-positioning booster seats are currently recommended for children from ages 4 years and 40 lbs, to at least age 8 years unless the child is greater than 4 feet, 9 inches (NHTSA, 2005). The purpose of the booster seat is to elevate a child in the automobile's chair so that the lap belt fits across the child's pelvis, not abdomen, and the shoulder belt does not override on the child's neck. Serious injuries have been documented when children in this age range are not restrained properly (Agran et al., 1987; Durbin et al., 2003; Winston et al., 2000). In fact, Durbin et al., (2003) have shown that the odds of injury are 59% lower if child is restrained appropriately in a belt-positioning booster seat compared with a vehicle safety belt (Durbin et al., 2003). In their study, no child experienced any abdominal, neck, back, or lower extremity injuries if he or she was properly restrained in a belt-positioning booster seat (Durbin et al., 2003).

Best Evidence Strategies to Increase Booster Seat Use

Evidence-based studies have revealed the effectiveness of CRS laws in increasing child restraint use and reducing the numbers of both fatal and nonfatal injuries (Ehiri et al., 2006; Zaza et al., 2001). Proper use and increased use of CRS has been achieved as a result of multiple strategies, including comprehensive legislation, enforcement of laws, education, media campaigns and public policies mandating use (NHTSA, 2002).

There are many reasons why parents choose not to buckle up their children. Many of these problems relate to the lack of knowledge parents have regarding the need for CRS (Spanier et al., 2002; Vaca et al., 2002). Other issues parents or caregivers face are the difficulties of correctly installing a CRS, securing the child in the seat correctly, and knowing which type of seat (infant, convertible, or booster) is the most appropriate. If there is comprehensive legislation which could clearly outline the best practices, this would help parents and caregivers to appropriately use and install the seats.

The NHTSA MV Occupant Safety Survey (2003) and research by Ramsey et al., (2000) and Ebel et al., (2003) reported 4 reasons why children 4 to 8 years of age are riding improperly restrained (i.e., not in booster seats):

1) parents lack knowledge about booster seats;
2) parents think booster seats are loose, unstable, or unsafe;
3) parents think their children are large enough to avoid them; and
4) pressure from their children and even other parents to quit treating their children like babies.

Importantly, the Australian public health community found that a lack of comprehensive legislation is a contributory factor for misuse. As demonstrated by Reeve et al., (2007) current Australian legislation specifies the use of CRS for infants up to the age of 1; however for older children the legislation is less definitive stating only that an "appropriate" restraint should be used. Accordingly, the responsibility of choosing appropriate restraint use largely rests upon the parents and caregivers. Similarly, an additional Australian study (Koppel et al., 2008) shows that in the absence of clearer guidelines and comprehensive legislation for CRS use, the responsibility largely rests upon parents and caregivers to determine which particular restraint is appropriate for their child. However, parents and caregivers' ability to determine appropriate restraint use for their older children and the factors that influence appropriate restraint use has received little attention.

Although belt-positioning booster seats reduce the risk for injury by nearly 59% for children 4 to 8 years of age compared with seatbelts (Dellinger et al., 2002; Durbin et al., 2003), in many countries booster seat requirements for children over 4 years of age are not required by law. However, the US, UK and more recently a few Canadian provinces are moving towards a more comprehensive policy including a compulsory requirement for children aged 4-8 years or children less than 135 cm in height, should be seated in booster seats.

Legislation has been shown to have dramatic positive effects on children's restraint wearing rates (Winston et al., 2004; 2007). For example, a recent longitudinal study from the US found that children in states with laws requiring booster seats for children up to the age of 7 years were 39% more likely to be appropriately restrained than were children in other states (Winston et al., 2007). Similarly, since the introduction of a mandatory CRS law in New Zealand for children up to the age of 5 years, the use of CRS increased by 15% to reach 89% in 2005 for this age group (New Zealand Ministry of Transport, 2006). In Japan evidence shows that after compulsory legislation, usage rates increased by approximately 6 fold (Desapriya et al., 2004, 2008).

Comprehensiveness of Current CRS Laws and Booster Seat Legislation: Dissemination of Evidence-based Practices to Parents and Caregivers

Impact of Current CRS Law

We have highlighted the various CRS laws in the US.

Table 4. Child Restraint Laws

February 2008			
	Who is covered?		
State	Must be in child restraint	Adult safety belt permissible	Maximum fine 1st offense
Alabama	younger than 1 or less than 20 pounds in a rear-facing infant seat; 1 through 4 years or 20-40 pounds in a forward-facing child safety seat; 5 but not yet 6 in a booster seat.	6 through 14 years	$25[1]
	law states no preference for rear seat		
Alaska	3 years and younger	4 through 15 years	$50[1]
	law states no preference for rear seat		
Arizona	4 years and younger	not permissible	$50
	law states no preference for rear seat		
Arkansas	5 years and younger and less than 60 pounds	6 through 14 years or 60+ pounds	$100
	law states no preference for rear seat		
California	5 years and younger or less than 60 pounds[2]	6 through 15 years or 60+ pounds	$100[1]
	children 5 years and younger or less than 60 pounds must be in the rear seat[2]		
Colorado	younger than 1 year and less than 20 pounds in a rear-facing infant seat; 1 through 3 years and 20-40 pounds in a forward-facing child safety seat; 4 through 5 years and less than 55 inches in a booster seat[3]	6 through 15 years or 55 inches or more	$50
	law states no preference for rear seat		
Connecticut	younger than 1 year or less than 20 pounds in a rear-facing restraint system; 1 through 6 years who is less than 60 pounds in a child restraint system (booster seats may only be used in a seating position with a lap and shoulder belt)	7 through 15 years and 60+ pounds[4]	$60[4]
	law states no preference for rear seat		
Delaware	7 years and younger and less than 66 pounds[5]	8 through 15 years or 66+ pounds[5]	$25
	law states no preference for rear seat		
District of Columbia	7 years and younger	8 through 15 years	$75[1]
	law states no preference for rear seat		
Florida	3 years and younger	4 through 5 years	$60[1]
	law states no preference for rear seat		
Georgia	5 years and younger and 57 inches or less[6]	more than 57 inches	$50[1]
	5 years and younger must be in rear seat if available		
Hawaii	3 years and younger in a child safety seat; 4 years through 7 years must be in a booster seat or child restraint	4 through 7 years who are taller than 4'9"; 4 through 7 years who are at least 40 pounds seated in a rear seat where if there are no available lap/shoulder belts, may be restrained by a lap belt	$100[7]
	law states no preference for rear seat		

Table 4. Child Restraint Laws (Continued)

February 2008			
	Who is covered?		
State	**Must be in child restraint**	**Adult safety belt permissible**	**Maximum fine 1st offense**
Idaho	6 years and younger	not permissible	$100
	law states no preference for rear seat		
Illinois	7 years and younger	8 through 15 years; children who weigh more than 40 pounds seated in the rear where only a lap belt is available	$50
	law states no preference for rear seat		
Indiana	7 years and younger when driver holds an Indiana license[8]	8 through 15 years	$25[1]
	law states no preference for rear seat		
Iowa	younger than 1 year and less than 20 pounds in a rear-facing child seat; 1 through 5 years	6 through 10 years	$25
	law states no preference for rear seat		
Kansas	all children 3 and younger must be in a child restraint; children 4 through 7 who weigh less than 80 pounds and children 4 through 7 who are less than 57 inches tall must be in a child restraint or booster seat	all children 8 through 13 years; children 4 through 7 years who weigh more than 80 pounds, and children 4 through 7 years who are taller than 57 inches	$60
	law states no preference for rear seat		
Kentucky	40 inches or less	not permissible	$50
	law states no preference for rear seat		
Louisiana	younger than 1 year or less than 20 pounds in a child safety seat; 1 through 3 years or 20-39 pounds in a forward-facing child safety seat; 4 through 5 years or 40-60 pounds in a child booster seat	6 through 12 years or greater than 60 pounds	$50
	law states no preference for rear seat		
Maine	less than 40 pounds in a child safety seat; 40-80 pounds and less than 8 years in a safety system that elevates the child so that an adult seat belt fits properly	8 through 17 years or less than 18 years and more than 4'9"	$50
	11 years and younger and less than 100 pounds must be in rear seat if available		
Maryland	5 years and younger or 40 pounds or less[2]	6 years and more than 40 pounds through 15 years	$25
	law states no preference for rear seat		
Massachusetts	4 years and younger or 40 pounds and less	5 through 11 years	$25
	law states no preference for rear seat		
Michigan	3 years and younger	not permissible	$10
	law states no preference for rear seat		
Minnesota	3 years and younger	not permissible	$50
	law states no preference for rear seat		
Mississippi	3 years and younger	not permissible	$25
	law states no preference for rear seat		

Missouri	3 years and younger must be in a child restraint; all children who weigh less than 40 pounds must be in a CR; 4 through 7 years who weigh at least 40 pounds but less than 80 pounds and who are 4'9" or shorter must be in either a CR or booster seat; children 4 years	all children 8 through 16 years; all children 4 years and older who weigh 80 pounds or more or who are taller than 4'9"	$50; $10 for violations involving children taller than 4'9" or who weigh 80 pounds or more
	law states no preference for rear seat		
Montana	5 years and and younger and less than 60 pounds	not permissible	$100
	law states no preference for rear seat		
Nebraska	5 years and younger	6 through 17 years[10]	$25[1]
	law states no preference for rear seat		
Nevada	5 years and younger and 60 pounds or less	not permissible	$500[11]
	law states no preference for rear seat		
New Hampshire	5 years and younger who are less than 55 inches tall	6 through 17 years; younger than 6 who are at least 55 inches tall	$25
	law states no preference for rear seat		
New Jersey	7 years and younger and less than 80 pounds	not permissible	$25
	children 7 years and younger and less than 80 pounds must be in the rear seat if available		
New Mexico	younger than 1 year in a rear-facing infant seat; 1 through 4 years or less than 40 pounds in a child safety seat; 5 through 6 or less than 60 pounds in a booster seat	7 through 17 years	$25
	children younger than 1 year in a rear-facing infant seat must be in the rear seat if available		
New York	3 years and younger in all seats; 6 years and younger in all seats	4 through 15 years; 7 through 15 years	$100[1]
	law states no preference for rear seat		
North Carolina	7 years and younger and less than 80 pounds[12]	8 through 15 years + children 40-80 pounds in seats without shoulder belts	$25[1]
	law states no preference for rear seat		
North Dakota	6 years and younger and less than 57 inches or less than 80 pounds	7 through 17 years; 6 years and younger and at least 57 inches tall and at least 80 pounds; 6 years and younger and at least 40 pounds, if there are no available lap/shoulder belts, may be restrained by a lap belt	$25[1]
	law states no preference for rear seat		
Ohio	3 years and younger or less than 40 pounds	4 through 14 years[13]	$25[1]
	law states no preference for rear seat		
Oklahoma	5 years and younger[14]	6 through 12 years	$25
	law states no preference for rear seat		

Table 4. Child Restraint Laws (Continued)

February 2008			
	Who is covered?		
State	**Must be in child restraint**	**Adult safety belt permissible**	**Maximum fine 1st offense**
Oregon	younger than 1 year or 20 pounds or less must be in a rear facing child safety seat; 40 pounds or less must be in a child safety seat; more than 40 pounds but 4 feet and 9 inches or less must be in a safety system that elevates the child so that an adult seat belt fits properly	taller than 4 feet and 9 inches	$90
	law states no preference for rear seat		
Pennsylvania	7 years and younger[15]	not permissible	$100
	law states no preference for rear seat		
Rhode Island	6 years and younger and less than 54 inches and less than 80 pounds	6 years and younger who either weigh 80 pounds or more or who are at least 54 inches tall; 7 through 17	$75
	children 6 years and younger must be in rear seat if available		
South Carolina	younger than 1 year or less than 20 pounds in a rear-facing infant seat; 1 through 5 years and 20-39 pounds in a forward-facing child safety seat; 1 through 5 years and 40-80 pounds in a booster seat secured by lap-shoulder belt (lap belt alone is impermissible)	1 through 5 years and 80+ pounds or any child 5 years and younger if the child's knees bend over the seat edge when sitting up straight with his/her back firmly against the seat back	$150
	children 5 years and younger must be in rear seat if available		
South Dakota	4 years and younger and less than 40 pounds	5 through 17 years; all children 40+ pounds, regardless of age	$20
	law states no preference for rear seat		
Tennessee	younger than 1 year or 20 pounds or less in a rear-facing infant seat; 1 through 3 years and 20+ pounds in a forward-facing infant seat; 4 through 8 years and less than 4'9" in a booster seat	9 through 15 years or any child 12 or younger who is 4'9" or taller	$50
	children 8 years and younger and less than 4'9" must be in rear seat if available; rear seat recommended for children 9 through 12		
Texas	4 years and younger and less than 36 inches	not permissible	$200
	law states no preference for rear seat		
Utah	4 years and younger	5 through 15 years	$45
	law states no preference for rear seat		
Vermont	younger than 1 year or less than 20 pounds in a rear-facing infant seat; 2 through 7 and more than 20 pounds	8 through 15 years and more than 20 pounds	$25
	law states no preference for rear seat		
Virginia	7 years and younger unless they have a physician	8-15 years[16]	$50

	exemption		
	children in rear-facing devices must be in a rear seat if available; if not available, they may be placed in front only if front passenger airbag is deactivated		
Washington	younger than 8 and less than 4'9"	8 through 15; younger than 8 and 4'9" or taller	$86
	12 years and younger must be in rear seat if practical		
West Virginia	7 years and younger and less than 4'9"	7 years and younger and 4'9" or taller	$20
	law states no preference for rear seat		
Wisconsin	children younger than 1 and all children who weigh less than 20 pounds are required to be in a rear-facing infant seat; children 1 through 3 years who weigh at least 20 pounds but less than 40 pounds are required to be in a forward-facing child safety seat; children 4 through 7 who both weigh at least 40 pounds but less than 80 pounds and who are less than 57 inches tall are required to be in a booster seat	8 years and younger and more than 80 pounds and 57 inches or taller	$75
	children 3 and younger must be in a rear seat, if available		
Wyoming	8 years and younger	not permissible	$50
	children 8 years and younger must be in the rear seat if available		

[1]This state assesses points for violations.

[2]In California, children weighing more than 40 pounds may be belted without a booster seat if they are seated in the rear seat of a vehicle not equipped with lap/shoulder belts. The California rear seat requirement does not apply if: there is no rear seat; the rear seats are side-facing jump seats; the rear seats are rear-facing seats; the child passenger restraint system cannot be installed properly in the rear seat; all rear seats are already occupied by children under 12 years; or medical reasons necessitate that the child not ride in the rear seat. A child may not ride in the front seat of a motor vehicle with an active pasenger airbag if the child is under 1 year of age, or weighs less than 20 pounds or is riding in a rear-facing child restraint system.

[3]In Colorado, if a child 4-5 years and less than 55 inches is being transported in a vehicle equipped with a lap belt only, then the child must be restrained with the lap belt. The law is secondary for children ages 4-5 years who must be in booster seats.

[4]The fine in Connecticut is $15 if the child is 4-16 years and 40 pounds or more. Connecticut also requires a mandatory child restraint education program for first or second violation.

[5]In Delaware, children younger than 12 years/65 inches or less must be restrained in a rear seat if a vehicle has a passenger airbag unless the airbag has been either deactivated or designed to accommodate smaller people. Exceptions: no rear seat or rear seat occupied by other children younger than 12 years/65 inches or less.

[6]In Georgia, children weighing more than 40 pounds are permitted to be restrained in the back seat of a vehicle by a lap belt if the vehicle is not equipped with lap and shoulder belts or when the lap and shoulder belts are being used by other children who weigh more than 40 pounds.

[7]Hawaii drivers are charged $50 for a mandatory child restraint education program and a $10 surcharge deposited into a neurotrauma special fund.

[8]In Indiana, children younger than 8 years must be restrained in adult belts if it's reasonably determined they cannot fit in child restraints. If the driver does not hold an Indiana driver's license, then children under 16 must be restrained by either a child restraint or a safety belt. Children weighing more than 40 pounds are permitted to be restrained by a lap belt if the vehicle is not equipped with

lap and shoulder belts or if all lap and shoulder belts other than those in the front seat are being used to restrain other children who are younger than 16.

[9]In Maryland, vehicles registered out of state are required to restrain children under the age of 4 years or 40 pounds or less in a child restraint system.

[10]Nebraska's law is secondary for those children who may be in safety belts and standard for those who must be in a child restraint device.

[11]In Nevada, the minimum fine is $100. An alternative to the fine is at least 10 but not more than 50 hours of community service.

[12]In North Carolina, children younger than 5 who weigh less than 40 pounds must be restrained in a child safety seat in the rear seat if the vehicle has a passenger airbag, unless the child restraint system is designed for use with airbags.

[13]In Ohio, the law is secondary for children 4 through 14 years.

[14]In Oklahoma, children weighing more than 40 pounds are permitted to be restrained in the back seat of a vehicle by a lap belt if the vehicle is not equipped with lap and shoulder belts or when the lap and shoulder belts are being used by other children who weigh more than 40 pounds.

[15]In Pennsylvania, the law is secondary for children ages 4 through 7 years who must be in booster seats.

[16]Children at least 4 years but less than 8 years may be belted if any licensed physician determines that use of a child restraint system by a particular child would be impractical by reason of the child's weight, physical fitness, or other medical reason, provided that any person transporting a child so exempted shall carry on his person or in the vehicle a signed written statement of the physician identifying the child so exempted and stating the grounds for the determination.

(Source: http://www.iihs.org/laws/childrestraint.aspx reprinted with permission from Insurance Institute of Highway Safety (2008).)

Booster Seat Legislation in Canadian Provinces

Booster seat legislation is crucial to closing the gap for children who are too large for child car seats, yet too small for seatbelts. Between 1997 and 2001, the death rate from car crashes in Canada dropped by 52% among children under 5 years of age and by 25% among children between the ages of 10 and 14 years, but did not drop at all for children aged between 5 and 9, those in the booster seat range. The Canadian province of Quebec passed and implemented legislation in June 2002 and Ontario implemented a booster seat law in September 2005. Nova Scotia recently passed a booster seat law and it came into effect in 2007. The British Columbia booster seat law came into effect July 2008. Similarly, the Prince Edward Island booster seat law was enacted in January 2008. Effective from July 2008 the Newfoundland and Labrador booster seat law came into effect. Meanwhile New Brunswick's booster seat law came in to effect May 2008. The Government of Alberta has indicated a commitment to adding booster seat requirements to the legislative agenda. The remaining provinces and territories are without specific booster seat legislation (NORP 2010, Annual Monitoring Report, 2004).

Ontario

Children less than 1.45 m in height (approximately 57 inches) and under 80 pounds (36 kg) or under 8 years of age will be required to use appropriate CRS in MV.

British Columbia

Children less than 1.45 m in height (approximately 57 inches) and under 9 years of age will be required to use appropriate CRS in MV.

New Brunswick

Children less than 1.45 m in height (approximately 57 inches) and under 9 years of age will be required to use appropriate CRS in MV.

Newfoundland and Labrador

Children less than 1.45 m in height (approximately 57 inches), and under 80 pounds (36kg) and under 8 years of age will be required to use appropriate CRS in MV.

Nova Scotia

Children less than 1.45 m in height (approximately 57 inches) and fewer than 9 years of age will be required to use appropriate CRS in MV.

Prince Edward Island

Children less than 1.45 m in height (approximately 57 inches), if they are over 40 pounds (18kg) and under 10 years of age will be required to use appropriate CRS in MV.

Quebec

Children must ride in a booster seat until they have a minimum seated height (measured from the seat to the top of the head) of 25 inches (63 cm) (NORP 2010, Annual Monitoring Report, 2004).

European Booster Seat Legislation

According to New Programme for the Assessment of Child-restraint Systems (NPACS), European Directive 2003/20/EC on compulsory seatbelt wearing/CRS use effectively requires all children who are under 12 years of age and are less than 150 cm tall, to be sitting in an appropriate CRS (albeit a booster cushion for the taller child) as a condition should the child be traveling in a vehicle (NPACS, 2007).

UK Booster Seat Legislation

From September 2006, all children under 1.35 m in height (approximately 53 inches) and under 12 years of age are required to use appropriate CRS in cars, vans and other good vehicles on UK roads. For children under 3 years of age, there are virtually no exceptions to the requirement. Further changes requiring everyone, adults and children, to be correctly restrained will come into force in 2009 (UK Department of Transportation, 2006).

US Booster Seat Legislation

Only half of the states that have child safety laws include guidance for children 4 to 8 years of age in booster seats. Importantly, 44 states have laws pertaining to booster seats. State legislation largely varies, such as permitted unrestrained travel for different specific circumstances or the inability of drivers to receive a citation from the police unless they have been stopped for another traffic violation. States with primary enforcement have increased restraint use by 12 to 23% and have decreased MV-related fatality rates (US Preventative Task Force, 2007). The other important limitation of those 44 states that currently have laws that apply to children in booster seats is that most of these do not cover all children up to 8 years of age.

Japanese Booster Seat Legislation

It appears that Japanese policy-makers and parents have a weak awareness and understanding of the safety benefits related to CRS (Desapriya et al., 2008). A survey conducted by Japan Automobile Faderation (JAF) in 1998 revealed that a mere 8.5% of children under the age of 14 years used safety seats (JAF, 2006). In April 2000, Japanese law required that CRS must be used for all children aged 0-5 years. The introduction of the compulsory CRS law resulted in an increase in CRS use from approximately 8.5% in 1998 to 47% by 2004 (Desapriya et al., 2005). This law was not comprehensive because it only covered children up to 5 years of age and missed children who typically require booster seats (typically up to 8 years old). The law in Japan portrays the wrong safety message to children, parents and caregivers, basically indicating that only seatbelts are required to restrain children over 5 years of age. Hence, the child's safety is highly compromised if they graduate prematurely to seatbelts without booster seats as shown in the literature.

Belgium's Booster Seat Legislation

One recent study conducted in Belgium found that more than half of children are not appropriately restrained, according to their age, weight or height, even though there has been a law since September 2006 which states that children under 1.35 m in height require appropriate booster seats (Vesentini and Willems, 2007). Much like the UK's booster seat law, Belgium's law falls short of the European booster seat directives for a height requirement of 1.50 m. More importantly, many studies have demonstrated that children do not achieve an adequate seated height for good adult seatbelt fit until they reach a standing height of 145 cm (Cameron et al., 2006; Ramsey et al., 2000; Rivara et al., 2001; Weber, 1993; Winston et al., 2005; Winston and Durbin, 1999).The use of adult seatbelts by children under this height is likely to result in suboptimal levels of protection for child MV occupants.

Gaps in Current Child Restraint Laws

While all 50 states have some sort of occupant restraint law to protect children, as Dellinger et al., (2002) point out, these laws vary dramatically, and most have serious gaps in coverage. Similarly, the international laws that were cited above have serious gaps. In short, current child restraint laws only inform parents and caregivers on how to restrain their children, and do not necessarily reflect the safest way for children to ride in MV. Still few states has no provision in child restraint legislation that requires booster seats for children aged 4-8 years. In addition there are no provisions in the legislation to prohibit children from sitting in the front seat as all of the above mentioned international child safety laws do not emphasize and encourage parents and caregivers to restrain their children in the back seat. Height requirement varies widely by state, province, and country.

The UNECE (2003) report outlines the laws related to child occupant front seating (UNECE, 2003). Front seating is permitted without provision in Denmark, France, Germany, Sweden, UK, and US. Front seating is completely prohibited in Finland; and it is prohibited until the age of 12 years in Azerbaijan, Belgium, Croatia, Czech Republic, Portugal, Romania, Slovenia, and Yugoslavia. It is prohibited until the age of 10 years in Monaco, and Turkey. Front seating is permitted if appropriate child restraints are in place in several countries including Austria, Bulgaria, Hungary, Israel, Italy, Latvia, Lithuania, Luxembourg, Netherlands, Norway, Poland, Russian Federation, Slovakia, and Spain. In Iceland, front seating is permitted, but children under 6 years of age must be in a CRS, and in Switzerland, front seating is prohibited until the age of 7 years. Varying conditions are in place, such as permission to be seated in the front seat if a child's height is over 150 cm (Austria, Hungary, Luxembourg, and Slovakia). Interestingly, in Norway children are permitted to sit in the front seat in cars without seatbelts (registered before 1971). As demonstrated, there are various types of legislation regarding this issue in several countries (UNECE, 2003).

Rear Seating for Child Occupants

One simple, low-cost method to improve children's safety in the vehicle is to restrain them in the rear seat rather than in the front, whenever this is possible (Lennon, 2007). One of the inherent weaknesses of several of the booster seat laws are that they do not have the provision to promote rear seating of young children. Scientific evidence shows that all children up to the age of 13 are safest in the back seat. It is best to ensure that child occupants are properly restrained for their age and size and stay in the proper position at all times. This will help prevent the air bag from injuring child occupants. It is estimated that 25-40% of vehicles with child occupants, carry them in the front seat in the US, while estimates for Australia suggest that the proportion is even higher (Lennon, 2005; 2007).

One recent in-depth Australian study that examined injuries to children aged 2-8 years of age presenting at 2 hospital ED after a crash found that children sitting in the front seat of the vehicle were 2.5 times more likely to have incurred serious injury than those occupying the rear seats (Brown et al., 2005). Greenberg-Seth and colleagues (2004) reported the efficacy of an intervention for increasing rear seating for child occupants over a short time period. The objective of their project was to increase child rear seating in conjunction with proper CRS use by reinforcing educational messages with incentives for positive behavior. Special attempts were made to reach lower-income segments of society through collaboration with existing community organizations.

Increasing Appropriate CRS/Booster Seat Use In Our Communities

The underlying reasons for such high rates of misuse of this important child safety device are not well understood. Partners in Child Passenger Safety (PCPS) have recently emerged as a major force in misuse prevention efforts in North America and world. Many countries have used PCPS evidence based information on child occupant safety research to enact new legislation to protect older child occupants. Their scientifically sound research has shown that 83% of children are being inappropriately graduated to an adult seatbelt and as a result, children are suffering from significant head, spine and brain injuries in crashes (Winston et al., 2000). Although advocates have promoted booster seat use to protect children who are involved in vehicle crashes (Winston and Durbin, 1999), use still remains low.

These findings are also consistent with previous research reporting relatively high proportions of children who commonly graduate to seatbelts too early. That is, children are moved from a booster seat into a seatbelt or moved directly into a seatbelt from a forward facing toddler seat rather than using a booster seat as an interim step. A possible explanation for the premature graduation to seatbelts could be that parents are either uncertain or unaware of the appropriate thresholds for transition from a booster seat to a seatbelt (Cameron et al., 2006; Ramsey et al., 2000; Rivara et al., 2001).

However, extensive literature shows that uncertainty is a key issue for many parents when deciding to use or not to use booster seats. Therefore, comprehensive guidelines by manufactures should include the upper end of the height range, rather than weight criteria alone, since height, particularly seated height, plays a significant role in determining the correct alignment of the seatbelt on the child's pelvis and shoulder.

Comprehensive Legislation and a Combination of Interventions

As Rivara and colleagues (2001) pointed out, one of the most effective enabling factors for injury control is legislation. All 50 states currently have laws requiring infants and young children to be restrained in approved restraint devices. Comprehensive child restraint legislation could effectively guide parents as to the type of restraint to use as well as draw their attention to critical safety dimensions. Unfortunately, the age, height and weight coverage for current child restraint laws vary substantially. It is necessary to reduce this gap between states.

Based on the findings from this study (Rivara et al., 2001), a key recommendation is the need for more definitive child restraint legislation pertaining to child restraint use for older children, supplemented by an extensive educational and awareness strategy for parents and children. A uniform booster seat law is preferred. It should include all children under 9 years of age and less than 145 cm in height. Booster seat legislation has been shown to have a dramatically positive effect on children's restraint usage rates (Winston et al., 2004; 2007) thereby reducing parents' uncertainty concerning when to graduate their child into a seatbelt. As a veteran pediatric injury prevention experts like Simons-Morton and Winston (2006) have shown, laws change parents' and caregivers' perceptions of the value of the safety practices. Importantly, laws are capable of communicating public health safety values of the CRS.

A 2006 Cochrane Collaboration systematic review and meta-analysis (Ehiri et al., 2006) demonstrated that a combination of interventions is more effective than a single strategy approach to increase booster seat use in our communities. It is well known that the causes of traffic crashes are multifactorial; therefore, preventive interventions should not focus solely on a single factor. In addition, periodic monitoring of CRS use/misuse involves the current status of occupant protection laws in many countries. Various countries have either recently passed booster seat laws or policy proposals are under consideration. With such variation in legislation, it will be important to continue to monitor public compliance and Child-restraint Systems use or misuse.

Unaffordability is a Disincentive for CRS Use

As mentioned earlier, affordability seriously affects usage. Therefore, a cost reduction of CRS and a government subsidy may be necessary to facilitate increased CRS purchasing and use (Desapriya et al., 2005). This process would be expensive, but when viewed against greater public health and traffic safety benefits in the long-term, it may be economical.

Strategies to Increase Adult Restraint Use

As shown in the previous literature, child occupant restraint use is highly correlated with adult and caregivers' seatbelt use (Ehiri et al., 2006, Ramsey et al., 2000; Rivara et al., 2001, US Preventative Task Force, 2007; Vesentini and Willems, 2007). Therefore, an effort to increase adult seatbelt use is a viable policy to increase appropriate CRS usage in communities. A meta-analysis by Dinh-Zarr et al., (2001), has identified 6 main interventions to increase seatbelt use by MV occupants as follows: (a) safety belt laws; (b) primary enforcement of these laws (that is, allowing police to stop and charge drivers for not wearing seatbelts, as opposed to charging them only if they are stopped for other offences); (c) enhanced enforcement programs; (d) incentives; (e) mass media; and, (f) education programs (Dinh-Zarr et al., 2001). Perceptions of traffic laws and penalties are more important in influencing behavior than the actual laws and penalties. For this reason, some countries aggressively publicize their seatbelt usage laws and penalties (Shults et al., 2001).

Evaluation studies, summarized by Salzberg et al., (2004) show conclusively that high-visibility enforcement increases seatbelt use and decreases vehicle occupant injuries. In addition, driver's license demerit points may be a more severe sanction than fines, and may be effective in increasing seatbelt use among vehicle drivers and occupants. The principle of a demerit point system is that points accumulate and can lead to additional sanctions for repeated convictions, and ultimately license suspension or revocation. As a strategy to increase vehicle restraint use, we could add additional provisions for the accumulation of demerit points which could have implications for the cost of insurance.

Extensive Law Enforcement

As cited by Dellinger et al., (2002), a recent Harris poll demonstrated that 90% of the public favor stronger enforcement of laws that require all children to be buckled up. A systematic review of the scientific literature by the *Guide to Community Preventive Services Task Force* found that CRS use laws are an effective intervention, and therefore the task force has strongly recommended passage and enforcement of comprehensive child restraint laws (Dellinger et al., 2002; Zaza et al., 2001). Deterrence theory indicates that the most effective combination is tough, fair laws, vigorous enforcement, and intensive and targeted educational campaigns (DeJong and Hingson, 1998; Desapriya et al., 2004; Shults et al., 2001). However, a law on its own is often not sufficient; the key factor in the effectiveness of a traffic law is motorists' perceptions that they run a high risk of being detected and punished for violating the law. The perception of likelihood of apprehension is a much stronger deterrent than the severity or the swiftness of the penalty. However, an incentive or education program to support enforcement may be necessary.

Effective Coordination of Policy and Practice

The effective combination of engineering, education and enforcement (the 3 E's) are required to ensure optimal protection of child occupants (Simons-Morton and Winston, 2006). Therefore, continuous effective approaches are needed to coordinate policy-makers, researchers, engineers, educators, enforcement officers, child safety advocates and the public health community.

Multifaceted Community Campaigns

A multifaceted community campaign conducted by Ebel et al., (2003) was able to significantly increase booster seat use among eligible children. The community campaign shared elements with other successful community-based injury strategies since it had a strong theoretical basis, involved coalition-building, relied on a focused public health message, used focus groups to guide campaign messages and development, and included an evaluation of campaign effectiveness. This program might serve as a useful model for other communities working on increasing the use of appropriate CRS.

In addition, the CDC-P Guide to Community Preventive Services (2001) recommends CRS use laws, community-wide information and enhanced enforcement campaigns, and distribution or incentive programs plus education programs to increase CRS use for infants and children.

The NHTSA recommends education, training, enforcement, outreach, and legislation to increase proper restraint use for children. Effective interventions included education, demonstration of correct use, and child safety seat distribution programs and were tested during a time of growing cultural support and increased regulatory requirements for CRS use. NHTSA currently has 'Ease of Use' ratings on their website to help parents who are unsure of which type of CRS to use (NHTSA, 2008). They use a 5-star rating system to evaluate certain car seat features in 4 basic categories including, evaluation of instructions, evaluation of labels, vehicle installation features, and securing the child. On this site, they also offer free inspections by a certified child passenger safety technician (NHTSA, 2008).

Data from primary care studies were lacking interventions to increase the use of belt-positioning booster seats for children 4 to 8 years of age, an area where interventions are needed because of lower use and gaps in current CRS legislation (CDC-P, 2001).

Emergency Department (ED) Interventions

A recent review by Gittelman and Durbin (2005) suggested that an ED visit for an injury represents a "teachable moment" for the patient and their family, which may make the injured and their caregiver more receptive to educational information.

Based on this, Gittelman and colleagues (2006) conducted a prospective, randomized study that focused on families with children aged 4 to 7 years, weighing 40 to 80 lbs who presented at a pediatric ED without a booster seat and resided in lower socioeconomic communities. Subjects were randomly assigned to one of three groups: 1) received standard discharge instructions, 2) received five-minute booster seat training, and 3) received five-minute booster seat training and a free booster seat with installation (Gittelman et al., 2006).

Gittelman et al., (2006) obtained the participants' automobile restraint practices by telephone after 1 month. A total of 225 children were enrolled in this program. Before randomization in the study, 79.6% of parents reported that their child was usually positioned in the car with a lap/shoulder belt and 13.3% with a lap belt alone. Some parents (16.4%) had never heard of a booster seat, and 44.9% believed a lap belt was sufficient restraint. A total of 147 parents (65.3%) were contacted for follow-up at 1 month. Only 1 parent (1.3%) in the control group and 4 parents (5.3%) in the education group purchased and used a booster seat after their ED visit, while 55 parents (98.2%) in the education and installation group reported using the booster seat; 42 (75%) of these parents reported using the seat 100% of the time (Gittelman, et al., 2006). This study showed that an ED education intervention is a highly successful intervention for increasing booster seat use among lower socioeconomic communities. As reported earlier, another study has found that simply providing a booster seat is not sufficient for CRS use (Brixey et al., 2008). The combination of a free booster seat and the education component appears to be the best strategy.

Enhanced Knowledge for Parents and Caregivers: Correct Use of CRS and LATCH

Educational and awareness programs to improve parents' and caregivers' knowledge on safety benefits of the correct fit or use of a CRS using lower anchor attachment for CRS installation (instead of vehicle safety belts) needs to be developed and evaluated. Future educational interventions should also target older children as well as other individuals who drive with children, not just parents and caregivers (Durbin et al; 2004). Educational materials can be placed in a variety of settings where parents and caregivers have customary access. These settings include primary care offices, community-based organizations, child care settings, emergency departments, and commercial outlets and point of sale locations

Clear Instruction Manuals

Evidence from the above research shows that one possible contributor is poor comprehension of installation instructions. Poor comprehension often occurs when the required reading level of a particular text exceeds the reading capacity of the target population. In this situation, people often become fatigued and discouraged, which may affect compliance (Doak et al., 1996). Even though this factor is highly influential for CRS use and misuse in communities, it is identified as a factor in very few articles that have dealt with child safety and literacy (Powell et al., 2000; Powers, 1988). Parents' and caregivers' education has been cited as highly correlated to proper use of CRS in many countries including the US. Available literature indicates that CRS instructions in the US are currently written at a reading level that is too high. Experts in the area of health literacy recommend that materials such as CRS instruction and installation manuals should be targeted to a 5th or 6th grade reading level (Doak et al., 1996; Lane et al., 2000; Wegner and Girasek, 2003; Weiss and Coyne, 1997; Wil and Geller, 2004).

Given the high rate of CRS misuse and the associated risk of injury, as well as the large variety of CRS and vehicles available for purchase, each requiring slightly different methods of installation, parents and caregivers should receive adequate training to install CRS properly. The complexity of CRS use makes it difficult for parents, in the absence of expert instruction, to use CRS properly. However, few studies have examined how and where parents learn about CRS use, and whether hands-on training is more effective than non-demonstrated educational methods. Ruffin and Kantor (1992) determined that prenatal instructors and manufacturers' information were the best sources of general knowledge about CRS use, but ability to install CRS using this knowledge was not evaluated.

Media Campaigns

A recent study in Canada assessed the effectiveness of a national 1 week media campaign promoting booster seat use and reported that this 1 week campaign promoting the traffic safety benefits of booster seat use, substantially increased self-reported use of booster seats (Howard, 2006). However, the study reported that parents did not remember details of the campaign content, but did remember implications for their own child.

Future Directions For CRS-Related Research

As Simons-Morton and Winston (2006) have shown, research in effective child occupant safety and protection methods requires better coordination between occupant safety social and engineering theories, and laboratory and field research. This will enable better CRS product design. Coordination among researchers, engineers and policy makers is essential if innovations in child occupant injury prevention are to be developed and disseminated. A more comprehensive research agenda is required in rapidly motorizing countries to assess the prevalence of use/misuse barriers for the appropriate use of CRS and explore possibilities to enhance their use. Collaboration of researchers from motorized and rapidly motorizing countries may be beneficial so that information may be shared.

Feasibility of Integrated Child Restraint Systems in Vehicles

To improve the safety of CRS as well as reduce the misuse associated with their installation, the integration of CRS into the rear seats of vehicles has been increasingly recommended by safety advocates and traffic safety researchers. Comprehensive legislation reinforced by consumer education that is directed to parents and caregivers should emphasize that children should always travel in the rear seat of the vehicle. Integrated CRS are currently the safest known child safety systems for children between the ages of 4 and 8 years. The development of simple, improved availability, and a safer CRS, such as those integrated in the vehicle, is one initiative that can significantly improve current child occupant protection (Grantz, 2002).

Effective Communication and Dissemination of Best Practices

A study by Will and Geller (2004) found that there are 4 million CRS under manufacturer recall, that have yet to be returned by parents or caregivers for repair or replacement (Will and Geller, 2004). This shows that communication between parents, caregivers and government agencies responsible for child occupant injury prevention is lacking or ineffective. It is necessary to have improved communication strategies in place between CRS manufacturers, agencies that oversee safety aspects of CRS, and parents and caregivers in order to prevent child occupant injuries in the future. Importantly, the study by Will and Geller (2004) found that the greatest hurdle in educating for the proper use of CRS to parents and caregivers is the lack of parental participation in available safety intervention programs. Parental participation rates in these safety-seat checkpoints are disappointingly low; far below the needs shown by high rates of misuse. This portrays parental lack of interest in readily available information on the correct use of CRS.

Design Modifications

Since the correct use of CRS represents such a complex behavior, the best strategy to address the widespread problem of CRS use may be through the design of a less complicated CRS (Margolis et al., 1992). As Simons-Morton and Winston (2006) have discussed, we need to remove the barriers before any innovative CRS that can simplify the appropriate use and give maximum safety benefits and comfort to the child occupants, can be developed and utilized. Adoption of many potential safety enhancements to CRS have been precluded by a litigious US society (Simons-Morton and Winston, 2006). According to a study by Hummel which was quoted by Gantz, (2002), regardless of the type of restraint used, children are far more likely to be injured in side impact crashes than head-on crashes (Gantz, 2002). Therefore, urgent design enhancement is required to appropriately protect children in side impact crashes. This should address design enhancements of the CRS in the possible rotation during lateral impacts (Arbogast et al., 2004). In addition childhood obesity has safety implications beyond the obvious health risks. A study by Trifiletti et al., (2006) projects that there are nearly 300,000 children whose age-weight combinations place them at risk for difficulty in obtaining an appropriate CRS. The design changes of CRS have now become very important as conventional CRS and related guidelines will not provide maximum safety to obese children in a crash.

European Transport Safety Council (ETSC) Directives

The recent ETSC document (2003) has highlighted that frontal and side impact directives and consumer information from the European New Car Assessment Program (EuroNCAP) have led to the most rapid developments in vehicle occupant protection that Europe has experienced, but a great deal more can be achieved. In addition, ETSC has highlighted that even though research and development activity necessary for improvements in other areas of vehicle safety has been introduced, this now requires the political will to bring about its adoption in legislation. In addition, collaboration between European and North American countries would be mutually beneficial as a collaborative partnership would be cost-effective.

Dissemination of CRS Information to Rapidly Motorizing Countries

The information related to traffic safety benefits of CRS should be disseminated to rapidly motorizing countries to help reduce child occupant injuries and fatalities in the future. As we have mentioned previously in this chapter, CRS use in these countries remains dangerously low. As a first step, motorized countries could invest in training child safety seat technicians in these countries. There should be some mechanism in place to manufacture low cost CRS utilizing local resources. In addition, CRS loan programs should be implemented at selected sites in rapidly motorizing countries. We can select cities from Asia, Africa, Europe and South America to conduct a pilot study to examine the effectiveness of this program on the appropriate restraint use of vulnerable children in these economically disadvantaged countries. This project could yield long-term traffic safety benefits for motorized countries. Previous literature shows that some groups of children are at greater risk than others. CRS use is lower among First Nations populations, other ethnic groups (Mexican-Americans, African-Americans), rural populations and low income families (NHTSA, 2001). The proposed pilot projects in developing countries will enhance our knowledge concerning the ways and effective means of reaching these vulnerable ethnic and socially deprived groups in our societies, in an effort to influence them to make the right decision for their children's traffic safety.

Research Agenda Priorities

Based on the available evidence, we have determined some priorities for future research. This will require research teams from various industries to collaborate with the common goal of a reduction in CRS misuse through improved research and action. Some future research priorities include:

- Performance concerns regarding side air bags (with the growing number of side protection airbags fitted in cars, MVC research is required to identify their benefits and any associated problems)
- Side impact compatibility and advanced child occupant protection
- Development of advanced intelligent restraints
- More comprehensive biomechanical data, injury performance criteria, sophisticated child crash-test dummies
- Protection in side impacts at higher severities and for non-struck side occupants
- How to better improve the structural integrity of the vehicle and the CRS

REFERENCES

American Academy of Pediatrics (AAP) (2002). TIPP: the injury prevention program. A guide to safety counseling in office practice. Age-related safety sheets. Available at: http://www.aap.org/family/tippmain.htm. (Accessed Feb. 9, 2008).

Arbogast KB, Ghati Y, Menon RA, Tylko S, Tamborra N, Morgan RM. Field investigation of child restraints in side impact crashes. *Traffic Inj Prev.* 2005;6(4):351-60.

Arbogast KB, Durbin DR, Cornejo RA, et al. An evaluation of the effectiveness of forward facing child restraint systems. *Accid Analysis Prev.* 2004;36:585-589.

Arbogast KB, Jermakian JS. Field use patterns and performance of child restraints secured by lower anchors and tethers for children (LATCH). *Accid Anal Prev.* 2007;39(3):530-5.

Agran R, Dunkle D, Winn D. Effects of legislation on motor vehicle injuries to children. *Am J Dis Child.* 1987;141:959-64.

Blue Ribbon Panel on Child Restraint and Vehicle Compatibility (BRPCRVS, 1995). Blue Ribbon Panel on Child Restraint and Vehicle Compatibility: Recommendations. Washington, DC: Blue Ribbon Panel on Child Restraint and Vehicle Compatibility; May 30, 1995.

Britax/GMTV Campaign. Available at: http://www.britax.co.uk/gmtv_news.htm.

Brixey S, Ngui E, Guse C. Disparities in child passenger restraint use in an urban center. Paper presented at the 9th World Conference on Injury Prevention and Safety Promotion, Merida, Mexico, March 15-18th, 2008.

Brown J, Bilston L, MacCaskill M, Henderson M. Identification of injury mechanisms in children aged 2-8 years in motor vehicle accidents. Report for the Motor Accidents Authority, NSW. 2005.

Bull MJ, Stroup KB, Gerhart S. Misuse of car safety seats. *Pediatrics.* 1988;81:98-101.

Cameron LE, Segedin G, Nuthall A, Thompson J. Safe restraint of the child passenger. *J. Paediatr Child Health.* 2006;42:752–7.

Centers for Disease Control and Prevention (CDC-P). (2000). Occupant fatalities and restraint use among children aged 4-8 years. *MMWR.* 2000;49(7):135-7.

CDC-P. (2001). Motor-vehicle occupant injury: Strategies for increasing use of child safety seats, increasing use of safety belts and reducing alcohol-impaired driving – a report on recommendations of the Task Force on Community Preventive Services. Morbidity and Mortality Weekly Report, 50 (RR-7).

CDC-P. (2002). Child passenger safety. Updated March 25, 2002. Available at: http://www.cdc.gov/safeusa/move/childpassenger.htm

CDC-P. National Child Passenger Safety Week--February 14-20. *MMWR.* 1999;48;83-4.

Charlton JL, Fildes B, Laemmle R, Smith S, Douglas F. A preliminary evaluation of child restraints and anchorage systems for an Australian car. *Annu Proc Assoc Adv Automot Med.* 2004;48:73-86.

Charlton JL, Koppel S, Fitzharris M, Congiu M, Fildes B. Factors that influence children's booster seat use. Report No. 250 April 2006. MONASH University Accident Research Centre.

Cody BE, Mickalide AD, Paul HP, Colella JM. *Child passengers as risk in America: A national study of restraint use.* National SAFEKIDS Campaign, Washington, D.C., February 2002.

Cohen LR, Runyan CW, Downs SM, Bowling JM. Pediatric injury prevention counseling priorities. *Pediatrics.* 1997;99;704-10.

Decina LE, Knoebel KY. Child safety seat misuse patterns in four states. *Accid Anal Prev.* 1997;29(1):125-32.

Decina LE, Lococo K. Misuse of Child Restraints (USDOT/NHTSA Publication Number DOT HS 809 671), U.S. Department of Transportation, Washington, DC (2004).

Decina LE, Lococo K, Block A. Misuse of Child Restraints: Results of a Workshop to Review Field Data Results. USDOT/NHTSA Publication Number DOT HS 809 851, Traffic Safety Facts — Research Note, U.S. Department of Transportation, Washington, DC (2005).

Decina LE, Lococo KH. Observed LATCH use and misuse characteristics of child restraint systems in seven states. *J Safety Res.* 2007;38(3):273-81.

DeJong W, Hingson R. Strategies to reduce driving under the influence of alcohol. *Annual Review Public Health.* 1998;19:359–78.

Dellinger AM, Groff PC, Mickalide AD, Nolan PA. Kids in cars: closing gaps in child occupant restraint laws. *J Law Med Ethics.* 2002;30:150-6.

Desapriya EBR, Iwase N, Pike I, Brussoni M, Papsdorf M. Child motor vehicle occupant and pedestrian casualties before and after enactment of child restraint seats legislation in Japan. *Injury Control and Safety Promotion.* 2004;11(4):225-30.

Desapriya EB, Pike I, Joshi P. Prices and affordability in child restraint seats in Japan. *Inj Prev.* 2005;11(2):125.

Desapriya E, Pike I, Raina P. Severity of alcohol-related motor vehicle crashes in British Columbia: case - control study. *Int J Inj Contr Saf Promot.* 2006;13(2):89-94.

Desapriya E, Fujiwara T, Pike I, Babul-Wellar S, Scime G. Evaluation of compulsory child restraint seat law to reduce motor vehicle child occupant death in Japan 1994-2005. *Injury Prevention and Safety Promotion* 2008;15(2) 93-97.

Dinh-Zarr T, Sleet D, Shults RA, et al. Reviews of evidence regarding interventions to increase the use of safety belts. *American Journal of Preventive Medicine.* 2001;21:48–65.

Doak LG, Doak CC, Meade CD. Strategies to improve cancer education materials. *Oncol Nurs Forum.* 1996;23(8):1305-12.

Durbin DR, Arbogast KB, Moll EK. Seat belt syndrome in children: a case report and review of the literature. *Pediatric Emergency Care.* 2001;17(6):474-7.

Durbin DR, Elliott MR, Winston FK. Belt-positioning booster seats and reduction in risk of injury among children in vehicle crashes. *JAMA.* 2003;289(21):2835-40.

Durbin, DR., Chen, I., Elliott, M., Winston, FK. Factors associated with front row seating of children in motor vehicle crashes. Epidemiology. 2004;15, 345-349.

Ebel BE, Koepsell TD, Bennett EE. Too small for a seatbelt: predictors of booster seat use by child passengers, *Pediatrics.* 2003;111:323-7.

Eby DW, Kostyniuk LP. A statewide analysis of child safety seat use and misuse in Michigan. *Accid Anal Prev.* 1999;31(5):555-66.

Ehiri, J.E. Ejere, H.O., Magnussen, L., Emusu, D., King, W., Osberg, J.S. Interventions for promoting booster seat use in four to eight year olds traveling in motor vehicles. Cochrane Database Syst Rev. 2006;25;(1):CD004334.

European Transport safety Council (ETSC) 2003-Priorities for EU motor vehicle safety designs- Available at: http://www.etsc.be/oldsite/rep_road7.htm (Accessed 21 Jan. 2008).

Garrett JW, Braunstein PW. The seat belt syndrome. *J Trauma.* 1962;2:220-38.

Gittelman MA, Durbin D. Injury prevention: Is the pediatric Emergency Department the appropriate place? *Pediatric Emergency Care.* 2005;27(7):460-7.

Gittelman MA, Pomerantz WJ, Laurence S. An emergency department intervention to increase booster seat use for lower socioeconomic families. *Acad Emerg Med.* 2006;13:396-400.

Gantz T. (2002). Child occupant protection –current issues-Prevention Institute. Available at: http://www.preventioninstitute.org/traffic_child.html (Accessed 7th Feb. 2008).

Greenberg-Seth, J., Hemenway, D., Gallagher, SS., Ross, JB., Lissy , KS., 2004. Evaluation of a community-based intervention to promote rear seating for children. Am J Public Health. 2004; 94 (6), 1009-1013.

Grossman DC, Garcia CC. Effectiveness of health promotion programs to increase motor vehicle occupant restraint use among young children. *Am J Prev Med.* 1999;16:12-22.

Howard A. Automobile restraints for children: a review for clinicians. *CMAJ.* 2002;167(7):769-73.

Howard AW. Injury in childhood: a vexingly simple problem. *CMAJ.* 2006;10;175(8):899.

Insurance Institute of Highway Safety (IIHS). American Child Restraint Laws. http://www.iihs.org/laws/childrestraint.aspx reprinted with permission from Insurance Institute of Highway Safety (2008).

Insurance Institute of Highway Safety (IIHS) (2003). Latch systems for child restraints aren't always a snap; not every child restraint will work in every vehicle. Available at: http://www.highwaysafety.org/news_releases/2003/pr061103.htm.

Japanese Automobile Federation (JAF) (2006). JAF Chronology. Available at: http://www.jaf.or.jp/e/chrono.htm.

Johnston C, Rivara FP, Soderberg R. Children in car crashes: analysis of data for injury and use of restraints. *Pediatrics.* 1994;93:960-5.

Klinich K, Pritz H, Beebe M, Welty K. Survey of older children in automotive restraints. In: 38th Stapp Car Crash Conference; 1994; Fort Lauderdale, FL, USA: Society of Automotive Engineers; 1994; 245-64.

Kohn M, Chausmer K, Flood MH. Anticipatory guidance about child safety seat misuse: lessons from safety seat "checkups. *Arch Pediatr Adolesc Med.* 2000;154:606-9.

Koppel SK, Charlton JL, Fitzharris MC, Fildes B. Factors associated with the premature graduation of children into seatbelts. *Accid Anal Prev.* 2008;40(2):657-66.

Kulowski K, Rost W. Intra-abdominal injury from safety belts in auto accidents. *Archives of Surgery*. 1956;73:970-1.

Kunkel NC, Nelson DS, Schunk JE. Do parents choose appropriate automotive restraint devices for their children? *Clin Pediatr*. 2001;40(1):35-40.

Lane WG, Liu GC, Newlin E. The association between hands-on instruction and proper child safety seat installation. *Pediatrics*. 2000;106:924-9.

Lee KC, Shults RA, Greenspan AI, Haileyesus T, Dellinger AM. Child passenger restraint use and emergency department-reported injuries: a special study using the National Electronic Injury Surveillance System – All Injury Program, 2004. *J Saf Res*. 2008;39:25-31.

Lennon A. Where do children sit in Australian passenger vehicles? Results of an observational study. Proceedings of the Australasian Road Safety Research, Policing and Education Conference, Nov. 26-27, Wellington, NZ. 2005.

Lennon A. A risky treat: exploring parental perceptions of the barriers to seating their children in the rear seats of passenger vehicles. *Inj Prev*. 2007;13(2):105-9.

Margolis LH,, Wagenaar AC, Molnar LJ. Use and misuse of automobile child restraint devices. *American Journal of Disabled Child*. 1992;146:361-6.

Miller TR, Zaloshnja E, Hendrie D. Cost-outcome analysis of booster seats for auto occupants aged 4 to 7 years. *Pediatrics*. 2006;118(5):1994-8.

Mock C, Arreola Rissa C, Trevino Perez R, Almazan Saavedra V, Enrique Zozaya J, Gonzalez Solis R, Simpson K, Hernandez Torre M. Childhood injury prevention practices by parents in Mexico. *Inj Prev*. 2002;8(4):303-5.

National Highway Traffic Safety Administration (NHTSA). (1995). Research Note: Revised Estimates of Child Restraint Effectiveness. Report 96.855. Washington, DC: National Highway Traffic Safety Administration; 1995.

NHTSA. (1996). Third report to Congress on the effectiveness of occupant protection systems and use. Available qat http://www.nhtsa.dot.gov/people/injury/ airbags/ 208con2e.html.

NHTSA. (2000). A chronology of NHTSA actions and activities on child safety seats. Washington (DC): National Highway Traffic Safety Administration.

NHTSA. (2001). Seat Belts and Hispanics: 2000 Report. DOT HS 809 045. Washington, DC: National Highway Traffic Safety Administration; 2001.

NHTSA. (2002). A national strategy: increasing booster seat use for 4-to 8-year old children. Available at: http://www.nhtsa.dot.gov/CPS/booster_seat/National_Strategy/

NHTSA.(2003). Motor Vehicle Occupant Safety Survey. Washington (DC): National Highway Traffic Safety Administration.

NHTSA. (2004). NHTSA child safety seat information. Washington (DC): National Highway Traffic Safety Administration. Available at: http://www.alavoices.org/Files/AlaVoices/documents/CPS-NHTSA%202004%20Child%20Safety%20Seat%20Guidelines.pdf.

NHTSA. (2005). NHTSA Vehicle Safety Rule Making and Supporting Research Priorities: Calendar Years 2005-2009. at: http://www.nhtsa.gov/cars/rules/rulings/PriorityPlan-2005.html#VI. (Accessed 19th October 2007).

NHTSA. (2006). Programs and activities to increase booster seat use. Washington (DC): National Highway Traffic Safety Administration. Available at:

http://www.nhtsa.dot.gov/people/injury/childps/BoosterSeatProgress/pages/NHTSALed.htm (Accessed 4th February 2008). (US Library of Congress, S.980 "Anton's Law", passed December 2003).

NHTSA. (2007). Recommends that children should be at least 145 cm tall to make the transition from a booster seat to a seatbelt (NHTSA, 2007) Booster Seat web page. Available at: www.BoosterSeat.gov. (Accessed May 2007).

NHTSA. (2008). Ease of Use Ratings. Child Passenger Safety web page. Available at: http://www.nhtsa.gov/portal/site/nhtsa/template (Accessed 10th March 2008).

National Occupant Restraint Program (NORP) 2010, Annual Monitoring Report- 2004. National Occupant Restraint Program (NORP) Annual Monitoring Report- 2004

Available at: http://www.ccmta.ca/english/norp/pdf/norp_report03.pdf (Accessed 12th January 2008).

New Programme for the Assessment of Child-restraint Systems (NPACS). Available at:http://www.npacs.com/ (Accessed 11th October 2007).

New Zealand Ministry of Transport, 2006. Safety belts. Available at: http://www.transport.govt.nz/belts-index. (Accessed 7th February 2008).

Ong MEH, Ooi SBS, Manning PG. Reviews of 2517 childhhood injuries seen in a Singapore Emergency department in 1999. Mechanism and injury prevention suggestion. *Singapore Med J.* 2003;44(1):12-9.

Pampers. Available at: http://www.pampers.com/en_GB/home/jhtml/index.jhtml?_requestid=30657.

Peden MM, Scurfield R, Sleet DA, Mohan D, Hyder AA, Jarawan E, Mathers C, Eds, 2004. World Report on Road Traffic Injury Prevention, Health Organization, Geneva.

Powell EC, Tanz RR, Uyeda A, Gaffney MB, Sheehan KM. Injury prevention education using pictorial information. *Pediatrics.* 2000;105(1):e16.

Powers RD. Emergency Department Patient Literacy and the Readability of Patient-Directed Materials. *Ann of Emergency Medicine.* 1988;17:124-6.

Ramsey A, Simpson E, Rivara FP. Booster seat use and reasons for nonuse. *Pediatrics.* 2000;106(2):20-5.

Reeve KN, Zurynski YA, Elliott EJ, Bilston L. Seatbelts and the law: how well do we protect Australian children? *Med J Aust.* 2007;186(12):635-8.

Rivara FP, Bennett E, Crispin B, Kruger K, Ebel E, Sarewitz A. Booster seats for child passengers: lessons learned for increasing their use. *Inj Prev.* 2001;7(3):210-3.

Roadmap to State Highway Safety Laws. Roadwork Ahead: The Unfinished Safety Agenda. HS-043 742, 1-69. Washington, DC: Advocates for Highway and Auto Safety; 2004.

Rothenstein J, Howard A, Parkin P, Khambalia A, Macarthur C. Community pediatricians' counseling patterns and knowledge of recommendations relating to child restraint use in motor vehicles. *Inj Prev.* 2004;10(2):103-6.

Rowling M. Japan approves child safety seats. *Lancet.* 1999;353:9168.

Ruffin MT, Kantor R. Adults' knowledge about the use of child restraint devices. *Fam Med.* 1992;24(5):382-5.

Salzberg PM, Moffat JM. Ninety-five percent: An evaluation of law, policy, and programs to promote seat belt use in Washington State. *J Safety Research.* 2004;35(2):215–22.

Scottish In-Car Child Safety Initiative. Available at: www.protectchild.co.uk/protectchild03/pages/03pages/frame03.html www.scotland.gov.uk/library5/transport/sbws-09.asp.

Sehgal A, Jain S, Jyothi MC. Parental awareness regarding childhood injuries. *Indian J Pediatr.* 2004;71(2):125-8.

Shults RA, Elder RW, Sleet DA, et al. Review of evidence regarding interventions to reduce alcohol-impaired driving. *Am J Preventive Med.* 2001;21(Suppl. 4): 66–88.

Sibbald B. Canada ranks in middle in child-injury mortality rate, report indicates. *CMAJ.* 2001;15;164(10):1483.

Simons-Morton BG, Winston FK. Translational research in child and adolescent transportation safety. *Eval Health Prof.* 2006;29:33 DOI: 10.1177/0163278705284442.

Snowden A. (2006). Canadian children in vehicle safety-Available at:
http://www.auto21.ca/news/Canadian_Children_In_Vehicle_Safety_Study.pdf (accessed 11 Feb 2008).

Spanier AJ, Mercante D, Barkemeyer B. Child safety seat knowledge among postpartum mothers in an urban setting. *South Med J.* 2002;95:1017-21.

Stapp JR. Human tolerance to deceleration. *American Journal of Surgery.* 1957;93(4):734-40.

Stewart D, Lang NJ, Emery S. LATCH- Lower Anchors and Tethers for Child Restraints (Fourth Edition) (2005) Seattle, WA.

Taft CH, Mickalide AD, Taft AR. *Child passengers at risk in America: A national study of car seat misuse.* National SAFEKIDS Campaign, Washington, D.C., February, 1999.

Transport Canada (1998). Child restraint use in Canada: 1997 survey data. Road Safety Leaflet CL 9804, TP 2436: Transport Canada available on Transport Canada website – www.tc.gc.ca/roadsidesafety/tp2436/cl9804/menu_e.htm.

Trifiletti LB, Shields W, Bishai D, McDonald E, Reynaud F, Gielen A. Tipping the scales: obese children and child safety seats. *Pediatrics.* 2006;117(4):1197-202.

Turner C, McClure R, Nixon J, Spinks A. Community-based programs to promote car seat restraints in children 0-16 years – a systematic review. *Accid Anal Prev.* 2005;37(1):77-83.

United Nations Children's Fund (UNICEF). (2001). Innocenti Report Card No.2: A league table of child deaths by injury in rich nations. UNICEF Innocenti Research Centre Florence, Italy.

United Nations Economic Commission for Europe (UNECE). (2003). Working party on Road Traffic Safety. Available at: http://www.unece.org/trans/main/wp1/wp1fdoc/TRANS-WP1-80r2e.pdf (Accessed 7th Feb. 2008).

United Nations Economic Commission for Europe (UNECE). (2005) Synthesis of replies to questionnaire on child restraint use (2005) Available at: http://www.unece.org/-trans/doc/2005/wp1/WP1-47-INF1cr1e.pdf (Accessed 7th Feb. 2008).

United Kingdom (UK) Department of Transportation (2006). Available at: http://www.-thinkroadsafety.gov.uk/campaigns/childcarseats/faq21.htm (accessed 21 Jan. 2007).

United States (US) Preventative Services Task Force. Counseling about proper use of motor vehicle occupant restraints and avoidance of alcohol use while driving: US Preventive Services Task Force recommendation statement. *Ann Intern Med.* 2007;7;147(3):187-93.

Vaca F, Anderson CL, Agran P, Winn D, Cheng G. Child safety knowledge among parents utilizing emergency services in a Level 1 Trauma Center in Southern California. *Pediatrics.* 2002;110(5):e61.

Valent F, Barbone F. Automotive child restraint systems in Northeastern Italy. *Acta Paediatrica.* 2003;92(8):958–64.

Vesentini L, Willems B. Premature graduation of children in child restraint systems: an observational study. *Accid Anal Prev.* 2007;39(5):867-72.

Weber K. Child passenger protection. In: Nahum AM, Melvin JW, eds. Accidental Injury: Biomechanics and Prevention. New York, NY: Springer-Verlag; 1993:493-511.

Wegner M, Girasek D. How readable are child safety seat instructions? *Pediatrics & Child Health.* 2003;111(3):588-91.

Weiss BD, Coyne C. Communicating with Patients Who Cannot Read. *New Engl J Med.* 1997; 337(4):272-4.

Will KE, Geller ES. Increasing the safety of children's vehicle travel: from effective risk communication to behavior change. *J Safety Res.* 2004;35:263–74.

Williams AF. Comment on occupant and licensing interventions. *Am J Prev Med.* 1999;16:6-8.

Williams SB, Whitlock EP, Edgerton EA, Smith PR, Beil TL. US Preventive Services Task Force. Counseling about proper use of motor vehicle occupant restraints and avoidance of alcohol use while driving: a systematic evidence review for the US Preventive Services Task Force. *Ann Intern Med.* 2007;7;147(3):194-206.

Winston FK, Baxt C, Kassam-Adams NL, Elliott MR, Kallan MJ. Acute traumatic stress symptoms in child occupants and their parent drivers after crash involvement. *Arch Pediatr Adolesc Med.* 2005;159(11):1074-9.

Winston FK, Chen IG, Elliott MR, Arbogast KB, Durbin DR. Recent trends in child restraint practices in the United States. *Pediatrics.* 2004;113(5):e458-64.

Winston IG, Chen R, Smith MR. Parent driver characteristics associated with sub-optimal restraint of child passengers, *Traffic Injury Prev.* 2006;7;373–80.

Winston FK, Durbin DR. Buckle up! Is not enough. *JAMA.* 1999;281(22):2070-2.

Winston FK, Durbin DR, Kallan MJ, Moll EK. The danger of premature graduation to seat belts for young children. *Pediatrics.* 2000;105:1179–83.

Winston FK, Kallan MJ, Elliot MR, Xie D, Durbin DR. Effect of booster seat laws on appropriate restraint use by children 4 to 7 years old involved in crashes. *Arch Pediatr Adolesc Med.* 2007;161:270-5.

Winston FK, Rineer C, Menon R, Baker S. The carnage wrought by sudden economic change: Traffic-related mortality and the unification of Germany. *BMJ.* 1999;318:1647-50.

World Health Organization (WHO). (2002). A 5-year WHO Strategy for road traffic injury prevention. Available at: http://www.who.int/violence_injury_prevention.

Zaza S, Sleet DA, Thompson RS, Sosin DM, Bolen JC. Regarding interventions to increase use of child safety seats. *Am J Prev Med.* 2001;4:31-47.

In: Consumer Product Safety Issues
Editors: P. A. Varga, M. D. Pintér

ISBN: 978-1-60456-826-4

Chapter 2

SAFER STORAGE OF FIREARMS AT HOME AND RISK OF SUICIDE: A CONSUMER PRODUCT SAFETY ISSUE

Edmond D. Shenassa[1,*], Michelle L. Rogers[1] and Kirsten L. Spalding[1]

[1]Department of Community Health, Brown Medical School, Providence, RI, USA

ABSTRACT

Despite the ubiquity of firearms in American homes, and the abundance of evidence linking access to firearms at home with the risk of suicide by firearm, information on consumer product safety issues and policies, such as placement of triggers locks, that may protect household members against risk of suicide by firearm is conspicuously sparse. Here we report results from the first study to estimate the protective effect of storing firearms locked and/or unloaded (i.e., safer storage practices). We analyzed data from the latest National Mortality Followback Survey (1993 NMFS), a nationally representative survey of next-of-kin of 22,957 decedents representing 2.2 million individuals. In our analysis we partially control for respondents' intention to die and focus on suicides that are likely to have been impulsive. Relative to decedents who did not practice safer storage of firearms, risk of suicide by firearms was less 67% among those who store their firearms locked and/or unloaded. Our findings strongly support the utility of consumer product safety devices and regulations intended to reduce the likelihood of unauthorized or impulsive use of firearms.

---

* Correspondence to: Edmond D. Shenassa, Sc.D., Epidemiology Section, Department of Community Health, Brown Medical School, 121 S. Main St., Providence, RI 02912 , Email: Edmond_Shenassa@Brown.edu

Safer Storage of Firearms at Home: A Consumer Product Safety Issue

Approximately 45% of American households own a firearm; approximately 75% of handgun owners and 38% of firearm owners claim protection of themselves and their property as the primary reason for storing firearms at home [1-3]. Access to firearms at home is linked with household members' risk of violent death [4-9], particularly suicide. Ecological studies consistently link state, regional, and national levels of firearm availability with *firearm-related* suicide rates [10-27] as well as *overall* suicide rates [13, 16, 17, 21, 25-27], particularly among people under age 25 [10, 27, 28] (Table 1). Individual-level studies commonly, but not universally [29, 30] , link firearm ownership with an elevated risk of suicide (Table 2). Even more convincingly, studies of firearm-related suicides uniformly report that among adults, risk of suicide by firearm is 3 to 12 times higher among handgun purchasers as compared to controls [31-33]. Similarly, adolescents with access to firearms at home are 4 times more likely to commit suicide by firearm than other adolescents [34]. Most persuasively, in a nationally representative sample, access to firearms at home was associated with a 17-fold increase in the risk of suicide by firearm relative to those without such access [9].

Despite the ubiquity of firearms in American homes, and the abundance of evidence linking access to firearms at home with the risk of suicide by firearm, information on consumer product safety issues and policies, such as placement of trigger locks, that may protect household members against risk of suicide by firearm is conspicuously sparse, and the sum of the available information is inconclusive [34-40] (Table 3). The inconsistency of findings from these studies can be attributed to their focus on small geographic areas, small sample sizes, and in some instances, under-representation of racial and ethnic minorities (see [41, 42] for reviews).

Here we report findings from the first study to address these shortcomings by analyzing data from a nationally representative sample to estimate the protective effect of storing firearms locked and/or unloaded and to consider, in the analysis, respondents' intention to die. Furthermore, we focus on suicides that are likely to have been impulsive. Evidence suggests that as many as half of all suicides are contemplated for less than five minutes prior to the suicide attempt [43-45]. In this study we estimate the protective effect of safer firearm storage practices (i.e., storing firearms locked and/or unloaded) on the risk of suicide by firearms among people with relatively low intention to die.

Methods

Sample

We analyzed data from the latest round of the National Mortality Followback Survey (1993 NMFS). This survey, conducted by the National Center for Health Statistics, uses a sample of US residents who die in a given year to supplement information from the death certificate with information from the next-of-kin (hereafter, informant) or another person

Table 1. Ecological Studies of the Link between Firearm Availability and Suicide.

Authors	Unit of analysis/ population and measure of suicide	Measure of firearm availability	Findings
Birckmayer & Hemenway, 2001	Suicide and firearm suicide rates among those ages 15-24, 25-44, 45-64, and 65-84 in 9 U.S. census regions – 1979-1994 (NCHS MF)	Proportion of households with firearms (NORC)	Prevalence of household firearm ownership predicted overall suicide rate among those 15-24 (p<.05) and 65-84 (p<.05) years old, and firearm suicide rate among those 15-24 (p<.05), 45-64 (p<.05), and 65-84 (p<.05) years old
Clarke & Jones, 1989	U.S. suicide and firearm suicide rates –1959-1984 (*Vital Statistics*)	Proportion of households with handguns (Gallup; NORC)	Prevalence of handgun ownership predicted the rate of firearm suicides (p<.01), but not the overall rate of suicide
Hemenway & Miller, 2002	Suicide and firearm suicide rates in 9 U.S. census regions – 1988-1997 (NCHS MF)	Proportion of households with handguns (NORC)	Across the 9 regions, household handgun ownership rates predicted firearm suicide (p<.01) and overall suicide rates (p<.05)
Kaplan & Geling, 1998	Firearm suicide and homicide rates in 9 U.S. census regions – 1989-1991 (NCHS MF)	Prevalence of firearm ownership (NORC)	Prevalence of firearm ownership predicted firearm suicide among whites (p<.05) and black males (p=.02); gun ownership has a stronger impact on firearm suicides than homicides
Kellermann & Reay, 1986	Number of firearm suicides relative to all firearm deaths that occurred at home in King County, Washington – 1978-1983	Whether a firearm was kept in the home	For every self-protection homicide involving a firearm kept in the home, there were 37 suicides involving firearms
Kellermann et al., 1998	Number of firearm-related attempted and completed suicides relative to all fatal and nonfatal shootings in/around a home in Memphis, Tennessee; Seattle, Washington; and Galveston, Texas – Nov. 1992–May 1994	Whether the gun involved was kept in the home	For every self-defense/legally justifiable shooting, there were 11 attempted or completed suicides
Killias, 1993	Proportion of suicides with a firearm, rate of firearm suicide, and overall suicide rate in 14 countries – 1983-1986	Proportion of households reporting gun ownership (International Crime Survey, 1989)	Prevalence of household gun ownership predicted proportion of suicides with a gun (p<.01), rate of firearm suicide (p<.01), and overall suicide rate (p<.05)
Kleck & Patterson, 1993	Suicide and firearm suicide rates in 170 U.S. cities – 1979-1981	Proportion of firearm related homicides (1979-82); aggravated assaults (1979-80), and robberies (1979-80); the dollar value of stolen property reported to the police which was due to firearms thefts, 1979-81	Gun prevalence predicted total suicide rate (p<.05) and firearm suicide rate (p<.01)
Authors	Unit of analysis/ population and measure of suicide	Measure of firearm availability	Findings

Table 1. (Continued)

Authors	Unit of analysis/ population and measure of suicide	Measure of firearm availability	Findings
Lester, 1988	Suicide rates by firearm and other methods in 9 U.S. census regions – 1970 (Vital Statistics of the United States, 1970)	Proportion of homicides and suicides committed with firearms; accidental death rate from firearms; strictness of state handgun control laws; and subscription rates to firearm magazines	States with a higher availability of firearms had higher firearm suicide rates
Lester, 1989	Suicide and firearm suicide rates in the 48 continental U.S. states – 1980 (NCHS)	Per capita circulation of 3 firearm magazines, Shooting Times, Guns and Ammo, and American Handgunner	States with a higher per capita circulation of firearm magazines had higher suicide and firearm suicide rates ($p<.01$)
Lester, 1990	Suicide rates by firearm and other methods in 20 countries – 1980	Proportion of homicides by gun	Proportion of homicides using guns predicted firearm suicide rate ($p<.05$)
Lester, 2000	Firearm suicide rates in 36 nations – circa 1993 (76)	Accidental firearm mortality rate; proportion of homicides involving firearms	Accidental firearm mortality rate was not associated with the firearm suicide rate; proportion of homicides involving firearms predicted the firearm suicide rate ($p<.05$)
Lester, 2001	Percentage of suicides using firearms among total Canadian population and those aged 0-11, 12-17, 18-34, 35-54, and 55+ years – 1970-1995	Rate of accidental death from firearms	Rate of accidental death from firearms predicted percentage of suicides using firearms for all age groups ($p<.05$) except 55+
Bridges, 2002	Replicated Lester, 2001, using 3 additional years of data (1996-1998)	Rate of accidental death from firearms	Rate of accidental death from firearms predicted percentage of suicides using firearms for all age groups ($p<.01$) except 55+
Miller et al., 2002	Suicide and firearm suicide rates for women in all U.S. states – 1988-1997 (NCHS MF)	Household gun ownership rates (BRFSS; NORC General Social Surveys); Cook's Index; and fraction of all suicides that involved a gun	IRRs for firearm suicide in states with high gun levels to states with low gun levels: Cook's Index 1.64 (1.39-1.94) FS/S 2.01 (1.83-2.22) BRFSS 1.88 (1.44-2.44); MRR of high to low gun states: 6.5 for gun suicides, 1.5 for all suicides
Miller et al., 2002	Suicide and firearm suicide rates for 5-14 year olds in all U.S. states – 1988-1997 (NCHS MF)	Household gun ownership rates (BRFSS; NORC); Cook's Index; and FS/S	IRRs for firearm suicide in states with high gun levels to states with low gun levels: Cook's Index 1.48 (1.23-1.79) FS/S 1.64 (1.40-1.92) BRFSS 1.67 (1.29-2.17); MRR of high to low gun states: 6.7 for gun suicides, 2.0 for all suicides
Miller et al., 2002	Suicide and firearm suicide rates for all U.S. states, 9 census regions – 1988-1997 (NCHS MF)	Proportion of households with guns and handguns (NORC); fraction of all suicides that involved a gun	IRRs, state level: suicide 1.13 (1.06-1.19), firearm suicide 1.43 (1.34-1.53)

Table 2. Individual-Level Studies of the Link Between Firearm Availability & Suicide.

(a) Adult Individual-Level Studies

Authors	**Number of suicide victims (cases)**[1]	**Comparison group**	**Outcome assessed/ Exposure**	**OR (CI)*/p-value**
Bailey et al., 1997	120 females	120 matched community-dwelling controls	RS among women/ ≥ One gun in home	4.6 (1.2-17.5)
Beautrais et al., 1996	197	1028 community-dwelling controls	RS/ Access to a firearm at home	1.4 (0.96-1.99)
Conwell et al., 2002	86, ages 50 and over	86 matched community-dwelling controls	RS/ ≥ One gun in home	3.23 (1.15-11.20)
Cummings et al., 1997	353	1756 matched community-dwelling controls	RS/ History of family handgun purchase	1.9 (1.4-2.5)
Grassel et al., 2003	887 suicide victims and 1546 gun suicide victims	208,738 non-injury deaths	-RS -RFS/ Handgun purchase	6.8 (5.7-8.1) 12.5 (10.4-15.0)
Kellerman n et al., 1992	438	438 matched community-dwelling controls	RS/ ≥ One gun in home	4.8 (2.7-8.5)
Wiebe, 2003	1959	13,535 community-dwelling controls	-RS -RFS/ Having a gun in home	3.44 (3.06-3.86) 16.89 (13.26-21.52)
Wintemut e et al., 1999	238,292 handgun purchasers in California – observed 1991-1996	General adult population of California	-RS in first year after handgun purchase -RFS in first year after handgun purchase	SMR 4.31 SMR 7.12

[1] Suicide victims in all studies cited in this table were community dwellers

Table 2. (Continued).

(b) Adolescent Individual-Level Studies

Authors	**Number of suicide victims (cases)**	**Comparison group**	**Outcome assessed/ Exposure**	**OR (CI)*/p-value**
Brent et al., 1988 (37)	27	56 suicidal psychiatric inpatients (18 who had seriously considered and 38 who had attempted suicide)	RS / Firearms available in home	2.7 (1.1-6.4)
Brent et al., 1991 (36)	47	(a) 47 suicide attempters (b) 47 never-suicidal psychiatric controls	RS / Gun available in home	a: OR 2.1 (1.2-3.7) b: OR 2.2 (1.4-3.5)
Brent et al., 1993 (38)	7, all with no apparent psychopathology	(a) 60 adolescent suicide victims with psychiatric disorder (b) 38 community controls with no psychiatric disorder	RS / Gun available in home	a: none reported b: p=.04
Brent et al., 1993 (33)	67	67 matched community-dwelling controls	RS / Any gun available in the home	3.3 (1.4-7.7)
Brent et al., 1994 (39)	63, all with a history of affective illness	23 community-dwelling controls with a history of affective illness	RS / Handgun available in the home	p<.01
Brent et al., 1999 (35)	140	131 community-dwelling controls	RS / Gun in home	4.00 (1.30-14.70)
Shah et al., 2000 (34)	36 firearm suicide victims	36 matched community-dwelling controls	RFS / Household access to firearms	3.91 (1.11-13.80)

RS: Risk of suicide
RFS: Risk of firearm suicide
OR: Odds ratio
CI: Confidence interval
SMR: Standardized mortality ratio

Table 3. Studies of Firearm Storage Practices and Suicide Risk.

Authors	Cases	Controls	Measures of firearm storage	Odds Ratio (CI)*/p-value
Brent et al., 1988	27	56 suicidal psychiatric inpatients (18 who had seriously considered and 38 who had attempted suicide)	Firearms stored loaded Firearms stored locked	NS** NS**
Brent et al., 1991	47	(a) 47 suicide attempters (b) 47 never-suicidal psychiatric controls	Guns stored locked Guns stored together with ammunition Guns stored loaded	NS** NS** NS**
Brent et al., 1993	7, all with no apparent psychopathology	(a) 60 adolescent suicide victims with psychiatric disorder (b) 38 community controls with no psychiatric disorder	Loaded gun in the home	Cases vs. a: p<.01 Cases vs. b: p<.01
Brent et al., 1993	67	67 matched community-dwelling controls	Loaded gun in the home	NS**
Conwell et al., 2002	86, ages 50 and over	86 matched community-dwelling controls	≥ 1 gun kept unlocked ≥ 1 gun kept loaded	9.52 (1.52-58.82) 6.41 (1.17-35.71)
Kellermann et al., 1992	438	438 matched community-dwelling controls	Any gun kept loaded All guns kept unloaded Any gun kept unlocked All guns kept locked up	9.2 (4.1-20.1) 3.3 (1.7-6.1) 5.6 (3.1-10.4) 2.4 (1.0-5.7)
Shah et al., 2000	36 firearm suicide victims	36 matched community-dwelling controls	≥ 1 unlocked gun in the home	p=0.05 Crude OR 2.57 (0.98-6.70)

*Odds ratios = 95% confidence interval

**NS indicates not significant; p-values not reported

familiar with the decedent's life history. The 1993 NMFS is based on a sample of 22,957 deaths representing 2,215,000 individuals.The death certificates were sampled from the 1993 Current Mortality Sample, a 10% random sample of death certificates from all states in the union (except South Dakota), and independent vital registration areas of the District of Columbia and New York City. The NMFS sample was stratified by broad age groups, two racial groups (Black and non-Black), and gender within 12 causes of death (including suicide; homicide; various types of injuries; chronic and infectious diseases; and substance abuse) [46.] Decedents below 35 and over 99 years of age, as well as female and Black decedents were oversampled.

Informants were identified either on the death certificate as having provided information or were identified by the funeral director. When an informant was not identified in this manner or could not be located, another person familiar with the decedent's life history was recruited. Informants were contacted by personalized letter and asked to participate in the survey [41]. Informants for the survey included children (17.7%), parents (17.5%), spouses (15.3%), siblings (10.1%), other relatives (8.9%), in-laws (3.4%), friends (3.2%), staff persons at institution where decedents resided (1.4%), neighbors (1.2%), and step-parents (0.8%). Eighty-three percent of informants participated in the survey. All questions in the survey refer to the last year of life, except where, as we will note, the question refers to the last *month* of life.

To reduce the likelihood of misinformation provided by respondents not well acquainted with the decedent, we included in the analyses only information reported by the spouse, sibling or offspring (if the decedent was age 18 or older), or a parent, sibling or offspring (if the decedent was age 25 or younger). As an extra precaution, we also required the informant to have resided with the decedent for at least a year. Finally, the study sample included only those who had no missing information on any variables of interest.

We further focused our analyses on decedents who, according to the informant, had not expressed a wish to die during the last month of life. Recognizing that there may be residual confounding by intention to die, we identified two subgroups within the study-sample. The first group is comprised of decedents who did not have any of the three strongest correlates of suicide: 1) consumed alcohol in the last year of life, 2) experienced an employment change in the last year of life, and 3) visited a mental health professional in the last year of life. The second group is comprised of those who died of suicide. We reason that, *on average*, intention to die is relatively lower among decedents who did not have any of the three strongest correlates of suicide, and relatively higher among those who died of suicide. Thus, for the current analysis, we make the simplifying assumption that if the risk associated with unsafe firearm storage is highest among those with the highest intention to die, and lowest among those with the lowest intention to die, then we have reasonably succeeded to group decedents by, and partially control for, their intention to die within limitations imposed by the dataset.

Measurement

The dependent variable in this study is an indicator of whether the decedent committed suicide with a firearm according to the ICD-9 codes for external cause of injury (i.e., E-codes) stated on the death certificate: E955.0 (suicide by handgun) and E955.1-E955.4 (suicide by all other and unspecified firearms[†]. The primary exposure of interest is whether the decedent kept any firearms in or around home during the last year of life, and whether the firearms were stored unloaded or locked in the decedent's home. We consider firearms stored in a locked drawer, cabinet, or closet and those stored with a trigger lock or other locking mechanism as locked; disassembled firearms were considered to be unloaded.

The regression models included the following variables: 1) *Socio-demographic characteristics* – gender, age, education level, and race/ethnicity; 2) *Correlates of mental well-being and stress* – three categorical variables reflecting the frequency with which the decedent engaged in religious activities, participated in moderate and/or vigorous physical activities (e.g., heavy housework, jogging, tennis, aerobics), and visited or talked on the telephone with family or friends; and a categorical measure of average alcohol consumption during the last year of life.

This variable accounts for both the frequency and the amount of alcohol consumption.

The reference group, non-drinkers, includes both lifetime abstainers and former drinkers. Those who were reported to have consumed alcohol in their last year of life are classified as either infrequent light drinkers (<3 drinks, 1-2 times per week or less often), infrequent heavy drinkers (3+ drinks, 1-2 times per week or less often), frequent light drinkers (<3 drinks, 3+ times per week), or frequent heavy drinkers (3+ drinks, 3+ times per week). We also included indicators of whether the decedent experienced a demotion or a job loss, quit a job, or retired during the last year of life; whether the decedent avoided or refused needed health care during the last year of life; and a variable indicating whether the decedent saw a mental health professional during the last year of life. 3) Number of depressive symptoms during the last month of life. These symptoms include: 1) seeming drowsy or sluggish, 2) seeming unresponsive or withdrawn, 3) seeming impatient or annoyed, 4) expressing feelings of worthlessness, 5) crying for long periods for no apparent reason, 6) having trouble sleeping or sleeping more or less than usual, 7) eating more or less than usual, and 8) having trouble concentrating or difficulty making decisions. The alpha reliability for this scale is 0.77. We categorized the scale as 0, 1-2, 3-4, and 5 or more symptoms.

Analysis

Our analyses followed the study aims. Using logistic regression, we first estimated the risk of suicide by firearm associated with having access to firearm(s) at home (Model A in tables), and then we estimated the protective effect of safer storage practices among firearm owners (Model B in tables). Next, we estimated the risk of suicide by firearm associated with having access to firearm(s) at home, and the protective effect of safer storage practices among the two groups of decedents with relatively low and high intention to die. All observations are weighted to account for the over-sample of certain types of death certificates and for the non-response of proxy respondents, to adjust for the sampling design as well as survey non-response [47, 48] using SUDAAN [49] and the weights provided with the NMFS [46].

[†] In the NMFS, suicide by airgun (E955.6) was included in a category containing E-codes E954, E955.5-E959, and was therefore, not able to be classified with suicides by firearms.

Results

The study-sample consists of 4,996 decedents representing 721,436 individuals aged 15 and older (Table 4). Compared with the original NMFS sample, decedents in our study are slightly younger and somewhat over representative of males, individuals with a college education, and non-Hispanic whites. Table 5 shows that, relative to those who died from other causes, victims of suicide by firearm were significantly more likely to be younger, male, and Hispanic. Females who did not own firearms were most likely to commit suicide by drugs, whereas males who did not own firearms were most likely to commit suicide by hanging. Both male and female firearm owners were most likely to commit suicide by firearms (data not shown). Individuals who committed suicide by firearm were also significantly more likely to have: never engaged in religious activities; exercised frequently; been frequent and/or heavy drinkers; experienced an employment change in the past year; avoided or refused needed health care; exhibited two or fewer depressive symptoms; and visited a mental health professional in the past year. Additionally, those who committed suicide by firearm were more likely than others to have a firearm kept in/around their home and were less likely to have kept it unloaded.

Among all decedents in the study-sample (Table 6A), risk of committing suicide is a function of age and gender, but independent of level of education or race/ethnicity. Analysis of the original NMFS sample (analyses not shown) yielded similar effects for age, gender, and education. Other key predictors of suicide by firearms include decedents' alcohol consumption, employment change, and visit(s) to a mental health professional during the last year of life. In particular, infrequent heavy drinkers are 3.1 times more likely than non-drinkers to commit suicide by firearm. Those who visited a psychiatrist or psychologist for a personal, emotional, behavioral, or mental problem were 3.3 times more likely to commit suicide by firearm. Among decedents with access to firearms (Table 6B), similar associations were found between the risk of suicide by firearm and alcohol consumption, employment change, and visit(s) to mental health professionals.

Table 4 Decedent's Demographic Characteristics by Cause of Death and Sample Inclusion

Decedent's Characteristics	All deaths in NMFS (n=19,018)	All deaths in study (n=4,996)	All suicides in NMFS (n=1616)	All suicides in study (n=304)	Firearm suicides in NMFS (n=977)	Firearm suicides in study (n=190)	All other deaths in study (n=4806)
Mean age, y (SE)	72.1 (0.3)	70.9 (0.4)	46.7 (1.0)	46.8 (1.1)	47.3 (0.5)	50.3 (1.4)	71.0 (0.4)
Male, %	51.1	57.1	78.3	82.4	86.9	87.3	56.9
Education, %							
Less than high school [a]	39.7	38.0	28.5	32.3	30.8	35.8	38.0
High school	37.6	35.6	42.0	40.2	41.0	40.0	35.6
College	22.7	26.3	29.5	27.5	28.1	24.1	26.3
Race/ethnicity, %							
White Non-Hispanic and other [a]	84.7	87.1	87.3	85.9	87.9	84.7	87.1
Black Non-Hispanic	12.0	10.7	7.0	9.3	6.6	9.4	10.7
Hispanic	3.3	2.2	5.5	4.8	5.3	6.0	2.2

Source: 1993 National Mortality Followback Survey (NMFS).

[a] Reference group

Table 5. Descriptive Statistics by Cause of Death

Decedent's Characteristics	Study Sample (n=4,996)	Committed Suicide by Firearm (n=190)	Died from Other Causes (n=4,806)	
Correlates of Mental Well-Being and Stress				
Frequency engaged in religious activities, %				
Never [a]	37.0	57.2	36.9	*
Less than once per month, 1-3 times per month, 1-2 times per week	56.2	36.5	56.3	*
3-6 times per week, every day	6.8	6.3	6.8	
Frequency engaged in moderate/vigorous physical activities, %				
Never [a]	45.1	22.2	45.2	*
Less than once per month, 1-3 times per month, 1-2 times per week	19.9	22.7	19.9	
3-6 times per week, every day	35.0	55.1	34.9	*
Frequency of contact with family or friends, %				
Never [a]	5.3	3.9	5.3	
Less than once per month, 1-3 times per month, 1-2 times per week	24.6	23.0	24.6	
3-6 times per week, every day	70.1	73.1	70.1	
Alcohol consumption, %				
Non-drinker [a]	75.8	41.1	76.0	*
Infrequent light	11.3	15.8	11.3	
Infrequent heavy	6.4	13.6	6.3	*
Frequent light	2.0	10.1	2.0	*
Frequent heavy	4.6	19.4	4.5	*
Experienced employment change, %	1.6	17.1	1.5	*
Avoided or refused needed health care, %	14.2	22.5	14.2	*
Visited a psychiatrist/psychologist, %	4.9	85.5	4.8	*
Depression				
Number of depressive symptoms, %				
None [a]	23.3	32.1	23.2	*
1-2	34.9	27.6	34.9	*
3-4	25.0	24.5	25.0	
5+	16.9	15.9	16.9	
Possession and Storage of Firearms				
Firearm(s) kept in/around the home, %	35.7	91.3	35.4	*
Firearm(s) kept locked, % [b]	43.3	29.1	43.5	
Firearm(s) kept unloaded, % [b]	81.2	61.8	81.4	*

Source: 1993 National Mortality Followback Survey (NMFS).

[a] Reference group

[b] Statistics are presented only for those who own firearms (n=1,608).

* $p < .05$ – Test of difference between those who committed suicide by firearm and those who died from other causes.

Table 6. Logistic Regression Models of the Risk of Suicide by Firearm Relative to Death from Other Causes

Decedent's Characteristics	A. Study Sample OR (95% CI)	B. Only Firearm Owners OR (95% CI)
Sociodemographic Characteristics		
Age	0.95 (0.94 to 0.96)	0.96 (0.94 to 0.97)
Gender		
Male	2.01 (1.28 to 3.15)	2.21 (1.27 to 3.86)
Female[a]	1.00	1.00
Education		
Less than high school [a]	1.00	1.22 (0.69 to 2.17)
High school	0.93 (0.59 to 1.47)	1.00
College	0.91 (0.54 to 1.54)	1.09 (0.58 to 2.03)
Race/ethnicity		
White Non-Hispanic and other [a]	1.00	1.00
Black Non-Hispanic	1.01 (0.62 to 1.64)	0.58 (0.30 to 1.14)
Hispanic	2.07 (0.96 to 4.46)	1.75 (0.63 to 4.85)
Correlates of Mental Well-Being and Stress		
Frequency engaged in religious activities		
Never [a]	1.00	1.00
Less than once per month, 1-3 times per month, 1-2 times per week	0.28 (0.18 to 0.43)	0.26 (0.16 to 0.45)
3-6 times per week, every day	0.60 (0.29 to 1.23)	0.64 (0.27 to 1.52)
Frequency engaged in moderate/vigorous physical activities		
Never [a]	1.00	1.00
Less than once per month, 1-3 times per month, 1-2 times per week	1.18 (0.63 to 2.21)	0.94 (0.41 to 2.16)
Decedent's Characteristics	A. Study Sample OR (95% CI)	B. Only Firearm Owners OR (95% CI)
3-6 times per week, every day	1.73 (0.95 to 3.17)	2.81 (1.38 to 5.72)
Frequency of contact with family or friends		
Never [a]	1.00	1.00
Less than once per month, 1-3 times per month, 1-2 times per week	1.10 (0.28 to 4.37)	0.63 (0.16 to 2.48)
3-6 times per week, every day	0.85 (0.22 to 3.28)	0.44 (0.12 to 1.60)
Alcohol consumption		
Non-drinker [a]	1.00	1.00
Infrequent light	1.16 (0.65 to 2.06)	0.98 (0.49 to 1.97)
Infrequent heavy	3.08 (1.64 to 5.79)	3.49 (1.63 to 7.48)
Frequent light	1.68 (0.79 to 3.57)	1.55 (0.62 to 3.85)
Frequent heavy	2.55 (1.30 to 5.01)	4.65 (2.13 to 10.13)
Experienced employment change	3.89 (2.25 to 6.72)	4.23 (2.00 to 8.92)
Avoided or refused needed health care	1.19 (0.71 to 1.98)	0.95 (0.51 to 1.77)
Visited a psychiatrist/psychologist	3.35 (1.69 to 6.65)	4.93 (2.32 to 10.48)
Depression		
Number of depressive symptoms		
None [a]	1.00	1.00

Decedent's Characteristics	A. Study Sample OR (95% CI)	B. Only Firearm Owners OR (95% CI)
1-2	0.67 (0.39 to 1.14)	1.12 (0.59 to 2.13)
3-4	0.98 (0.54 to 1.78)	1.01 (0.48 to 2.13)
5+	0.78 (0.41 to 1.51)	1.18 (0.54 to 2.58)
Possession and Storage of Firearms		
Firearm(s) kept in/around the home	18.60 (10.70 to 32.33)	
Firearm(s) kept locked		0.33 (0.20 to 0.56)
Firearm(s) kept unloaded		0.30 (0.18 to 0.49)
N	4,996	1,608

Source: 1993 National Mortality Followback Survey (NMFS).

[a] Reference group

Table 7A shows the risk of suicide by firearm relative to death from other causes among decedents in the study-sample, as well as among the two groups with relatively lower and higher intention to die. The risk of suicide by firearms is highest among those with the highest intention to die (i.e., subsample consisting entirely of suicide victims; OR=30.3, medium among those with relatively moderate intention to die (i.e., study-sample, OR=18.6), and lowest among the sample with the lowest intention to die (i.e., subsample excluding those with any of the 3 strongest correlates of suicide; OR=16.0).

Next, we estimated the preventive effect of safer firearm storage practices among the study-sample as well as among the two groups with lower and higher intention to die (relative to the study sample). The protective effect of safer firearm storage is strongest among those with lowest intention to die (locked: OR=0.19; unloaded: OR=0.22) and medium among those with relatively moderate intention to die (locked: OR=0.33; unloaded: OR=0.30), among those with the strongest intention to die there is no association between safer firearm storage and risk of suicide.

Table 7. Comparison of groups with relatively low, medium, and high intention to die

	A. Risk of suicide by firearm at home relative to death from other causes	**B. Protective effect of safer firearm storage on risk of suicide by firearm at home**	
		Locked	**Unloaded**
Level of intention to die	OR (95% CI)	OR (95% CI)	OR (95% CI)
Group with relatively low intention to die [a]	15.98 (6.61 to 38.64)	0.19 (0.06 to 0.54)	0.22 (0.08 to 0.59)
Study sample (medium intention to die) [b]	18.60 (10.70 to 32.33)	0.33 (0.20 to 0.56)	0.30 (0.18 to 0.49)
Group with relatively high intention to die [b]	30.27 (12.02 to 76.21)	0.86 (0.25 to 2.92)	0.46 (0.16 to 1.36)

Source: 1993 National Mortality Followback Survey (NMFS).

[a] Model includes age; gender; education; frequency of religious activities, physical activities, and contact with family/friends; avoidance/refusal of health care; and depressive symptoms.

[b] Model includes age; gender; education; frequency of religious activities, physical activities, and contact with family/friends; alcohol consumption; employment change; avoidance/refusal of health care; depressive symptoms; and visit(s) to a mental health professional.

Discussion

In line with previous research [e.g.] [50, 51], we found an elevated risk of suicide among males, heavy drinkers, and persons with recent employment changes. Distribution of suicide methods by gender also replicated previous work [52]. Also in accord with previous works [4, 9, 31-40, 53, 54], we found a 19-fold elevation in the risk of suicide by firearm among decedents with access to firearms at home relative to others. This points to the importance of consumer product safety practices that can reduce the risk of suicide among the substantial proportion of the US population with access to firearms at home. In the first study to analyze the protective effect of firearm storage practices in a nationally representative sample, we found that firearm owners who practiced safer storage of their firearms were on average 67% less likely to die from firearm-related suicide than those who did not store their firearms safely.

These findings must be considered in light of the study's limitations and strengths. One shortcoming of our study pertains to the assumption that household firearm ownership is a reliable proxy of firearm access. Evidence suggests that considering firearm ownership alone is likely to miss those with ready access to firearms outside of home [55]. Thus, we may have underestimated the proportion of participants with ready access to firearms. This underestimation was probably more prevalent among those who did not die of a firearm-related injury. This would result in a degree of overestimation of our odds ratios, as it is likely that some members of the comparison groups have had easy access to firearms. However, it is highly unlikely that the large effect size can be explained away by this source of misclassification.

We partially addressed an important limitation of existing literature on suicide and firearms by using proxies for subjects' intention to die. We examined three groups which, on average, had relatively high, medium, and low intention to die. As expected, access to firearms at home was most lethal among those with highest intention to die, for whom safer firearm storage practices was not protective. For the group with the moderate intention to die, access to firearms at home was moderately lethal, as was the protective effect of safer storage practices. The most striking results regard the group with the lowest relative intention to die. For this group, access to firearms at home is associated with relatively low risk of suicide by firearms, while the protective effect of storing firearms locked and/or unloaded was greatest. Considering that all those with stated intention to die were excluded from the study, and considering that this group excludes those with any of the three strongest correlates of suicide, it appears that the decedents in this group had a fairly low intention to die and may have engaged in an impulsive act. This suggests that safer firearm storage may be most effective in reducing impulsive suicidal acts.

We recognize the single-mindedness by which some repeatedly pursue self-annihilation [56, 57]. In the absence of firearms these people are likely to simply use the next most lethal suicide method. Nonetheless, even if all of those who committed suicide by firearm simply switched to the next most lethal method, a significant reduction in suicide mortality is still realized. This is because, irrespective of individuals' intention to die, suicide attempts involving firearms are 2.6 times more lethal than those involving suffocation, the second most lethal suicide method (see [52] for a detailed examination of relative lethality of suicide attempt by firearms).

Our findings lend further support to the utility of consumer product guidelines or products intended to reduce the likelihood of unintended or impulsive use of firearms, including lock and personalization technology [58-63], and consumer product regulations to reduce firearm-related injuries and fatalities [58, 59, 61-68]. The importance of this approach is even more evident when considering the difficulty of changing individuals' firearm storage practices [58-63, 69], and the failure of otherwise safety conscious people to store their firearms safely[70-72]. A substantial proportion of adult firearm owners report unsafe storage practices [2, 73].

Firearm training has frequently been proposed as a means of promoting safer firearm storage. However, information on the effectiveness of these programs is sparse. At least two community-based programs have distributed gun locks or safes to firearm owners; both appear to have improved firearm storage practices among participants, particularly among parents [74, 75]. In contradiction, a nationally representative survey of firearm owners suggests that firearm training may actually be associated with reduced likelihood of safer storage practices [2] and at least one counseling intervention in a pediatric primary care setting failed to promote safer firearm storage [76]. Thus, the efficacy of firearm training in regard to firearm storage practices remains questionable.

In the US, legislation mandating minimum safety features on new firearms can be an essential part of the solution, one which is supported by the majority of American adults. In a recent national survey, 88 percent of respondents favored childproofing standards for new handguns, and 71 percent favored personalization [77]. At present, criteria exist which require that any handgun imported into the US meet a set of minimum safety standards [78]. However, the net effect of these criteria is modest given that they don't apply to firearms manufactured within the US, and they are, alone, not enough to ensure that safety features are consistently applied to new handguns [78]. Additionally, in recent years some states have passed legislation to hold firearm owners responsible when a child's access to firearms leads to injuries [79]. However, introduction of this legislation was not followed by significant decline in the rate of firearm-related suicides among children [79]. Clearly, current legislation pertaining to firearm safety is insufficient; implementing federally mandated safety standards may be necessary to reduce the risk of violent death associated with firearm ownership. To the extent that changes in the legal and social milieu are lasting, they are likely to reduce suicide risk of future generations [80].

Shenassa was supported by grant R40MC03600-01-00 from the Maternal and Child Health Bureau, Department of Health and Human Services.

References

[1] Blendon RJ, Young JT, Hemenway D. The American public and the gun control debate. *Jama.* Jun 12 1996;275(22):1719-1722.

[2] Hemenway D, Solnick SJ, Azrael DR. Firearm training and storage. *Jama.* Jan 4 1995;273(1):46-50.

[3] Young JT, Hemenway D, Blendon RJ, Benson JM. Trends:Guns. *Public Opinion Q.* 1996;60(4):634-649.

[4] Bailey JE, Kellermann AL, Somes GW, Banton JG, Rivara FP, Rushforth NP. Risk factors for violent death of women in the home. *Arch Intern Med.* Apr 14 1997;157(7):777-782.
[5] Kellermann AL, Reay DT. Protection or peril? An analysis of firearm-related deaths in the home. *N Engl J Med.* Jun 12 1986;314(24):1557-1560.
[6] Kellermann AL, Rivara FP, Rushforth NB, et al. Gun ownership as a risk factor for homicide in the home. *N Engl J Med.* Oct 7 1993;329(15):1084-1091.
[7] Kellermann AL, Somes G, Rivara FP, Lee RK, Banton JG. Injuries and deaths due to firearms in the home. *J Trauma.* Aug 1998;45(2):263-267.
[8] Miller M, Azrael D, Hemenway D. Rates of household firearm ownership and homicide across US regions and states, 1988-1997. *Am J Public Health.* Dec 2002;92(12):1988-1993.
[9] Wiebe DJ. Homicide and suicide risks associated with firearms in the home: a national case-control study. *Ann Emerg Med.* Jun 2003;41(6):771-782.
[10] Birckmayer J, Hemenway D. Suicide and firearm prevalence: are youth disproportionately affected? *Suicide Life Threat Behav.* Fall 2001;31(3):303-310.
[11] Bridges FS. Gun availability and use of guns for murder and suicide in Canada: a replication. *Psychol Rep.* Jun 2002;90(3 Pt 2):1257-1258.
[12] Clarke RV, Jones PR. Suicide and increased availability of handguns in the United States. *Soc Sci Med.* 1989;28(8):805-809.
[13] Hemenway D, Miller M. Association of rates of household handgun ownership, lifetime major depression, and serious suicidal thoughts with rates of suicide across US census regions. *Inj Prev.* Dec 2002;8(4):313-316.
[14] Kaplan MS, Adamek ME, Geling O, Calderon A. Firearm suicide among older women in the U.S. *Soc Sci Med.* May 1997;44(9):1427-1430.
[15] Kaplan MS, Geling O. Firearm suicides and homicides in the United States: regional variations and patterns of gun ownership. *Soc Sci Med.* May 1998;46(9):1227-1233.
[16] Killias M. International correlations between gun ownership and rates of homicide and suicide. *Cmaj.* May 15 1993;148(10):1721-1725.
[17] Kleck G, Patterson E. The impact of lgun control and gun ownership levels on violence rates. *Journal of Quantitative Criminology.* 1993;9(3):249-287.
[18] Krug EG, Powell KE, Dahlberg LL. Firearm-related deaths in the United States and 35 other high- and upper-middle-income countries. *Int J Epidemiol.* Apr 1998;27(2):214-221.
[19] Lester D. Gun control, gun ownership, and suicide prevention. *Suicide Life Threat Behav.* Summer 1988;18(2):176-180.
[20] Lester D. Firearm availability and the incidence of suicide and homicide. *Acta Psychiatr Belg.* 1988;88(5-6):387-393.
[21] Lester D. Gun ownership and suicide in the United States. *Psychol Med.* May 1989;19(2):519-521.
[22] Lester D. The availability of firearms and the use of firearms for suicide: a study of 20 countries. *Acta Psychiatr Scand.* Feb 1990;81(2):146-147.
[23] Lester D. Firearm availability and the use of firearms for suicide and homicide. *Percept Mot Skills.* Dec 2000;91(3 Pt 1):998.
[24] Lester D. Gun availability and use of guns for murder and suicide in Canada. *Psychol Rep.* Dec 2001;89(3):624.

[25] Miller M, Azrael D, Hemenway D. Household firearm ownership and suicide rates in the United States. *Epidemiology.* Sep 2002;13(5):517-524.

[26] Miller M, Azrael D, Hemenway D. Firearm availability and suicide, homicide, and unintentional firearm deaths among women. *J Urban Health.* Mar 2002;79(1):26-38.

[27] Miller M, Azrael D, Hemenway D. Firearm availability and unintentional firearm deaths, suicide, and homicide among 5-14 year olds. *J Trauma.* Feb 2002;52(2):267-274; discussion 274-265.

[28] Sloan JH, Rivara FP, Reay DT, Ferris JA, Kellermann AL. Firearm regulations and rates of suicide. A comparison of two metropolitan areas. *N Engl J Med.* Feb 8 1990;322(6):369-373.

[29] Beautrais AL, Joyce PR, Mulder RT. Access to firearms and the risk of suicide: a case control study. *Aust N Z J Psychiatry.* Dec 1996;30(6):741-748.

[30] Miller M. Geriatric suicide: the Arizona study. *Gerontologist.* Oct 1978;18(5 Pt 1):488-495.

[31] Cummings P, Koepsell TD, Grossman DC, Savarino J, Thompson RS. The association between the purchase of a handgun and homicide or suicide. *Am J Public Health.* Jun 1997;87(6):974-978.

[32] Grassel KM, Wintemute GJ, Wright MA, Romero MP. Association between handgun purchase and mortality from firearm injury. *Inj Prev.* Mar 2003;9(1):48-52.

[33] Wintemute GJ, Parham CA, Beaumont JJ, Wright M, Drake C. Mortality among recent purchasers of handguns. *N Engl J Med.* Nov 18 1999;341(21):1583-1589.

[34] Shah S, Hoffman RE, Wake L, Marine WM. Adolescent suicide and household access to firearms in Colorado: results of a case-control study. *J Adolesc Health.* Mar 2000;26(3):157-163.

[35] Brent DA, Perper J, Moritz G, Baugher M, Allman C. Suicide in adolescents with no apparent psychopathology. *J Am Acad Child Adolesc Psychiatry.* May 1993;32(3):494-500.

[36] Brent DA, Perper JA, Allman CJ, Moritz GM, Wartella ME, Zelenak JP. The presence and accessibility of firearms in the homes of adolescent suicides. A case-control study. *Jama.* Dec 4 1991;266(21):2989-2995.

[37] Brent DA, Perper JA, Goldstein CE, et al. Risk factors for adolescent suicide. A comparison of adolescent suicide victims with suicidal inpatients. *Arch Gen Psychiatry.* Jun 1988;45(6):581-588.

[38] Brent DA, Perper JA, Moritz G, Baugher M, Schweers J, Roth C. Firearms and adolescent suicide. A community case-control study. *Am J Dis Child.* Oct 1993;147(10):1066-1071.

[39] Conwell Y, Duberstein PR, Connor K, Eberly S, Cox C, Caine ED. Access to firearms and risk for suicide in middle-aged and older adults. *Am J Geriatr Psychiatry.* Jul-Aug 2002;10(4):407-416.

[40] Kellermann AL, Rivara FP, Somes G, et al. Suicide in the home in relation to gun ownership. *N Engl J Med.* Aug 13 1992;327(7):467-472.

[41] Brent DA. Firearms and suicide. *Ann N Y Acad Sci.* Apr 2001;932:225-239; discussion; 239-240.

[42] Miller M, Hemenway D. The relationship between firearms & suicide: a review of the literature. *Aggression and Violent Behavior.* 1999;4(1):59-75.

[43] Simon OR, Swann AC, Powell KE, Potter LB, Kresnow MJ, O'Carroll PW. Characteristics of impulsive suicide attempts and attempters. *Suicide Life Threat Behav.* 2001;32(1 Suppl):49-59.

[44] Williams CL, Davidson JA, Montgomery I. Impulsive suicidal behavior. *J Clin Psychol.* Jan 1980;36(1):90-94.

[45] Williams CL, Sale IM, Wignall AL. Correlates of impulsive suicidal behaviour. *N Z Med J.* Apr 27 1977;85(586):323-325.

[46] The National Mortality Followback Survey. ftp://ftp.cdc.gov/pub/health_Statistics/NCHS/Dataset_Documentation/NMFS.

[47] Brogan DJ. Software for sample survey data, misuse of standard packages. In: Armitage P, Colton T, eds. *Encyclopedia of Biostatistics.* Vol 5. New York: John Wiley and Sons, Ltd.; 1998:4167-4174.

[48] Johnson DR, Elliott LA. Sampling design effects: Do They affect the analyses of data from the National Survey of Families and Households? *J Marr Fam.* 1998;60:993-1001.

[49] SUDAAN. Version 8.0.1. *Research Triangle Park.*

[50] Self-directed Violence. In: Krug E, Dahlberg LL, Mercy JA, Zwi A, Lozano R, eds. *World Report on Violence & Health.* Geneva: World Health Organization; 2002:185-212.

[51] Blakely TA, Collings SC, Atkinson J. Unemployment and suicide. Evidence for a causal association? *J Epidemiol Community Health.* Aug 2003;57(8):594-600.

[52] Shenassa ED, Catlin SN, Buka SL. Lethality of firearms relative to other suicide methods: a population based study. *J Epidemiol Community Health.* Feb 2003;57(2):120-124.

[53] Brent DA, Baugher M, Bridge J, Chen T, Chiappetta L. Age- and sex-related risk factors for adolescent suicide. *J Am Acad Child Adolesc Psychiatry.* Dec 1999;38(12):1497-1505.

[54] Brent DA, Perper JA, Moritz G, Baugher M, Schweers J, Roth C. Suicide in affectively ill adolescents: a case-control study. *J Affect Disord.* Jul 1994;31(3):193-202.

[55] Ikeda RM, Dahlberg LL, Kresnow MJ, Sacks JJ, Mercy JA. Studying "exposure" to firearms: household ownership v access. *Inj Prev.* Mar 2003;9(1):53-57.

[56] Proulx F, Lesage AD, Grunberg F. One hundred in-patient suicides. *Br J Psychiatry.* Sep 1997;171:247-250.

[57] Shah AK, Ganesvaran T. Inpatient suicides in an Australian mental hospital. *Aust N Z J Psychiatry.* Apr 1997;31(2):291-298.

[58] Frattaroli S, Webster DW, Teret SP. Unintentional gun injuries, firearm design, and prevention: what we know, what we need to know, and what can be done. *J Urban Health.* Mar 2002;79(1):49-59.

[59] Meadows O. Firearm Safety Devices. *Journal of Pediatric Health Care.* 2003;17(3):157-158.

[60] Milne JS, Hargarten SW. Handgun safety features: a review for physicians. *J Trauma.* Jul 1999;47(1):145-150.

[61] Siebel BJ. The case against the gun industry. *Public Health Rep.* Sep-Oct 2000; 115(5):410-418.

[62] Teret SP, Culross P. Product-oriented approaches to reducing youth gun violence. *The Future of Children.* 2002;12(2):119-131.

[63] Vernick JS, Teret SP. A public health approach to regulating firearms as consumer products. *University of Pennsylvania Law Review.* 2000;148(4):1193-2011.

[64] Freed LH, Vernick JS, Hargarten SW. Prevention of firearm-related injuries and deaths among youth. A product-oriented approach. *Pediatr Clin North Am.* Apr 1998;45(2):427-438.

[65] Hemenway D. Regulation of firearms. *N Engl J Med.* Sep 17 1998;339(12):843-845.

[66] Hemenway D. The public health approach to motor vehicles, tobacco, and alcohol, with applications to firearms policy. *J Public Health Policy.* 2001;22(4):381-402.

[67] Teret SP, Lewin NL. Policy and technology for safer guns: an update. *Ann Emerg Med.* Jan 2003;41(1):32-34.

[68] Wintemute GJ. The relationship between firearm design and firearm violence. Handguns in the 1990s. *Jama.* Jun 12 1996;275(22):1749-1753.

[69] Brent DA, Baugher M, Birmaher B, Kolko DJ, Bridge J. Compliance with recommendations to remove firearms in families participating in a clinical trial for adolescent depression. *J Am Acad Child Adolesc Psychiatry.* Oct 2000;39(10):1220-1226.

[70] Connor SM, Wesolowski KL. "They're too smart for that": predicting what children would do in the presence of guns. *Pediatrics.* Feb 2003;111(2):E109-114.

[71] Coyne-Beasley T, McGee KS, Johnson RM, Bordley WC. The association of handgun ownership and storage practices with safety consciousness. *Arch Pediatr Adolesc Med.* Aug 2002;156(8):763-768.

[72] Farah MM, Simon HK, Kellermann AL. Firearms in the home: parental perceptions. *Pediatrics.* Nov 1999;104(5 Pt 1):1059-1063.

[73] Schuster MA, Franke TM, Bastian AM, Sor S, Halfon N. Firearm storage patterns in US homes with children *Am J Public Health.* Apr 2000;90(4):588-94.

[74] Coyne-Beasley T, Schoenbach VJ, Johnson RM. "Love our kids, lock your guns": a community-based firearm safety counseling and gun lock distribution program. *Arch Pediatr Adolesc Med.* Jun 2001;155(6):659-664.

[75] Horn A, Grossman DC, Jones W, Berger LR. Community based program to improve firearm storage practices in rural Alaska. *Inj Prev.* Sep 2003;9(3):231-234.

[76] Oatis PJ, Fenn Buderer NM, Cummings P, Fleitz R. Pediatric practice based evaluation of the Steps to Prevent Firearm Injury program. *Inj Prev.* Mar 1999;5(1):48-52.

[77] Teret SP, Webster DW, Vernick JS, et al. Support for new policies to regulate firearms. Results of two national surveys. *N Engl J Med.* Sep 17 1998;339(12):813-818.

[78] Milne JS, Hargarten SW, Kellermann AL, Wintemute GJ. Effect of current federal regulations on handgun safety features. *Ann Emerg Med.* Jan 2003;41(1):1-9.

[79] Cummings P, Grossman DC, Rivara FP, Koepsell TD. State gun safe storage laws and child mortality due to firearms. *Jama.* Oct 1 1997;278(13):1084-1086.

[80] Shenassa ED. Society, physical health and modern epidemiology. *Epidemiology.* Jul 2001;12(4):467-470.

In: Consumer Product Safety Issues
Editors: P. A. Varga, M. D. Pintér
ISBN: 978-1-60456-826-4

Chapter 3

Unconscionable Activism

Henry I. Miller[*,†]
The Hoover Institution, Stanford University, Stanford, CA, U.S.A

Abstract

Anti-technology lobbyists continue to promulgate and perpetuate various misapprehensions about a wide range of products and technologies, from pesticides and other chemicals to fat substitutes. Recombinant DNA technology (also known as "gene splicing" or "genetic modification" (GM)) applied to agriculture to create plants with new or selectively enhanced traits has been among the most abused -- accused repeatedly of being untested, unproven, unsafe, unwanted and unregulated. However, anti-biotechnology activists ignore our vast experience and the scientific consensus that gene-splicing is an extension, or refinement, of less precise, less predictable techniques; that the acreage increases each year because farmers in two dozen countries are so pleased with the results; and that Americans have consumed more than a *trillion* servings of foods that contain gene-spliced ingredients, with not a single untoward reaction.

What makes false alarms about biotech – or any new technology – hard to expose is the virtual impossibility of demonstrating the absolute safety of any activity or product: There is always the possibility that we haven't yet gotten to the nth hypothetical risk or to the nth dose or the nth year of exposure, when the risk will finally be demonstrated. It is logically impossible to prove a negative, and all activities pose some nonzero risk of adverse effects.

Anti-technology, anti-business activists – whether they are found in NGOs, government or the media – claim to fear a world in which exploitative, rapacious, multinational corporations conspire to strip away individual choice from the world's farmers and consumers. Yet it is *they* who are guilty of the mendacity and manipulation they imagine they see in others; *they* who are guilty of stripping away the freedom of researchers to research, doctors to doctor, and consumers to consume vaccines and drugs that can be life-saving. Were anti-technology, anti-business activists to have their way, unwise, regressive public policy could reduce significantly the pursuit of knowledge and production of wealth worldwide.

---

* Correspondence to: Phone (650) 725-0185; Home (650) 368-1221; Mobile (650) 906-9014

† Henry I. Miller, M.D., is a fellow at The Hoover Institution, Stanford University, Stanford, CA 94305-6010, U.S.A. From 1977 to 1994, he was an official at the U.S. National Institutes of Health and Food & Drug Administration. He is the co-author (with Gregory Conko) of "The Frankenfood Myth: How Protest and Politics Threaten the Biotech Revolution," chosen by Barron's as one of the 25 Best Books of 2004.

Introduction

Activism can be a good thing. Libertarians and civil rights advocates lobby for constraints on undue government intrusion into our lives, and professional associations further the interests of its members. We all benefit from getting to shop in the marketplace of ideas.

However, all is not good-faith, constructive activism. Some of the goods in the marketplace are shoddy, and they poison both commerce and civil discourse.

Activists from NGOs and the media have targeted a panoply of products, technologies and industries that they dislike – pesticides, food additives, chemicals in general, pharmaceuticals, nuclear power and biotechnology, among others – for opprobrium, over-regulation, and where possible, elimination. It seems that no stratagem, no misrepresentation, no outright lie is too outrageous for them. Several particularly egregious, pernicious examples are discussed below.

Baseless Buffoonery Bedevils Biotech

Biotechnology has been especially victimized by irresponsible activism. A prototypic example is Jeremy Rifkin's relentless, decades-old antagonism toward recombinant DNA technology, or gene-splicing, applied to the production of innovative new drugs, gene therapy for life-threatening diseases, agriculture, or anything else. Thirty years ago, he and his followers disrupted a public meeting, chanting, "We shall not be cloned," and displaying signs proclaiming, "Don't Xerox Life." That was hardly radical by the standards of the 1970's but Rifkin's statements about biotechnology threatening "a form of annihilation every bit as deadly as nuclear holocaust" are extreme and baseless, variations on The Big Lie – that biotech is untested, unsafe, unproven, unwanted and unregulated -- which seems increasingly to be a mainstay of radical activism.

A broad scientific consensus long has held that the newest techniques of biotechnology are no more than an extension, or refinement, of earlier ones applied for centuries -- and that gene transfer or modification by gene-splicing techniques does not, per se, confer risk. Rifkin's assertions about biotechnology ignore the seamless continuum that exists between old and new biotechnology and the monumental contributions that both have made to medicine, agriculture and innumerable scientific disciplines.

The late Harvard evolutionary biologist Stephen Jay Gould, by his own admission, tried to be sympathetic to Mr. Rifkin's views but was overwhelmed by his "extremism" and "lack of integrity" and by his showing "no understanding of the norms and procedures of science." Professor Gould characterized Mr. Rifkin's anti-biotechnology book, "Algeny," as "a cleverly constructed tract of anti-intellectual propaganda masquerading as scholarship," concluding that he had not "ever read a shoddier work." But Professor Gould had not read Rifkin's later literary efforts on a variety of topics, which are at least as bad.

And then there is Greenpeace. During a House of Lords Select Committee hearing in 1999, Lord Melchett, then director of Greenpeace, was asked "Your opposition to the release of [gene-spliced plants], that is an absolute" and definite opposition? It is not one that is dependent on further scientific research?" He replied: "It is a permanent and definite and complete opposition."

Greenpeace International may have attained the nadir of anti-biotechnology activism when, in 1995, the organization announced that it had "intercepted a package containing rice seed genetically manipulated to produce a toxic insecticide, as it was being exported ... [and] swapped the genetically manipulated seed with normal rice."

The rice seeds stolen by Greenpeace had been genetically improved for insect resistance and were en route to the International Rice Research Institute in the Philippines from the Swiss Federal Institute of Technology in Zurich. The modified seeds were to be tested to confirm that they would grow and produce high yields of rice with less chemical pesticide. In the Philippines and many developing countries in Asia where rice is a staple food, disease-resistant and insect-resistant rice are desperately needed.

Greenpeace also has invented tales for inhabitants of developing countries about gene-spliced crops causing homosexuality, impotence and baldness. In Africa, they have promulgated the myth that these improved crops cause impotence and increase the spread of HIV/AIDS. Doreen Stabinsky, a "science advisor" to Greenpeace International (an oxymoron if there ever was one), has claimed that cotton fiber (as in underwear), animal feed and cotton-seed oil from Bt-cotton plants can lead to an increase in the occurrence of antibiotic-resistant bacteria, including those that cause tuberculosis and gonorrhea. There is no evidence whatever for such claims.

As bizarre as any of the biotech critics is Jeffrey Smith, who, unfortunately, was permitted to publish a commentary in this publication on November 1, 2007. The essence of his article was virtually entirely misstatements, misrepresentations and outright falsehoods. The piece cites a litany of wholly imaginary problems of toxicity, allergy, harm to animals and ecosystems, etc., etc., allegedly caused by gene-spliced plants and foods. It ignores agbiotechs manifest successes, including higher yields, less use of chemical pesticides, and more environment-friendly cultivation practices, to say nothing of the potential for drought-resistant plants and plant-derived pharmaceuticals. Smiths screed is a vile parody of serious commentary. He seems ignorant of the entire history of agriculture, both pre- and post-gene-splicing, but exhibits an *idée fixe* about biotech. And about other things as well. Smith considers himself to be a yogic flyer who is capable of levitation. (I am not making this up; see the photo at http://www.cgfi.org/cgficommentary/jeffrey-smith%20article.) And according to the *Daily Illini* newspaper (October 28, 1996),[Smith "explained the idea and effects of yogic flying, which relies on practicing transcendental meditation, to the small crowd. Smith presented charts with evidence of a correlation between the presence of yogic flyers and an increase in the quality of life and a decrease in crime. Smith cited limited yogic flying programs in Washington D.C. and near the Middle East that resulted in less crime and more harmony." Rather than being afforded an opportunity to opine in a prominent publication, Mr. Smith should probably be ensconced in some sort of assisted living arrangement, and on appropriate medication.

Then there are the supposedly more moderate groups such as the Pew Initiative on Food and Biotechnology, the Center for Science in the Public Interest (CSPI) and Environmental Defense, that pose as open-minded skeptics, rather than antagonists. They claim not to wish to eliminate biotechnology but only to want it "properly" regulated. However, beneath the rhetoric, their arguments and actions – which consistently ignore the context necessary to understand the potential risks and benefits of the new biotechnology – lead us to much the same place as biotech's declared enemies: efforts to create a groundswell of anxiety and to elicit unnecessary, hugely burdensome government regulation that will make biotech product testing and commercialization untenable. The more "moderate" groups are subtler, and therefore more insidious, than the anti-biotech players who show their colors unambiguously.

The reports published by the lavishly funded Pew Initiative on Food and Biotechnology, which in 2007 lost its funding and folded, are worth exploring. They received extensive media and government attention, largely because Pew touted itself as occupying the thoughtful, disinterested middle ground in the biotechnology debates.

But Pew's PR machine saying that didn't make it so.

Contrary to their remonstrations that they were non-partisan and agnostic about biotechnology, Pew's workshops, conferences, and publications invariably showed a pervasive pro-regulation bias and attempted to create a presumption of genuine controversy where none existed. (Activists understand that over-regulation is an effective tool to inhibit innovation and slow the diffusion of even a superior technology or product that they dislike.) Pew's 2003 report, "Public Sentiment About Genetically Modified Food," supposedly reflecting public sentiment toward gene-spliced foods, was a typically disingenuous pastiche of truisms, half-truths and sleight-of-hand. The survey found that "Americans' knowledge about [gene-spliced] foods remains low," with 54 per cent saying they had heard nothing or not much about them. Then, without enlightening the subjects or offering them any sort of context, the survey went on to pose leading questions about safety and regulation. Not surprisingly, 89 percent agreed with the statement, "Companies should be required to submit safety data to the FDA for review, and no genetically modified food product should be allowed on the market until the FDA determines that it is safe." Almost nine out of ten consumers say they want safe food: What a surprise.

The 2004 report was no better. Only 32 percent of those surveyed reporting they had heard "a great deal or some" about gene-spliced foods (a 12-point decline since 2001, in spite of biotech's expanding applications and successes!), which suggests that fully two-thirds of the sample were offering completely uninformed opinions.

Like the previous survey, the 2004 version revealed that consumers are in favor of safe food: eighty-five percent want regulators to ensure "that [gene-spliced] foods are safe before they come to market," and more than 90 percent favor the labeling of gene-spliced foods and food ingredients.

Activists Exploit Scientific Illiteracy

The Pew approach to polling described above is reminiscent of the example of Idaho junior high school student Nathan Zohner, who found that 86 percent of survey respondents thought the substance "dihydrogen monoxide" should be banned, after they were told that prolonged exposure to its solid form causes severe tissue damage, exposure to its gaseous form causes severe burns, and it has been found in excised tumors of terminal cancer patients. Only one in 50 of young Nathan's survey respondents correctly identified dihydrogen monoxide as *water*, or H_2O. As any pollster (or common sense) will tell you, it's not hard to design survey questions to elicit a desired response – a trick that Pew mastered.

Because public understanding of science is so meager, hoodwinking consumers on surveys isn't difficult. A study by the U.S. National Science Foundation found that fewer than one in four know what a molecule is, and only about half understand that the earth circles the sun once a year.

The public's muddled view of biotechnology in particular is reflected in the results of a survey of 1,200 Americans, released in October 2003 by the Food Policy Institute at Rutgers University. In an eleven-item true/false quiz that was part of the survey, more than half of the subjects received a failing grade (defined as less than 70 percent correct answers). Only 57 percent recognized that the statement "ordinary tomatoes do not contain genes, while genetically modified tomatoes do," is false. Perhaps most shocking of all, only two thirds knew that eating genetically modified fruit would not alter their own genes! One wonders whether the one-third who got this question wrong think that if they eat rabbit stew, they will begin to hop.

The Pew surveys took advantage of respondents' ignorance about key facts: 1) with the exception of wild berries and wild mushrooms, wild game, and fish and shellfish, virtually *all* the organisms—plants, animals, microorganisms—in our food supply have been modified by one genetic technique or another; 2) because the techniques of the new biotech are more precise and predictable than their predecessors, biotech foods are likely to be even *more* safe than other foods; 3) food producers are already legally responsible for assuring the safety of their products, and the FDA does not normally perform safety determinations, but primarily conducts surveillance of marketed foods and takes action if any are found to be adulterated or mislabeled; and (4) unwarranted, excessive regulation, including unnecessary labeling requirements, discourages innovation, imposes costs that are passed along to the consumer and are a disproportionate burden on the poor.

What "location, location, location" is to real estate, "context, context, context" is to public policy – and the anti-biotech NGOs avoid it like the plague, because they know that it undermines their fear-mongering. Surveys by Pew and other similar groups purposefully exploit consumers' (understandable) lack of familiarity with the nuances of both the new biotech and the way that food is currently regulated. Pew's questioning whether the FDA should assure the safety of gene-spliced foods before they're marketed is like asking whether repeat child molesters should be banned from teaching kindergarten.

Another example of unconstructive – and also disingenuous – activism is illustrated by letter to the editor in the July 2005 issue of *Nature Biotechnology* from Greg Jaffe, an attorney at the Center for Science in the Public Interest (which is notorious for its fringe views and extreme recommendations about food and nutrition), that is an extraordinary combination of duplicity and illogic. Jaffe compares the length of time taken by USDA and FDA to review and approve gene-spliced crops and foods during 1994-2000 and 2001-2005. He finds that "the time it took each agency to reach a regulatory decision more than doubled in the past five years for no explainable reason," and that this "trend should worry those who believe that genetic engineering can be used safely and can benefit farmers, consumers and the environment."

These findings are dubious and the conclusions are insupportable. For one thing, Jaffe doesn't look for data that offer "explainable reasons" for the longer review times. Did the workload increase? (It did: According to USDA http://www.isb.vt.edu/cfdocs/biocharts2.cfm -- during 2001-2005 the number of field trial applications approved per year was, on average, 22 per cent higher than during 1994-2000.) Did the workload/reviewer ratio at the agencies change? Did overworked reviewers put low priority, "me, too," applications on the back burner, allowing the average reviewing time to increase? Did the earlier applications represent "low lying fruit" – easier applications to review -- while the later ones were more complex and took longer? Did less experienced companies, which characteristically submit lower quality applications that take longer to review, enter the field? If you don't look for explanations, you don't find them.

For another thing, a trend is only a trend if it's real. The FDA data are completely uninterpretable because FDA review is only voluntary, and during the two time periods in question it is impossible to know what percentage of gene-spliced foods underwent such reviews, and how similar (or different) the products were. It is possible that in recent years, companies simply stopped consulting the FDA for "easy" cases, submitting only more difficult ones that require more intensive review. If so, both literally and figuratively, Jaffe may be comparing apples and oranges.

Without providing any foundation for the statement, Jaffe concludes, "The U.S. government needs to explain to the public why its . . . regulatory system is taking longer to come to decisions about the safety of [gene-spliced] crops. The public wants assurances that federal regulators are ensuring the safety of products..." Oh, *please*. The American public, which consumes about a *billion* servings of foods containing gene-spliced ingredients a day (that is no misprint), has more important things to worry about. Mr. Jaffe spends too much time hanging around with his fellow fruitcakes.

Finally, we come to the ultimate hypocrisy: Jaffe, the longtime antagonist of biotech and proponent of ever more stultifying regulation, bemoans the alleged "regulatory delay," because it "hurts developers by increasing uncertainty about the regulatory decision-making process and by increasing the cost of getting a product to market." But this is exactly what, for years, Jaffe has worked to achieve! He's like the teenager convicted of murdering his parents who asks for mercy from the court because he's an orphan.

In both flagrant and subtle ways, Pew, CSPI and other anti-biotechnology lobbyists continue to perpetuate various manifestations of The Big Lie – that gene-splicing applied to agriculture and food production is untested, unproven, unsafe, unwanted and unregulated. They ignore our vast experience and the scientific consensus that gene-splicing is an extension, or refinement, of less precise, less predictable techniques; that the acreage increases each year because farmers in two dozen countries are so pleased with the results; and that Americans have consumed more than a *trillion* servings of foods that contain gene-spliced ingredients, with not a single untoward reaction.

GOLDEN RICE: A STERLING EXAMPLE

The "Golden Rice" saga is one of the vilest examples in the annals of cynical, anti-technology activism. In 2000, a university research team based in Switzerland and Germany announced an extraordinary scientific tour de force that resulted in the addition of beta-carotene, or provitamin A, to rice grains. The creation of this "Golden Rice" (so called because of its yellow color) was widely hailed as an example of how gene-splicing can benefit society, especially the inhabitants of less developed countries. Vitamin A supplementation of the diet prevents blindness and can be life-saving to the millions of children who are deficient in the vitamin.

Astonishingly, activists lost no time in attacking even this beneficent innovation. First, they claimed that the rice itself would be unhealthy, because too much Vitamin A can be toxic. That claim was rapidly discredited by nutritionists, who explained that Golden Rice contains beta-carotene, the chemical *precursor* of vitamin A that is not toxic at any dose. Then, torturing the data and flip-flopping as they are wont to do, Greenpeace declared that Golden Rice had too *little* beta-carotene and that an adult "would have to eat around 9kg [19.8 pounds] of cooked rice daily to satisfy his/her daily need of vitamin A." Greenpeace's Benedikt Haerlin threatened "direct action" against test plants in the field. Greenpeace's radical media allies, including the UK's *Guardian*, rushed to support the Luddite cause, with Michael Pollan of the *New York Times Magazine* dubbing Golden Rice "the great yellow hype." They criticized the developers of Golden Rice for working with for-profit companies to make seed available to the poor.

These reactions are the most detestable sort of distortion and misrepresentation. Even small amounts of vitamin supplementation can have huge effects. Golden Rice and other similar products can be life-enhancing, life-saving adjuncts to persons with vitamin A deficiency—but only if their producers can overcome NGO opposition, media misreporting and regulatory excesses, and get the new varieties to the farmers who wish to grow it. (It is unlikely to be widely available until at least 2010.)

Activists Act in Bad Faith

Such blatant and rabid militancy might make those who "merely" demand stifling regulation appear temperate by comparison. But correspondence published in the journal *Science* in 2003 opened a window into the motivations of the so-called "moderate" wing of the anti-biotech lobby. Steven H. Strauss, a Professor of Forest Science at Oregon State University, proposed in an article in that journal a very modest streamlining of the regulation of negligible-risk genetic constructions of gene-spliced plants. The reform that he suggested would remedy, in a small way, the irreconcilable paradox in the current federal oversight of plant biotechnology: that the use of the most precise and predictable techniques is far more stringently regulated than techniques that are less precise and predictable. In other words, Strauss was lobbying for regulatory proportionality, recognition of the basic principle that the degree of oversight should be commensurate with the degree of risk.

Jerry Cayford, of the Washington DC-based Resources for the Future, responded with a letter published in *Science*: "Steven H. Strauss makes a plea for less onerous field trial regulations for less radical genetic modifications . . . thereby helping smaller companies and public-sector investigators to be able to afford to try out crop variants. Unfortunately, his pleas ignore the politics of the genetically modified (GM) food debate . . . Strauss' proposal, reasonable as it may be, asks critics to surrender a major bargaining chip—strict regulation of field trials—but offers them nothing in return." In other words, although it would favor consumers, researchers, and the public interest, sensible regulatory policy is not a goal in itself but is merely a bargaining chip to be held or given up in a negotiation among radical groups, business interests, academic researchers, and government regulators!

Strauss's response to Cayford deplored this "hostage-taking" attitude, observing that "the costs to people and environment of effectively losing genetic engineering from most agricultural sectors as a result of excess regulation are too great for so simple-minded a political approach." He added that there are few practices more "'democratizing' than protecting and promoting the ideas and work of society's innovators when applied to improve food quality, dependability, and affordability."

The *coup de grace* in Strauss's response serves as a worthy epilogue to the unworthy efforts and venal motivation of biotech's antagonists, whether they are blatantly belligerent or subtly shifty: "[W]ith the high level of regulation and stigma successfully implanted in places such as Europe, policies and attitudes may take a generation or more to change course. The opportunity costs in dollars, and costs to human health and environment, will be incalculable."

In both flagrant and subtle ways, Jeremy Rifkin, Andew Kimbrell, Margaret Mellon, Mae-Wan Ho, Jeffrey Smith and other anti-biotechnology zealots continue to perpetuate various elements of The Big Lie, ignoring our vast experience and the scientific consensus that gene-splicing is an extension, or refinement, of less precise, less predictable techniques. To say nothing of the fact that North Americans have consumed more than a *trillion* servings of foods that contain gene-spliced ingredients, with not a single documented untoward reaction.

No one should mistake the anti-biotech NGOs' misdemeanors for naive exuberance or excessive zeal in a good cause. Their motives are self-serving and their tactics vicious, an ongoing example of the sentiments expressed by Linus van Pelt, a character in the *Peanuts* comic strip, "I love humanity; it's people I can't stand."

What makes false alarms about biotech – or any new technology – hard to expose is the virtual impossibility of demonstrating the absolute safety of any activity or product: There is always the possibility that we haven't yet gotten to the *n*th hypothetical risk or to the *n*th dose or the *n*th year of exposure, when the risk will finally be demonstrated. It is logically impossible to prove a negative, and all activities pose some nonzero risk of adverse effects. In any case, no responsible person would say that any and every manifestation of agricultural biotech will be of intrinsically low or negligible risk: Once again, context is important, and if were to modify a food plant to express the toxin of the bacterium *Clostridium botulinum* – deadly botulinum toxin, the active agent in the drug Botox – we would need to take precautions to prevent the introduction of harmful amounts of the toxin into food.

YELLOW JOURNALISM TINTS – AND TAINTS – BIOTECH COVERAGE

Unconstructive, anti-social activism comes not only from NGOs, but also from the mainstream media. The consistently biased, inaccurate reporting of former *New York Times* environmental reporter Keith Schneider induced some in the biotech community to devise an apocryphal Keith Schneider Yellow Journalism Prize. The *Times* seems committed to keeping it in the family; the current contenders include Andrew Pollack, who writes for both the business and science sections; Denise Caruso, who writes frequently on biotechnology (always antagonistically and inaccurately); and Michael Pollan, who pens anti-technology, anti-business screeds for the Sunday magazine.

Pollack's "Biotech's Sparse Harvest" on February 14, 2006, is a strong contender. His thesis is: "At the dawn of the era of genetically engineered crops, scientists were envisioning all sorts of healthier and tastier foods, including cancer-fighting tomatoes, rot-resistant fruits, potatoes that would produce healthier French fries and even beans that would not cause flatulence. . . Resistance to genetically modified foods, technical difficulties, legal and business obstacles and the ability to develop improved foods without genetic engineering have winnowed the pipeline."

While Mr. Pollack misses many of the nuances about biotechnology applied to agriculture and food production, he devotes ample ink to the anti-biotech crowd, including the Pew Initiative on Food and Biotechnology and the radical Friends of the Earth. Memo to Mr. Pollack: All points of view on scientific and technological issues are not created equal. Good journalism is not served by a "balance" that creates a kind of moral equivalence between those who hold ideological, anti-biotech views and those with supportable, legitimate viewpoints -- not unlike equating creationism theory with Darwinian theory.

Reflecting the views of biotech's antagonists, Pollack approaches the subject as though the genetic engineering of plants was fundamentally new and its risks unknown. How convenient – but hardly informative for readers – to ignore the context of new and conventional biotech applied to agriculture.

The use of gene-splicing to craft small, precise genetic changes that enhance or introduce desirable traits into plants has been a stunning technological success -- but excessive and unscientific regulation and the intractable opposition of activists have slowed its translation into consumer-friendly foods. Contrary to the implication in Mr. Pollack's article, gene-spliced "potatoes that would produce healthier French fries" (because they have a higher than usual starch content and, therefore, absorb less oil) have been developed – but they were never commercialized because anti-biotech activists bullied the fast-food chains into rejecting them. Like the Greg Jaffe saga above, how ironic and hypocritical that the same activists who have opposed agbiotech relentlessly for 20 years now decry the early "hype" and "over-selling" of its benefits.

Mr. Pollack's statement, "Developing non-allergenic products and other healthful crops has also proved to be difficult technically," is simply untrue. A vast spectrum of such plants has been crafted by laboratory scientists, but they cannot afford the gratuitously inflated regulatory costs to test the plants in the field. Excessive and unwise regulation, spurred by the drumbeat of activists' opposition, is a major reason that products in the development pipeline "do not include many of the products once envisioned," to quote Mr. Pollack. Unscientific and discriminatory EPA and USDA regulatory policies make field trials with gene-spliced plants ten to twenty times more expensive than a similar plant engineered with less precise, less predictable conventional genetic techniques. Unlike pharmaceutical development, agricultural R&D is a low-budget enterprise that leads to largely incremental advances, and unscientific regulation and hugely inflated regulatory costs make the development of many promising and valuable food products uneconomical.

Finally, Mr. Pollack's disparaging assertion that "industry. . . has been peddling the same two advantages – herbicide tolerance and insect resistance – for 10 years," is puzzling. (And "peddling" is an oddly disparaging word choice for someone writing what is supposedly a straight news article.) These traits have been of monumental importance -- not only to farmers' bottom line, but to occupational health and the natural environment. Enhanced pest resistance in plants has obviated the need for hundreds of millions of pounds of chemical pesticides (and thereby reduced environmental and occupational exposures), and herbicide tolerance has made possible a shift to more benign herbicides and to environment-friendly no-till farming.

Although the *Times* has a long history of reporters' overt antagonism toward biotechnology, it's hardly alone among major media outlets. In a "news" piece in August 2006, a *Wall Street Journal* article reporter Debbie Carlson managed to lend credence to virtually every aspect of activists' Big Lie. Carlson's inclusion of comments primarily from intransigent, ideological, long-time opponents of agricultural biotechnology -- instead of from bona fide experts on the technology and its regulation – represents either blatant bias or inexcusable reportorial carelessness.

The statements of the anti-biotech activists quoted in the article are predictably disingenuous. The claim of long-time anti-biotech campaigner Jane Rissler that USDA is "not required to do an ecological/biological implications review" is simply wrong. And the touting of "marker-assisted selection" for genetic improvement of plants – an old technique of limited usefulness – is another attempt by activists to disparage the superior technique of gene-splicing, which is, in fact, more precise, predictable and safe, as well as far more versatile than older methods.

Ms. Carlson's assertion that "laws regulating biotechnology will need to keep up with the work of scientists" is baseless. Regulation is already far *ahead* of the science, in the sense that it is excessively risk-averse and inhibitory: Case by case reviews are performed on every gene-spliced plant variety that is to be field-tested -- in defiance of the broad and long-standing scientific consensus that gene-splicing is essentially an extension, or refinement, of less precise and predictable genetic techniques. (Conventional genetic modification is essentially exempt from regulation.) Speculation by activists about the need for more regulation is merely a ploy to obstruct progress.

Ms. Carlson lacks perspective on the recent history of the genetic improvement of plants. If she had bothered to consult some genuine experts, she would have discovered that the scientific consensus holds that gene-splicing is an extension, or refinement, of less precise, less predictable genetic techniques. One such older technique, in use since the 1950s, is induced-mutation breeding, which involves exposing seeds or cells to ionizing radiation or toxic chemicals to induce random, desirable genetic mutations. Thousands of mutation-bred crop varieties have been commercialized in North America and Europe, and since the 1930's plant breeders have performed "wide cross" hybridizations in which large numbers of "alien" genes are moved from one species or one genus to another to create plant varieties that cannot and do not exist in nature. Common commercial varieties derived from wide crosses include tomato, potato, oat, rice, wheat, corn and pumpkin, among others.

Using these pre-gene-splicing technologies, plant breeders and food producers lack knowledge of the exact genetic changes that produced the desirable traits; and more important, they have no idea what other changes have occurred concomitantly in the plant, including those that could raise levels of toxins or alter the ability to cause allergic reactions. Greater precision is what makes gene-splicing superior.

Finally, Ms. Carlson is incorrect in stating that the commercialization of drought-resistant crops is "far away," or that "it is too early to predict what regulatory trials will be required." Their development is advanced, many are already in field testing in various parts of the world; moreover, their evaluation is, in fact, quite straightforward and not very different from the thousands of gene-spliced plant varieties possessing myriad new traits that have been reviewed and approved by regulators.

Enhanced drought-resistance in commercial crops will be an extraordinary advance, perhaps more important than any now in development. Fresh water is in increasingly short supply in much of the world. Irrigation for agriculture accounts for roughly 70 percent of the world's fresh water consumption – even more in areas of intensive farming and arid or semi-arid conditions – so the introduction of plants that grow with less water would allow much of that essential resource to be freed up for other uses. Especially during drought conditions – which currently plague much of Europe, Africa, Australia, and the United States -- even a small percentage reduction in the use of water for irrigation could result in huge benefits, both economic and humanitarian.

Ms. Carlson's article does not merely lack "balance." In fact, she could not have salvaged this article even with a few quotes from bona fide experts on biotechnology and agriculture: The activists' outright lies make it irredeemable. All points of view on scientific and technological issues are not created equal. Good journalism is not served by creating a kind of moral equivalence between those who hold ideological, anti-technology views and those with legitimate, evidence-based viewpoints -- not unlike equating creation theory with Darwinian theory.

Anti-Biotech Activism Is Only the Tip of the Iceberg

Biotechnology is not the only current victim of cynical, hypocritical, anti-social activism. Consider activists' intractable antagonism to the spraying of pesticides to kill insects that carry disease. The use of any pesticides – let alone the possible resurrection of the use of DDT, which was banned in the United States several decades ago -- has been greeted by near-hysterical resistance from environmental activists, who have attacked the killing of mosquitoes as "disrupting the food chain." And several years ago New York's Green Party literature declared, "These diseases only kill the old and people whose health is already poor."

Since the banning of DDT, insect-borne diseases such as malaria and dengue -- and now West Nile virus -- have been on the rise. The World Health Organization estimates that malaria kills about a million people annually, and that there are between 300 million and 500 million new cases each year.

Not only did government regulators underplay scientific evidence of the effectiveness and relative safety of DDT when they banned it, but they also failed to appreciate the distinction between its large-scale use in agriculture and more limited application for controlling carriers of human disease. Although DDT is a (modestly) toxic substance, there is a world of difference between applying large amounts of it on crops -- as American farmers did before it was banned -- and using it carefully and sparingly to fight mosquitoes and other disease-carrying insects. A basic principle of toxicology is that the dose makes the poison.

The regulators who banned DDT also failed to take into consideration the inadequacy of alternatives. Because it persists after spraying, DDT works far better than many pesticides now in use, some of which are toxic to fish and other aquatic organisms.

Another activists' "success" story was a 2007 statewide ban in California on children's toys that contain more than minuscule amounts of chemicals called phthalates, which are widely used to soften plastic toys and are found in scores of other common items, from shower curtains and traffic cones to phone cords. They also have critical applications in medicine, including surgical instruments and intravenous tubing and bags.

Undeterred by the judgment of experts but swayed by a few rat experiments and a single flawed epidemiological study, in late 2007 Sen. Diane Feinstein (D-Calif.) introduced a bill that would expand California's folly to the entire nation. Unfortunately, her legislation ignores the basic principles of toxicology. For starters, rodents are not humans. So while rat studies may be useful for suggesting what sorts of toxicity to look for in humans, often they do not predict harmful effects. Indeed, the toxicity of phthalates in rats appears not to be replicated in humans or other primates.

Second, as noted above, the dose makes the poison. This means that the mere presence of something in the body does not imply harm; one needs to know the dose and length of exposure, what it does (if anything) in the body, how it is disposed of, and so forth. Virtually any substance, including water, can be toxic at high enough levels. Consider an example taught to all medical students. Part of the workup for hypertension (high blood pressure) is to inquire whether the patient eats large amounts of licorice, because it contains glycyrrhizin, a chemical that promotes sodium and fluid retention and raises blood pressure.

Extremely versatile and cost-effective, phthalates are among the most rigorously studied additives in the United States and Europe over the last 50 years. A blue-ribbon panel of scientists chaired by former U.S. Surgeon General C. Everett Koop performed a careful review of the scientific literature on phthalate exposure and risk-assessment experiments. "Consumers can be confident that [phthalate-containing] toys and medical devices are safe," said Koop. "The panel's findings confirm what the U.S. Food and Drug Administration and the Consumer Product Safety Commission have been saying about these products all along. There is no scientific evidence that they are harmful to children or adults."

Numerous studies have shown that human exposure to phthalates under ordinary circumstances is low and harmless. In fact, the source of most phthalate exposure is food: According to a review performed by the National Institutes of Health, food constitutes approximately 85-90 percent of phthalate exposure in adults, and 44-60 percent in infants (depending on whether or not they are breast-fed). Nor are phthalates harmful even at *high* levels of exposure – in patients undergoing regular hemodialysis or extracorporeal oxygenation of their blood in an intensive care unit, for example.

The Consumer Product Safety Commission rejected a national ban on vinyl toys in 2003, after calculating likely exposure to diisononyl phthalate (DINP), the most common phthalate in children's toys. The total mouthing time for babies 3-12 months old is about 10 minutes per hour. Pacifiers (which do not contain phthalates) account for most of babies' sucking time, with their own body parts next. Soft vinyl toys containing DINP were sucked on for under 11 seconds per hour, or under 5 minutes a day. Even those whose sucking was in the 99th percentile were chewing on their DINP-containing toys for no more than 12 minutes per day. The CPSC concluded that a baby would have to suck for about ten times as long before he or she could consume enough DINP to have any potential adverse effects.

International scientists agree. In 2003, for example, the European Union's Institute for Health and Consumer Protection concluded in a risk assessment: "The end products containing DINP (clothes, building materials, toys and baby equipment) and the sources of exposure (car and public transport interiors, food and food packaging) are unlikely to pose a risk for consumers (adults, infants and newborns)." In spite of the reassuring risk assessments, politicians overruled them, and the EU instituted a "permanent" ban on phthalates in children's toys in 2005.

The public is harmed when lawmakers proscribe the use of a product that has been proven safe and useful. Inevitably, manufacturers will turn to -- and consumers will be exposed to -- alternatives that are likely to be less well tested.

Another example of a useful chemical in the crosshairs of activists is the fat substitute called olestra. After an eight-year review – one of the most intense evaluations of a food in history -- in 1996 the FDA concluded that product is safe for use in certain foods. From the outset, the Center for Science in the Public Interest (CSPI), a Washington-based Naderite lobbying group, has conducted a sleazy campaign of intimidation and disinformation against not only the product itself but against the regulators and independent advisors who evaluated and approved it.

Olestra is a potential boon to public health in the United States, where one person in three is obese and diets are dominated by fat. Americans have bought more than 9 billion servings of snacks cooked with olestra. If they had chosen to eat regular, full-fat chips instead, they would have consumed an additional half trillion calories and 100,000 tons of fat. Ironically, CSPI has consistently decried the amount of fat in Americans' diet, but since regulatory approval of olestra, the organization has abandoned objectivity and common sense in an all-out war on the product. Its executive director, Michael Jacobson, called it a "public health time bomb," adding that then-FDA Commissioner Kessler "has lit the fuse." Jacobson appeared on ABC's World News Tonight to tell the American public that "the only kind of label notice that would be acceptable to us would be . . . a skull and crossbones."

Inexplicably, even scientific and medical journals' activism at times contributes to public confusion about issues of safety and risk. In a May 2007 the journal *Cancer*, a publication of the American Cancer Society (ACS), ran a special online supplement which concluded that breast cancer is caused by trace chemicals in the environment, including pesticides, chemicals in cosmetics, and substances such as PCBs and DDT. Dr. Elizabeth Whelan, president of the New York-based American Council on Science and Health, observed that the paper contained several obvious, severe flaws and should not have survived peer review. She also pointed out the damage caused by the popular press' uncritical acceptance of such findings, and that the journal and the ACS performed a grave disservice. Instead of bolstering the notion that "allegedly inescapable, invisible, hostile chemical agents" are a major cause of cancer, she suggested that they should be reminding women to take prudent measures such as regular mammograms and being vaccinated against human papilloma virus.

Another prominent target of misplaced activism is the pharmaceutical industry, which for a half-century has been like the proverbial goose that lays gold eggs.

It is not unreasonable to want lower costs and wider availability of drugs (as I do), but some activists are demanding nothing less than an end to the protection of drug companies' intellectual property – although this is an industry whose innovators on average expend 12-15 years and upwards of a billion dollars in direct and indirect costs to bring a drug to the American market. Only about twenty percent of drugs that begin clinical testing proceed through trials and are eventually approved for marketing; and even more ominously, only three out of ten drugs that are finally marketed recoup their development costs.

The stock-in-trade of many of the drug industry's attackers is the same used by Rifkin, Greenpeace, Pew, CSPI, Union of Concerned Scientists, and the rest of the usual suspects: The Big Lie, repeated again and again, with endless variations and facets. These include accusations that innovator drug companies are really sponging off government-funded academic research, so, in effect, patients pay for drugs twice; fail to make expensive drugs available to developing countries; irresponsibly suppress the results of "negative" clinical trials; are insufficiently regulated; and that protection of intellectual property – primarily via patents – is an unfair "subsidy."

Many of these charges are demonstrably false. In 1999, for example, the NIH thoroughly investigated whether its research funding commonly leads to the development of pharmaceuticals, the profits from which taxpayers might be entitled to share. Of 47 drugs that had earned revenues of $500 million or more, NIH support had figured significantly in only four, two of which were actually the same drug. The NIH supports primarily pre-commercial, fundamental research into the biochemistry, physiology and molecular biology of cells and organisms, in health and disease.

Another fallacy is the notion that proprietary data submitted by innovative companies to regulatory agencies comprises a "public good" analogous to national defense -- something that cannot readily be withheld from one individual consumer without withholding them from all, and for which the marginal cost of an additional person consuming it, once it has been produced, is zero. The argument goes that information about drug development should be transferred from shareholders (including employees and pensioners) of innovator drug companies to manufacturers in developing countries. The rationale? To enable them to jump-start their commercial activities so that they can compete with the multinational companies that first discovered and developed the product. That only makes sense if you believe that the more than $50 billion spent on pharmaceutical R&D annually comes from the tooth fairy.

The ability of the developers to own their inventions and data, which is currently a contentious issue in bilateral and regional trade negotiations, is essential. It enables innovators to support sales, marketing and post-marketing surveillance activities -- not to mention providing the resources for R&D on new products. The compulsory transfer of proprietary inventions, technologies and data into the pubic domain constitutes a seizure of assets.

An egregious example is a proposal from a number of developing countries (led by Brazil) currently under discussion at a meeting of the Ad Hoc Intergovernmental Meeting on Genetic Resources and Disclosure Requirements under the auspices of the World Intellectual Property Organization (WIPO). It would create a mandatory international regime – a worldwide web of regulation – to ensure "Access (to) and Benefit Sharing" (ABS) of genetic resources. As part of a patent application, there would need to be disclosure of all flora, fauna, and traditional sources of knowledge that contributed to the development of a new product or technology. This sort of vague requirement is likely to benefit only the international lawyers who will argue, negotiate and litigate such issues endlessly.

Although it does not involve the production of a pharmaceutical (except in the sense that wine can be therapeutic) the story of a Pinot Noir grape variety brought to Northern California by Richard Peterson, a major figure in the wine industry there for four decades, is illustrative. In 1980, during a trip to England Peterson took cuttings from vines in a village called Wrotham. Subsequently, he experimented with these odd-looking, blue-gray grapevines in various parts of the California wine country, eventually planting them on two acres of the Peterson Family Christmas Tree Farm in Napa. In 2006, the fruit of those vines reached the market in the form of the 2000 Richard Grant Wrotham Pinot Reserve, at $60 a bottle.

Does Peterson owe more than a "thank you very much" to England or to the growers from whom he took the cuttings? And how about to the people who brought this Pinot Noir variety to Wrotham in the first place – namely, the Romans who invaded Britain 2000 years ago? How far back and how wide do we need to cast the net to resolve such issues of "ownership" of flora (or animals or microorganisms) used for commercial purposes?

This kind of situation illustrates that although ABS-oriented disclosure requirements may seem benign, and even equitable, they are a prescription for disaster. They would create a legal basis to challenge and overturn the validity and ownership of patents after they were approved, which could have significant negative economic impacts on industrialized and developing countries alike. A study by Tim Wolfe and Benjamin Zycher found that in ABS systems that focus on the confiscation of intellectual property, "companies would have greater incentive to avoid investment in any technology that might conceivably be linked back to a country or tribe's genetic resources or traditional knowledge. That includes not just drastically curtailing bioprospecting (and any potential benefits that would otherwise have been shared), but slashing investment and research into any new biotech or pharmaceutical product that might conceivably be threatened with patent revocation."

Wolfe and Zycher estimate the impacts in 27 individual countries of the imposition of ABS-oriented disclosure requirements on investment in biotech and pharmaceutical companies: "[I]n 20 years the biotech and pharmaceutical capital stock would be reduced $144 billion, or over 27%, relative to what it otherwise would have been. Interestingly, the United States is not the biggest loser. The nations included in the study from the European Union would lose a combined $79 billion in capital stock, compared to a US loss of $21.6 billion."

Patents are not the only form of intellectual property under attack. Activists similarly deride trademarks – the intellectual property represented by brand names – as a commercial ploy, ignoring that they differentiate products, help to avoid confusion for consumers, and identify the responsible manufacturers as the source of the drug if problems arise with side effects or quality control.

CONCLUSION

In his excellent book, "The March of Unreason," the British polymath Dick Taverne (aka Lord Taverne of Pimlico) analyzes the nature of irresponsible contemporary activism. He observes that "in the practice of medicine, popular approaches to farming and food, policies to reduce hunger and disease and many other practical issues, there is an undercurrent of irrationality that threatens the progress that depends on science and even [threatens] the civilized basis of our democracy," and that we ignore this trend at our peril. This irrationality emanates from a "new kind of fundamentalism" that has infiltrated many environmentalist campaigns -- an undiscriminating "Back-To-Nature" movement that views science and technology as the enemy and as a manifestation of an exploitative, rapacious and reductionist attitude toward nature. It is no coincidence, according to Lord Taverne, that eco-fundamentalists are strongly represented in anti-globalization and anti- capitalism demonstrations around the world.

What drives this ominous trend? A "politico-legal-media complex" that peddles fear in the guise of promoting safety, has effectively banished agricultural biotechnology from Europe and Africa, has the chemical industry on the run, and the pharmaceutical industry in its crosshairs.

Lord Taverne's views complement nicely historian Richard Hofstadter's classic analysis of religious and political movements in American public policy, "The Paranoid Style in American Politics." Hofstadter called the religious and political activists' obsession "paranoia," and observed that "the central image is that of a vast and sinister conspiracy, a gigantic and yet subtle machinery of influence set in motion to undermine and destroy a way of life." He identified a characteristic "leap in imagination that is always made at some critical point in the recital of events."

How ironic that these activists imitate the very conspiracies that they imagine are threats to society. Viewed from Hofstadter's model of the paranoid style, the "conspiracy" here is large, profitable, multinational companies deciding which products to pursue in order to maximize profits, and over-promoting them; while the "leap in imagination" resides in the conviction that stimulating competition and permitting the marketplace to decide on drugs' success or failure – that is, the free-market model – somehow is bad for patients, doctors, and, in particular, the inhabitants of less developed countries.

Anti-technology, anti-business activists – whether they are found in NGOs, government or the media -- fear a world in which exploitative, rapacious, multinational corporations conspire to strip away individual choice from the world's farmers and consumers. Yet it is *they* who are guilty of the mendacity and manipulation they imagine they see in others; *they* who are guilty of stripping away the freedom of researchers to research, doctors to doctor, and consumers to consume vaccines and drugs that can be life-saving. The machinations of anti-technology, anti-business activists impoverish us all, intellectually and economically. Were they to have their way, unwise, regressive public policy could reduce significantly the pursuit of knowledge and production of wealth worldwide.

Like cheap knock-offs of designer goods, some of the offerings in the marketplace of ideas may be attractive at first glance but not stand up to scrutiny. Only if we learn to distinguish the genuine from the fake will we be able to protect ourselves – and our supply of new pharmaceuticals, pesticides, food and other products – from the tyranny of irresponsible activists.

In: Consumer Product Safety Issues
Editors: P. A. Varga, M. D. Pintér
ISBN: 978-1-60456-826-4

Chapter 4

A Review: Analysing Pathogen Risks in Fresh-Cut Packaged Fruits and Vegetables

G.A. Francis*
Food Science Research Centre, Department of Life Sciences,
University of Limerick, Limerick, Ireland.

Abstract

As the consumption of fresh-cut produce increases, there has been an increasing number of serious food poisoning outbreaks linked to these products. The primary pathogens of concern are *Escherichia coli* O157:H7, *Salmonella* and *Listeria monocytogenes*; there are also important emerging threats from viral and protozoan pathogens. With no pasteurisation or kill step during processing, this food system relies on good agricultural and manufacturing practices, an incomplete decontamination process (dipping in chlorinated water or its equivalent), and refrigerated storage, as the only hurdles to prevent contamination and growth of human pathogens. These hurdles are ineffective, as demonstrated by the large number of foodborne outbreaks associated these products over the past 10 years.

This chapter describes the processing and packaging technologies involved in the production of fresh-cut produce. The effects of unit operations, such as antimicrobial washing and decontamination treatments, on the survival of pathogens are discussed. The implications of gas atmospheres and storage conditions on the growth of associated pathogens are reviewed in detail. While much progress has been made in the past decade in our understanding of the microbiological safety of these complex food systems, the need for further research and additional control measures are highlighted in order to assure a safer food supply and to restore confidence in the fresh-cut produce sector.

Keywords: Antimicrobial washing, foodborne pathogens, food safety, fresh produce, minimal processing, modified atmosphere packaging.

---

* Correspondence to: Food Science Research Centre, Department of Life Sciences, University of Limerick, Limerick, Ireland., E-mail: jillian.francis@ul.ie; Tel: +353.61.213427; Fax: +353.61.331490

Introduction

Fresh-cut produce may consist of peeled, sliced, shredded, trimmed and washed fruits or vegetables. These products are usually sealed within semi-permeable packages and stored at refrigeration temperatures. Fresh-cut, modified atmosphere packaged (MAP) produce provide substrates and environmental conditions conducive to the survival and growth of microorganisms. Preparation treatments such as peeling and shredding disrupt surface tissues, expose cytoplasm and provide a potentially richer source of nutrients than intact produce [14, 15, 40]. This, combined with high A_w and either close to neutral (vegetables) or low acid (many fruits) tissue pH, facilitate rapid microbial growth [28].

These products can harbour large and diverse populations of microorganisms. The main bacteria present are Gram-negative rods, predominantly *Pseudomonas*, *Enterobacter* or *Erwinia* species [44, 106, 163]. However, lactic acid bacteria have also been detected, and may predominate in salads held at abuse (30°C) temperatures [162]. Yeasts frequently isolated include *Cryptococcus*, *Rhodotorula*, and *Candida* [40], and commonly isolated moulds are *Aureobasidium*, *Fusarium*, *Mucor* and *Penicillium* [226]. The microorganisms and counts present vary depending on the product type and on environmental/agricultural, processing and storage conditions.

Traditionally, fresh and fresh-cut fruits and vegetables have had a good microbiological safety track record. However, human pathogens can contaminate fresh-cut produce and there has been an increase in the number of produce-linked foodborne outbreaks in recent years. Agricultural production and post-harvest handling are considered to be the prime sources of contamination [171]. Contamination may occur in the field from improperly composted animal manures, contaminated seed, soil, irrigation water and surface waterways, domestic and wild animals or birds, and during harvesting (e.g. from contaminated harvesting tools and equipment, poor handling practices). An investigation into a recent *E. coli* O157:H7 outbreak in the USA, linked to bagged fresh spinach, identified surface water, irrigation wells, cattle faeces and wild pig faeces as possible sources of contamination [166, 220]. Produce contamination can also occur during processing and packaging (e.g. cutting, shredding, washing, contact with contaminated work surfaces/packaging materials/equipment, poor hygiene practices of workers) and during transportation and distribution. The contamination problem may be further magnified as production volumes increase and distribution widens [220].

If contaminated in the field, the washing and sanitisation steps of minimal processing cannot be relied upon to ensure elimination of all pathogens. The risk of food poisoning is increased, as the majority of these products are consumed raw without a cooking treatment or any other step that effectively controls or eliminates pathogens prior to consumption. By extending shelf-life and protecting product quality, fresh-cut MAP produce can provide sufficient time for pathogens to grow to significant numbers on otherwise acceptable fresh foods [23, 24]. A comprehensive understanding of the implications of minimal processing and mild preservation technologies on pathogen survival and growth on produce is required. The effects of MAP technology on the survival and growth of spoilage organisms and on the competitive interactions between pathogens and spoilage organisms are also important [97]. While considerable progress has been made in the past decade in our understanding of the safety of these complex food systems, there are significant gaps in knowledge requiring further research in order to assure a safer food supply.

Foodborne Outbreaks

Contamination of fresh produce with human pathogens can occur anywhere in the farm-to-fork continuum, including agricultural production, harvesting, post-harvest handling, storage, processing, transport, distribution, retail display and final preparation. A range of pathogens have been isolated from produce (Table 1), [39, 100], and a number of food poisoning outbreaks have been linked to the consumption of fresh-cut produce and salads (Table 2). In 2006, several highly publicised foodborne outbreaks occurred in the USA including *E. coli* O157:H7 outbreaks linked to spinach and lettuce, *Salmonella* Typhimurium outbreaks associated with tomatoes and multiple outbreaks of botulism linked to carrot juice. In addition, *E. coli* O157:H7 in lettuce, *Salmonella* in lettuce and tomatoes and Norovirus associated with raspberries have caused several recent outbreaks in Europe [155]. Furthermore, botulism has been linked to a MAP dry coleslaw product [209] and *Salmonella* Newport has been traced to ready-to-eat salad vegetables [184]. The problem may be even greater as the short shelf-lives, complex distribution and universal consumption of fresh produce make produce-implicated outbreaks difficult to pin down; many food poisoning incidents related to fresh produce may go under-reported [178]. Contributory factors for the increase in the number of produce-linked foodborne outbreaks in recent years include increased consumption of fresh and fresh-cut produce, the greater range and diversity of products available to consumers and the elimination of seasonality by almost year-round availability of many commodities. This diversity and availability has been achieved by increased globalisation of the produce trade, but has brought with it new food safety risks and challenges. The pathogens of primary concern in fresh-cut MAP produce are pathogenic *E. coli* (e.g. *E. coli* O157:H7), *Salmonella* and *Listeria monocytogenes*, however, there are also important emerging threats from viral (Hepatitis A virus, Calicivirus, Norovirus) and protozoan (*Cyclospora*, *Cryptosporidium*, *Giardia*) pathogens.

Bacterial Pathogens

Salmonella is the most common cause of disease outbreaks linked to fresh fruit and vegetables. Salmonellae are abundant in faecal material and sewage-polluted water; consequently they may contaminate soil and crops with which they come into contact. Salmonellae from a range of produce, including sprouted seeds, cantaloupe melons, tomatoes, unpasteurised citrus juices, rocket and lettuce, have been responsible for several food poisoning outbreaks [135].

Fresh produce was not considered a significant vector for the transmission of *E. coli* O157:H7 until the mid-1990s, when a series of outbreaks associated with minimally processed horticultural products clearly showed that contamination can occur by indirect routes [70, 117]. The largest *E. coli* O157:H7 outbreak was in 1996 when >6,000 school children in Japan were infected with *E. coli* O157:H7 from white radish seed sprouts. Since

1993, 26 reported outbreaks of *E. coli* O157:H7 infection have been traced to contaminated lettuce and leafy green vegetables [161]. A recent multi-state outbreak in the USA linked to bagged fresh spinach caused approximately 205 confirmed illnesses, 31 cases of hemolytic uremic syndrome and three deaths [51, 220]. This outbreak was followed by two restaurant-associated outbreaks linked to the consumption of pre-washed lettuce [129]. These recent outbreaks have highlighted the dangers of centralised distribution and the great distances that fresh produce travels.

L. monocytogenes is widespread on plants and the agricultural environment generally, and is capable of surviving a variety of environmental stresses, including refrigeration temperatures [19], gas atmospheres commonly present within MAP produce and relatively low pH. Although there is a large amount of literature dedicated to researching the isolation, attachment, survival and growth of *L. monocytogenes* on produce [8, 76, 118, 160, 205, 223], only two fresh-cut produce related listeriosis outbreaks have been documented [124, 201]. However, vigilance is required due to the severity of the symptoms of disease caused by *L. monocytogenes* (meningitis, septicaemia, abortion and death in severe cases) and its relatively high mortality rate.

Viral and Protozoan Pathogens

The significance of viruses with respect to foodborne disease is clear with the inclusion of Norwalk virus, Hepatitis A virus and 'other viruses' within the top ten causes of foodborne disease outbreaks in the USA [62]. Outbreaks caused by Hepatitis A virus, Calicivirus and Norwalk-like viruses have been associated with the consumption of raspberries and strawberries, melons, lettuce, watercress, spring onions and diced tomatoes [28, 117, 119, 126, 135, 159, 195].

Viruses can be transmitted to produce by infected food handlers through the faecal-oral route during post-harvest processing and in food service establishments. In addition, viruses have been isolated from sewage and untreated water used for crop irrigation. A large outbreak in Pennsylvania associated with hepatitis A virus contamination of green onions was traced back to contamination in the field in Mexico [229]. Despite their significance, data regarding the effects of food processing and storage conditions on the survival and infectivity of viruses is limited. Nonetheless, the potential of several viruses to survive on vegetables for periods exceeding their normal shelf-life has been identified [11, 141, 200]. Survival appears to be dependent upon product type and pH, temperature and moisture content [11, 37, 67, 141, 147]. Viral properties, such as low infectious dose, resistance to chlorination, and resistance to commonly used produce washing regimes, have made it difficult to control enteric viruses on fresh produce [79, 140].

The protozoan parasites *Giardia lamblia*, *Cyclospora cayetanensis* and *Cryptosporidium parvum* have been the cause of serious foodborne outbreaks involving berries [120, 121], lettuce and onions [57] and raw sliced vegetables [168]. These organisms normally gain access to produce before harvest, usually as a result of contaminated manure or irrigation water and poor hygiene practices by food handlers [28, 219]. Chaidez *et al.*, [60] reported that *Cryptosporidium* and *Giardia* were detected in surface water and in wash water tanks at packing-houses and that the risk of contaminating fresh produce is high.

Table 1. Occurrence of pathogens on fresh-cut produce

Pathogen / Vegetable	Number (and %) of positive samples	Country and Comments	Reference
Listeria monocytogenes			
Cucumber slices	4/5 (80%)	Malaysia	8
Bean-sprouts	6/7 (85%)	Malaysia	8
Coleslaw	2/92 (2.2%)	Canada	201
	2/50 (4%)	Singapore	76
	3/39 (7.7%)	United Kingdom	160
Pre-packed salads	3/40 (7.5%)	Northern Ireland	118
	4/60 (6.7%)	United Kingdom	205
Chopped lettuce	5/39 (13%)	Canada	180
Cut and packaged lettuce	3/120 (2.5%)	Australia	211
Prepared mixed vegetables	8/42 (19%)	United Kingdom (contamination during processing suspected; <200/g present)	223
Fresh cut salad vegetables	11/25 (44%)	The Netherlands ($<10^2$/g present)	17
Chicory salads	(8.8%)	France (<1/g present)	172
Prepared vegetables	1/26 (3.8%)	United Kingdom	160
Processed vegetables and salads	(13%)	United Kingdom	167
Mung bean	19/102 (19%)	France	172
E. coli O157:H7			
Salad mix	0/63 (0%)	US	154
Celery	6/34 (18%)	Mexico	28
Salmonella spp.			
Salad mix	1/159 (0.6%)	Egypt	197
Endive	2/26 (7.7%)	Netherlands	216
Mung bean	30/344 (8.7%)	Thailand	134
Lettuce	82/120 (68%)	Italy	84
Fennel	64/89 (72%)	Italy	84
Aeromonas			
Cut lettuce	66/120 (55%)	Australia	211
S. aureus			
Salad greens	13/256 (5.1%)	UK	28
Cryptosporidium			
Lettuce	2/80 (2.5%)	Costa Rica	169
Carrots	1/80 (1.3%)	Costa Rica	169

Table 2. Foodborne outbreaks linked to the consumption of fresh fruits and vegetables

Pathogen	Product suspected	No. of cases	Location
Bacteria			
L. monocytogenes	Shredded cabbage in coleslaw	41	Canada
	Raw tomatoes, lettuce & celery	20	Boston, USA
Salmonella	Lettuce	96 & 56	UK & Finland
	Lettuce	386	UK
	Rocket	12	Norway
	Sliced watermelon	39	Michigan, USA
	Watermelon	17	Massachusetts
	Cantaloupe melon	22	Canada
	Cress sprouts	31	UK
	Mung sprouts	143	UK
	Alfalfa sprouts	124	Canada
	Tomatoes	84	Multi-state USA
	Tomatoes	174	Multi-state USA
	Tomatoes	561	USA & Canada
E. coli O157:H7	Spinach	205	USA
	Cantaloupe melon	9	Oregon, USA
	Radish sprouts	6561	Japan
	Alfalfa sprouts	108	US
	Lettuce	120	Sweden
	Lettuce	70	Montana
	Lettuce	23	Canada
	Lettuce	30	California
	Coleslaw	33	Indianapolis
	Fruit salad	47	Wisconsin
Cl. botulinum	Carrot juice		Multi-state USA & Canada
Parasites			
Cyclospora cayetanensis	Raspberries	1465	20 US states & Canada
	Raspberries	1012	Multi-state USA & Canada
	Blackberries	104	Canada
	Baby lettuce leaves	>91	Florida, USA
	Basil	>308	Multi-state USA
Cryptosporidium parvum	Green onions	54	Washington, USA
Giardia	Lettuce and onions	21	New Mexico
Viruses			
Norovirus	Raspberries	43	Sweden
	Raspberries	500	Denmark
	Raspberries	5	France
Norwalk virus	Melon	206	UK
Hepatitis A	Spring onions	400	USA
	Lettuce salad	103	Florida, USA
	Diced tomatoes	92	Arkansas
Calicivirus	Raspberries	>200	Canada

Adapted from: 100, 117, 135 and 172

The lack of sensitive methods for detecting and determining the survival or inactivation of oocysts has hampered incidence studies and research on the behaviour of protozoan parasites on refrigerated MAP produce. However, the increase in produce-linked outbreaks due to these organisms (see Table 2) indicates that research in this area is needed.

FACTORS AFFECTING PATHOGEN SURVIVAL AND GROWTH ON FRESH-CUT PRODUCE

Pathogen survival and growth on produce is influenced by a number of interdependent factors, principally storage temperature, product type / combinations, minimal processing operations (e.g. slicing, shredding, washing, disinfection and decontamination treatments), gas / package atmosphere and competition from the natural microflora present on produce.

Storage Temperature

Storage temperature is probably the single most important factor affecting survival and growth of pathogens on fresh-cut produce. Storage of produce at adequate refrigeration temperatures, will limit pathogen growth to those that are psychrotrophic/psychrophilic. While psychrotrophic organisms, such as *L. monocytogenes*, are capable of growth at low temperatures, decreasing the storage temperature (≤4°C) will reduce the rate of growth [32, 54]. *L. monocytogenes* populations remained constant or decreased on packaged vegetables stored at 4°C, while growth of *L. monocytogenes* was supported on all vegetables stored at 8°C, with the exception of coleslaw mix [95].

Mesophilic pathogens, such as *Salmonella* and *E. coli* O157:H7, are unable to grow where temperature control is adequate (i.e. ≤4°C). However, if temperature abuse occurs, they may then grow. Survival of *Salmonella* in produce stored for extended periods in chilled conditions may be of concern [185, 235]; *Salmonella* survived on a range of vegetables for more than 28 days at 2-4°C [128]. *E. coli* O157:H7 populations survived on produce stored at 4°C and proliferated rapidly when stored at 15°C [191]. Reducing the storage temperature from 8 to 4°C significantly reduced growth of *E. coli* O157:H7 on MAP vegetables; however, viable populations remained at the end of the storage period at 4°C [95]. Strict control of temperature throughout the chill-chain is crucial for maintaining microbiological safety [55, 65, 88, 105, 194], as even mild abuse temperature during storage can facilitate rapid proliferation of pathogens.

Product Type / Combinations

Each product type has a unique combination of compositional and physical characteristics and will have specific growing, harvesting, processing and storage conditions; pathogen survival and growth varies significantly with the type of product [10, 56, 131]. Dry coleslaw mix was largely unsuitable for *L. monocytogenes* and *E. coli* O157:H7 growth while significant growth of the pathogens occurred on shredded lettuce [95, 96]. Product factors that may affect pathogen survival and/or growth include: pH, presence of competitive spoilage microflora and/or natural plant antimicrobials and respiration rate/packaging interactions.

Product pH strongly influences the survival/growth of pathogens. Most vegetables have a pH of ≥5.0, and consequently support the growth of most foodborne bacteria. Many fruits have an acidic pH, however, a number of melons/soft fruits have pH values ≥5.0 and will support growth of many pathogenic bacteria [28, 85, 157, 171, 172]. *L. monocytogenes* survived and grew on apple slices, cantaloupe melon [65, 221], and tomatoes [31]. Acid tolerance is common in *E. coli* O157:H7 and *Salmonella* serotypes and these organisms can survive/grow in acidic produce [74, 152, 222, 227, 235]. Outbreaks of disease from *Escherichia coli* O157:H7 and *Salmonella* strains have occurred in apple cider and orange juice [58, 59], which have pH values between 3.5 and 4.0.

Some fruit and vegetable tissues have naturally occurring antimicrobials that provide varying levels of protection against pathogens [157, 208]. For example, the inhibitory effects of raw carrots and carrot juice on growth of *L. monocytogenes* have been reported [30, 33, 131, 174], and cooked cabbage and Brussels sprouts were inhibitory towards *Listeria* [33, 34, 131, 172]. Natural compounds from edible plants, such as oilseeds, herbs and spices, may have potential as effective antimicrobial agents against foodborne pathogens and spoilage bacteria on produce [9, 49, 77, 206, 207]. Nychas [176] reported antimicrobial activity of essential oils (EOs) from oregano, thyme, sage, rosemary, clove, coriander, garlic and onions against both bacteria and moulds. In addition, oils of oregano, thyme and clove were effective against *E. coli* [112, 207]. The antimicrobial activity of plant EOs is mainly attributed to the presence of high concentrations of phenol derivatives such as thymol, eugenol, and carvacrol, but there is evidence that minor components may also play a role in the overall antimicrobial effect [145, 183].

Fresh-cut produce harbour a large and diverse microflora. Effects of competition between the indigenous microflora and pathogens on fresh-cut packaged produce may impact on pathogen growth and therefore play an important role in product safety. Beansprouts did not support good growth of *L. monocytogenes* or *E. coli* O157:H7, due presumably to competition from high populations of background microflora, inhibition from the relatively high in-pack CO_2 levels (25-30%) and the more limited nutrient availability of intact vegetables [95].

Fresh-cut produce may be combined with cooked ingredients. Growth of *L. monocytogenes* on raw endive was probably limited by nutrient availability, but reached higher numbers when sweetcorn was added [53, 173]. Similarly, the addition of cooked products to raw vegetables supplied a source of nutrients and permitted rapid growth of both spoilage and pathogenic populations on such products [218].

Minimal Processing Operations

The unit operations employed during the production of minimally processed produce (i.e. peeling, slicing, shredding) cause the destruction of surface cells, affect product respiration rate and pH, and release nutrients and possibly antimicrobial substances from plant cells [15, 40], which will in turn affect the behaviour of pathogens. In general, pathogens do not grow well on uninjured surfaces of fresh intact produce; however, cutting or slicing facilitates microbial growth. For example, injuries to the wax layer, cuticle and underlying tissues increased bacterial adhesion and growth [113, 114, 202, 213, 215]. Peeling, cutting and shredding, should be carried out with equipment designed to cause the minimum of tissue disruption, as severe processing may facilitate more effective penetration and growth of pathogens [108, 150].

Shredding and slicing processes are important sources of contamination of fresh-cut produce. Pathogens can become attached to processing equipment (e.g. slicers, shredders) and once attached are very difficult to remove by chemical sanitisers [43, 101, 106, 172]. Indeed, *L. monocytogenes* has been recovered from the environment of processing operations used to prepare fresh-cut vegetables [233], highlighting the importance of strict hygiene during processing. Recommendations implemented to ensure quality and safety of produce relate to good agricultural and manufacturing practices, microbial specifications for the processed product, and proper storage conditions [172].

Washing / Disinfection / Decontamination Treatments

Washing in tap water removes soil and other debris, some of the surface microflora, and cell contents and nutrients released during slicing that help support growth of microorganisms [38]. However, while washing in tap water removed bacteria from exposed surfaces, substantial numbers still remained in hollows at the junction of epidermal cells and in folds in the epidermis [3, 29, 42, 130, 172]. In addition, due to the re-use of wash water in industry, washing may result in spread of bacteria and cross-contamination of products rather than decontamination [28, 35, 41, 106].

A variety of antimicrobial wash solutions have been used to reduce populations of microorganisms on fresh produce. The effectiveness of washing/antimicrobial dipping depends on a number of factors including: (1) type of treatment, (2) type, numbers, physiological growth phase and stress resistance of the target microorganism(s), (3) product type, (4) antimicrobial concentration, (5) pH of the solution, (6) contact time, (7) temperature of washing water and (8) general sanitation of plant and equipment [3, 25, 80, 81].

Chlorine is the most frequently used disinfectant for fresh fruits and vegetables; added to water as a solid, liquid or gas [3, 5, 35, 158]. Washing fresh produce with 50 to 200ppm of chlorinated water generally results in approximately 1 to 3-log_{10} reduction of microbial populations. Total microbial populations were reduced about 1000-fold when lettuce was dipped in water containing 300-ppm total chlorine, but no effect was seen against microbial populations on red cabbage or carrots [106]. The effects of chlorine in removing pathogens from produce have also been studied. The maximum log_{10} reductions of *L. monocytogenes*, after treatment with chlorine (200-ppm), were 1.7 for lettuce and 1.2 for cabbage [233]. Dipping coleslaw and lettuce in a chlorine solution (100-ppm) reduced initial *L. innocua* and *E. coli* populations, but resulted in enhanced growth during extended storage at 8°C [94]. Chlorine (100- to 200-ppm) was only marginally effective at reducing *E. coli* levels on lettuce tissue surfaces [26] and broccoli florets [18], while *Salmonella* populations on alfalfa sprouts were reduced by about 2-log_{10} CFU/g after treatment with 500-ppm chlorine [35].

Chlorine, used at concentrations currently permitted by the industry to wash fresh produce, cannot be relied upon to completely eliminate pathogens. The limited effectiveness of chlorine treatment on fresh-cut produce is due to a number of factors, but is largely attributed to the inability of chlorine sanitisers to reach pathogens attached to hydrophobic regions of cut plant tissue [3, 202]. The hydrophobic nature of the waxy cuticle on produce protects surface contaminants from exposure to chlorine which does not penetrate or dissolve these waxes [3]. The efficacy of disinfection treatments also depends on the nature of bacterial attachment [114, 115, 150], and microbial cells that become internalised or embedded in crevices, creases or injured tissues may be inaccessible to disinfection treatments [3, 143, 158, 202, 213, 214]. Biofilms occur naturally on field crops and bacterial cells within biofilms were shown to be more resistant to removal by washing and inactivation by sanitisers and disinfectants [135]. It is important to sanitise injured surfaces before cutting as once cut or injured surfaces are contaminated by pathogens, it is very difficult to remove attached microorganisms. The most useful effect of chlorine may be in inactivating vegetative cells in washing water and on equipment during processing as part of a HACCP system, thus avoiding build-up of bacteria and cross-contamination [230].

Another concern regarding the use of antimicrobial dips is that pathogens may not be fully eliminated by commercial treatments, while at the same time natural competitive organisms may be removed/reduced. *L. monocytogenes* inoculated onto disinfected (10% hydrogen peroxide) endive leaves grew better than on water-rinsed produce [53], and dipping lettuce in a chlorine (100-ppm) solution followed by storage at 8°C, significantly enhanced *Listeria* growth compared with undipped samples [99]. Disinfection before contamination with the pathogen occurs may increase growth of the pathogen because populations of competing microflora have been removed [20]. Therefore, strict hygiene practices and temperature management (i.e. ≤4°C) after reduction of microbial populations is crucial for safety.

Viruses and protozoan cysts on fruits and vegetables generally exhibit higher resistance to disinfectants than do bacteria or fungi [27]. Feline caliciviruses were very resistant to commercial disinfectants; however, peroxyacetic acid and H_2O_2 were effective at decontaminating strawberries and lettuce when used at four-fold higher concentrations than recommended by the manufacturers [110]. Treatment of *Cryptosporidium parvum* oocysts with 1-ppm ozone for 5 minutes resulted in <1 $\log_{10}$ inactivation [142].

Due to the ineffectiveness of chlorine in removing pathogens attached to hydrophobic regions of plant tissue and the increasing concern over the production of chlorinated organic compounds and their impact on human health and environmental safety, a variety of other disinfectants/sanitisers, including acidic electrolysed water [182], acidified sodium chlorite [210], peroxyacetic acid [181], chlorine dioxide [186, 233], hydrogen peroxide [199], organic acids [137], trisodium phosphate [233] and ozone [36, 48, 82] have been evaluated [26]. However, none of the disinfectant/antimicrobial treatments used during processing are likely to ensure elimination of all pathogens, and behaviour of surviving pathogens during subsequent storage remains unpredictable [18, 35, 181, 233].

The microbiological safety of fresh-cut produce is compromised, because of the limited efficacy of sanitiser-based washes, and due to the absence of any kill step that eliminates pathogens prior to consumption. Ideally, what is needed is a pasteurisation step, delivering a 6-log reduction in pathogen numbers. This can be achieved through the application of ionising irradiation [89], but this technology is currently unacceptable to consumers. In the absence of ionising irradiation, the sector must use an improved sanitisation / decontamination step and stronger antimicrobial hurdles during storage. Previous studies have shown that UV light was effective in reducing pathogenic bacterial loads and viruses on fresh fruits and vegetables [92, 232]. Dense phase carbon dioxide, or high pressure CO_2, is capable of penetrating into complex structures, and may be useful for inactivation of pathogens located within complex produce surfaces and internal tissues [234]. Novel decontamination techniques, including UV, pulsed electric fields, microwave, high intensity pulsed light and thermal destruction using condensing steam, may enhance pathogen destruction and these methods warrant further investigation [132].

Natural antimicrobials from edible plants have been studied for their potential as possible replacements for chemical additives because of their safety for human consumption and wide acceptance from consumers [9, 206]. Phenolic compounds present in plant essential oils (EOs) have been shown to possess broad antimicrobial activity and some are generally recognised as safe (GRAS). Application of EOs to fresh-cut produce would address the problem of unassailable hydrophobic surfaces and might help to reduce contamination and growth of spoilage and pathogenic bacteria [204]. Furthermore, use of combinations of plant antimicrobials/phytochemicals may result in additive or synergistic effects, increasing antimicrobial efficacy while at the same time reducing concentrations used and any negative sensory effects [111]. The susceptibility of bacteria to the antimicrobial effect of EOs appears to increase with a decrease in the pH, the storage temperature and the amount of oxygen within the package [49]. Therefore, the application of synergistic mixtures of plant antimicrobials/EOs, in combination with other mild preservation techniques (e.g. acid pH, refrigerated storage and modified atmosphere packaging), might help to improve the microbiological safety and quality of produce.

Modified Atmosphere Packaging (MAP)

When a fresh-cut product is packaged, it continues to respire thereby modifying the gas atmosphere inside the package, hence the term modified atmosphere packaging (MAP). Ideally, O_2 levels will fall from 21% in air to 2-5%, and CO_2 levels will increase to the 3-10% range. The gas mixtures control the product's biochemical and enzymatic reactions, inhibit the growth of microorganisms and extend the shelf-life. CO_2 inhibits the growth of bacteria by (i) affecting cellular enzymes and decreasing the rate of metabolic reactions [190], (ii) CO_2 product repression of carboxylases and decarboxylases [138], (iii) disrupting cell membrane structural integrity and/or specific functions [203], (iv) decreasing the substrate and intra-cellular pH [231], or by a combination of these mechanisms [75]. The extent of inhibition by CO_2 varies with the microorganism, CO_2 concentration, temperature of incubation, and type of food [68, 75, 83, 125].

Gas atmospheres within packages might be cause for public health concern in a number of ways. Firstly, the atmospheres and refrigeration temperatures employed may inhibit the development of some spoilage aerobic microorganisms [68, 86]. Consequently, their suppression may facilitate survival and growth of some pathogens, without the product showing obvious signs of spoilage. Furthermore, MAP increases the shelf-life of products, thus increasing the time available for pathogens to grow. Over-extending the shelf-life may allow development of significant populations, particularly if combined with exposure to even mild abuse temperatures. In addition to the target atmospheres described above, there is evidence of considerable incidence of 'unintended' atmospheres in commercial practice (e.g.

no O_2 or high levels CO_2 i.e. >20%), and these gas atmospheres have uncertain microbiological implications.

Of particular concern with refrigerated MAP produce is the growth of psychrotrophic, facultative anaerobic and microaerophilic microorganisms, which can tolerate refrigeration temperatures and low O_2 atmospheres [21], and a number of studies have indicated that MAP may select for such pathogens [32, 40, 123, 136]. Numerous researchers have shown that survival and growth of *L. monocytogenes* on produce is not reduced or affected by MAP [4, 24, 31, 32, 66, 131]. In other work, nitrogen flushing combined with storage at 8°C enhanced the growth of *L. monocytogenes* on shredded lettuce [99] and chicory salads [192]. Carlin *et al.* [52] examined the survival of *L. monocytogenes* on endive leaves stored at 10°C in air, or under 10%, 30% or 50% CO_2, with 10% O_2 and found that *L. monocytogenes* grew better as the concentration of CO_2 increased.

Salmonella and *E. coli* O157:H7 can grow under MAP conditions; however, there is insufficient information available on whether atmospheres inhibit or enhance their growth. CO_2 had little or no inhibitory effect on growth of *E. coli* O157:H7 on shredded lettuce stored at 13 or 22ºC, and growth potential was increased in an atmosphere of $O_2/CO_2/N_2$: 5/30/65, compared with growth in air [1, 72]. MAP did not influence the survival rate of hepatitis A virus on lettuce stored at 4°C, but an improvement in virus survival was observed on lettuce stored under high CO_2 levels (70% CO_2/30% N_2) at room temperature [37].

The effects of novel gas atmospheres including the use of high O_2 levels (i.e. >70% O_2) and noble gases [69] on pathogen growth have also been examined. In an agar-based study to investigate the effects of high O_2 (90%) and moderate CO_2 (10-20%) concentrations on foodborne pathogens at 8°C, Amanatidou *et al.* [4] noted inhibitory action against *L. monocytogenes*, *S.* Typhimurium, *S.* Enteritidis and *E. coli*. More research to determine the behaviour of bacterial as well as viral and protozoan pathogens on MAP produce is warranted.

Competition Between the Indigenous Microflora and Pathogens

Fresh-cut MAP produce harbour large populations of microorganisms including pseudomonads, lactic and bacteria (LAB) and *Enterobacteriaceae* [100, 172]. The background microflora provide indicators of temperature abuse largely by causing detectable spoilage, and levels can vary significantly for each product type and during storage.

LAB can exert antibacterial effects by one or more of the following mechanisms: lowering the pH [188]; generating H_2O_2 [187]; competing for nutrients [127]; and by producing antimicrobial compounds, such as bacteriocins [7, 116, 139]. A large portion of LAB isolates from beansprouts inhibited the growth of *L. monocytogenes* [50], and strains of LAB were reported to inhibit *A. hydrophila*, *L. monocytogenes* and *S.* Typhimurium on vegetable salads [225].

Various researchers have reported antagonism by the native microflora of vegetables against *Listeria* [97, 98, 153]. Reducing the background microflora of endive leaves [53] and shredded lettuce [99] resulted in enhanced growth of *Listeria*. A mixed bacterial population isolated from endive or lettuce reduced *Listeria* growth in vegetable media [53, 97, 98]. However, the inhibitory effects were dependent on gas atmosphere; in 3% O_2 (balance N_2)

growth of the mixed population was inhibited while *L. monocytogenes* proliferated [97]. Fluorescent pseudomonads have previously been shown to stimulate growth of *L. monocytogenes* in various foods, due to the release of potential nutrients by pseudomonads [153, 164, 172]. By contrast, strains of fluorescent pseudomonads slightly reduced final population densities of *L. monocytogenes* in an endive leaf medium [20]. *P. fluorescens* and *P. viridiflava* inhibited growth of *L. monocytogenes* on potato slices while *Erwinia carotovora* and *Xanthomonas campestris* did not effect its growth [153]. *Enterobacter* isolates (*Enterobacter cloacae, Enterobacter agglomerans*) significantly reduced *L. monocytogenes* growth during storage on a model lettuce medium; however, the inhibitory activities of *Enterobacter* spp. decreased as the concentration of CO_2 increased [97]. The native microflora of cantaloupe melon, especially the yeast and mould populations, might have out-competed *L. monocytogenes* for colonisable space and available nutrients, thus resulting in the decline of populations of *L. monocytogenes* [221].

Competitive microflora had a significant effect on the growth of *E. coli* O157:H7 in broth media; *Hafnia alvei* significantly inhibited the growth of *E. coli* O157:H7 at 37°C, whereas *Pseudomonas fragi* inhibited growth of the pathogen at 15°C [78]. A biocontrol strain of *P. syringae* inhibited growth of *E. coli* O157:H7 when co-inoculated onto apple tissue, and this inhibition was attributed to competition for nutrients or space [133]. *Enterobacter asburiae* decreased *E. coli* O157:H7 survival on lettuce foliage, while *Wausteria paucula* enhanced the survival of the pathogen [63]. Little is known about the mechanism(s) by which *Salmonella* manages to compete with natural microflora and survive on plant products [150]. *Pseudomonas* had an antagonistic effect on *Salmonella* [91, 151], while other work showed that Salmonellae grew better on vegetables when co-cultured with *Erwinia carotovora* or *P. viridiflava*, two major causes of bacterial soft-rot [228]. Competition with *Salmonella* was more effective with uncharacterised communities of bacteria than with individual strains, indicating additive and possibly synergistic activities [165].

Complex interactions with the indigenous microflora may have significant effects on survival and growth of pathogens. The importance of the indigenous microflora in controlling populations of pathogens, either by direct competition or through the production of antimicrobials, may provide biocontrol approaches for controlling pathogen contamination and growth. More research needs to be done to examine the influence of gas atmospheres, decontamination methods and background microflora on the survival/growth of pathogens on fresh-cut MAP produce in order to ensure that novel decontamination and preservation technologies can be applied safely.

Stress Response and Strain Variation Effects

Cumulative mild processing steps are employed during production of fresh-cut produce to improve microbiological safety. Pathogenic bacteria can respond or adapt to the mild or sub-lethal stresses encountered in these minimally processed foods in ways that increase their resistance to more severe treatments and enable better survival in foods [2, 46, 103]. Pathogens on plant surfaces are already stressed and stress may be increased during the multiple processing steps, potentially leading to very hardy bacteria geared towards survival

[135]. Cross-protection occurs because the overlapping stress responses enable bacteria exposed to one stress to become resistant to another stress. Apart from the enhanced survival in foods and increased resistance to subsequent food processing/preservation treatments, adapted or hardened pathogens may also have enhanced virulence [2, 6, 103, 196].

Two of the best studied adaptive tolerance responses are to acid (acid tolerance response) and heat (heat stress response). Acid adapted *L. monocytogenes*, *Salmonella* and *E. coli* O157:H7 survived significantly better in acidic foods such as fruit juices, when compared to non-adapted cells [104, 149]. Acid adaptation induces acid tolerance to more severe or normally lethal acid, but it can also induce cross-protection against other stresses such as thermal and osmotic stress [148, 156, 179]. Acid adaptation enhanced survival of *L. monocytogenes* during storage in packages of vegetables, which had relatively high in-pack CO_2 levels (25-30% in MAP coleslaw and bean sprouts) [96]. Exposure of *L. innocua* to acid and starvation stress decreased sensitivity to the quaternary ammonium compound cetrimide, whereas exposure to cold and heat stress increased sensitivity to this compound [170]. *E. coli* O157:H7 survived better in an acidic environment better at 4°C than at 10°C, which implies that induction of acid tolerance may enhance resistance to low temperature [64].

In addition to stress response effects, different strains of pathogens may respond differently to treatments including mild acid, low temperature and gas atmosphere [47]. There was significant variation among *L. monocytogenes* strains in their inherent stress resistance characteristics [13, 45, 71, 87, 93, 193, 212]; some strains may be more resistant to the stressful conditions encountered in fresh-cut produce. These stress response and strain variation effects should be considered when developing novel processing / preservation technologies for fresh-cut produce.

REDUCING PATHOGEN RISKS FROM FRESH-CUT PRODUCE

The microbiological safety of fresh produce is compromised because of the absence of an effective intervention step to eliminate field or process acquired contamination. Measures available to improve the microbiological safety of fresh-cut produce are listed below, based on current knowledge, highlighting research gaps and challenges.

To date a preventative approach to contamination of fresh produce, by the use of Good Agricultural Practices (GAPs), Good Manufacturing Practices (GMPs) and Hazard Analysis and Critical Control Points Programs (HACCP), has been the most effective means of ensuring fresh produce safety [198]. Pathogen risks from fresh-cut produce can be minimised by applying best practice at every stage throughout the supply chain, from agricultural production, pre-processing, processing, packaging, distribution to final preparation. At all stages, strategies to minimise contamination by pathogens, product storage at ≤4°C and education of workers and consumers are important recurring themes.

Agricultural production practices can have major implications for contamination of raw produce with pathogens [109], and producer awareness of their role in assuring food safety is vitally important. This is an extremely complex arena with great diversity in crop production methods and environmental factors [90]. The percentage of 'grower' implicated contamination incidents have been declining since 1996, and this trend is most likely due to implementation of GAPs by growers [198]. Improperly composted sewage or animal manure should not be applied to land where vegetables for processing are grown. Land subject to flooding or on which animals have grazed should be avoided [39]. Proximity to animal production facilities may also be a significant cross-contamination hazard. Irrigation should be carried out with clean water, and applied as trickle irrigation at ground level rather than as an overall spray [171]. However, serious deficits in water quality and availability exist globally, with water pollution from sewage and animal production facilities posing serious problems [178].

Birds such as gulls and pigeons, wild animals, domestic animals and insects should be excluded from packing houses and processing areas [171]. Wild birds are known to disseminate *Salmonella*, *Listeria* and *E. coli* O157, apparently picked up from feeding on garbage, sewage, etc; control of pre-harvest contamination of produce by wild birds and animals is particularly difficult [35]. Increasing globalisation of the fresh produce market and greater international trade pose serious new challenges [217], and knowledge of contamination levels in imported produce is minimal [35]. In the future, serious efforts will be made to develop and apply GAPs to wherever primary production takes place.

The microbiological quality and safety of fresh-cut produce is also greatly affected by the hygienic practices of employees. Poor handling practices during harvesting and processing can negate GAPs and HACCP programs implemented to ensure microbiological safety [166]. Workers involved in harvesting and handling should be trained in the principles of good handling and sanitation practices and provided with adequate hand-washing and toilet facilities in fields and packing houses [26]. In addition, how industry practices affect safety and what motivates workers to use safer practices are factors that warrant closer examination.

All processing should be geared to minimise opportunities for pathogen contamination and growth. Starting at harvest, bruising and cutting should be minimised prior to processing [150]. Immediately prior to processing, preliminary decontamination should be carried out by removing outer leaves, soil etc. from produce using sharp sanitised knives for any cutting. Peeling, cutting and shredding should be carried out with equipment designed to cause the minimum of tissue disruption as severe processing may facilitate more effective contamination and subsequent growth by pathogens [107]. GMPs should include effective work surface and machine sanitising to eliminate the risk of pathogen contamination from the processing environment or from machines used in processing [172, 233].

Although its benefits are questionable [39], antimicrobial dipping to an appropriate protocol is a valuable tool for reducing numbers of potential pathogens [35]. Strict hygiene and GMPs should be exercised to avoid contamination after dipping. Post-processing risks introduced by antimicrobial dips [20, 53, 99] should be addressed in HACCP protocols: the most important of these are measures to ensure that products are stored at ≤4°C at all times and the use of conservative use-by dates. More efficient washing and decontamination treatments are needed to replace or supplement current treatments. Future research in alternatives to chlorine dipping is likely to result in novel and more effective antimicrobial treatments (e.g. plant EOs / phytochemicals, antimicrobial peptides) and the application of additional interventions beyond the washing step. Where alternatives to chlorine are being introduced, any differences in their antimicrobial effects should be understood. Greater use of edible coatings (e.g. sucrose polyesters of fatty acids, cellulose derivatives, etc) to product surfaces can be expected [12, 144, 236]. Other additional novel biocontrol steps may be introduced, such as inoculation of produce with organisms inhibitory to one or more pathogens [224, 225]. Ionising radiation may also be used either alone or in combination with other treatments, as a means of improving microbial safety and extending the shelf-life of produce [73, 146, 175]. Doses in the range of <0.5 to 3 kGy have been shown to reduce or eliminate populations of pathogens and spoilage organisms on produce [89] and salmonellae were not recovered from alfalfa sprouts irradiated with 0.5 kGy [189]. However, despite the efficiency, safety and suitability to products with surface contamination [177], the use of irradiation will ultimately depend on its acceptance by consumers.

Packaging materials must be carefully selected to ensure that their gas permeability properties match the respiration rates of the products being packaged. This is necessary in order to achieve package atmospheres within the technically useful range of 2-5% O_2 and 3-10% CO_2 [16, 61], and to avoid the creation of unintended atmospheres with uncertain microbiological implications [22]. More reliable package atmospheres can be expected as a result of improved materials, temperature-responsive smart packaging and better software to define gas permeability requirements for individual products. There will be widespread commercial application of active packaging with antimicrobial and other properties. Other novel elements of MAP include the use of high oxygen and noble gas enriched atmospheres.

Ensuring that temperatures are kept at or below 4°C throughout the cold-chain is essential for microbial safety and requires considerable attention to detail. Refrigerated distribution requires suitably designed vehicles, properly loaded to allow for air movement [39]. Problems can also arise at consumer level where products are held for extended periods in cars or experience elevated temperatures in domestic refrigerators. Time-temperature indicators embedded in the packaging may have a significant role in ensuring safe storage temperatures are used. Distributors, retailers and consumers must be educated on the importance of low storage temperatures. General principles of good hygiene must be applied throughout the distribution and storage chain to avoid cross-contamination. Truck use needs to be monitored and vehicles appropriately sanitised. In the food service sector, training of operatives in food hygiene is important since many of these are teenagers that receive little food preparation experience in the modern home [35].

Many codes of practice have been published by national agencies [102] and industry sectors, and model farm-to-table HACCP protocols have been developed for some commodities (sprouted seeds, shredded lettuce and tomatoes). However, while HACCP principles are being applied by many growers, manufacturers, and distributors based on current knowledge, the US National Advisory Committee on Microbiological Criteria for Foods claim that there is insufficient evidence to put in place a comprehensive validated HACCP system for every fresh produce item [171].

Conclusion

As the consumption of fresh-cut products increases, there has been an increasing number of serious food poisoning outbreaks linked to these products. In the future, the rapid growth in the volume of produce consumption and in the globalisation of sourcing can be expected to continue. There will be greater emphasis on the development and application of GAP and GMP protocols, particularly for use of water and manure, for worker hygiene and for transportation of produce. Serious efforts will be made to apply these protocols to production in developing as well as industrialised countries. These initiatives should result in a safer, more reliable raw material stream.

Several critical gaps in knowledge are identified, notably uncertainty about the location of contaminating cells on or in plant tissues and the influence of environmental stresses on growth and infectivity [70]. Stress response and cross-protection must be considered when current food processing technologies are being modified or new control measures developed. These responses are particularly significant in minimal processing technology where the imposition of one sub-lethal stress may lead to the induction of multiple stress responses that reduces the efficacy of subsequent treatments [122]. More research is needed on how to use cumulative sub-lethal hurdles safely without inducing stress response. Furthermore, research on the mechanisms of resistance of pathogens to multiple stresses is required.

Other research trends driven by the needs of this sector include greater understanding of pertinent and emerging pathogens, particularly *E. coli* O157:H7, viruses and protozoan parasites; greater comprehension of processes of produce contamination generally, and of how to prevent them; development and application of more effective disinfection or decontamination technologies; application of active and intelligent packaging; and development of enhanced foodborne disease surveillance systems and molecular techniques for detection and typing of foodborne pathogens [220]. Recent outbreaks have highlighted the need for better traceability and an early warning system in the produce industry. These technologies can help establish the history of food items, sources of contamination and links between geographically isolated outbreaks of food poisoning with a common source [171].

REFERENCES

[1] Abdul-Raouf, U.M., Beuchat, L.R. & Ammar, M.S. (1993). Survival and growth of *E. coli* O157:H7 on salad vegetables. *Appl. Environ. Microbiol. 59*, 1999-2006.

[2] Abee, T. & Wouters, J.A. (1999). Microbial stress response in minimal processing. *Int. J. Food Microbiol. 50*, 65-91.

[3] Adams, M.R., Hartley, A.D. & Cox, L.J. (1989). Factors affecting the efficacy of washing procedures used in the production of prepared salads. *Food Microbiol. 6*, 69-77.

[4] Amanatidou, A., Smid, E.J. & Gorris, L.G.M. (1999). Effect of elevated oxygen and carbon dioxide on the surface growth of vegetable-associated microorganisms. *J. Appl. Microbiol. 86*, 429-438.

[5] Anonymous. (1973). Determination of chlorine in water. *Chem. Proc. 19*, 13-15.

[6] Archer, D.L. (1996). Preservation microbiology and safety: evidence that stress enhances virulence and triggers adaptive mutations. *Trends Food Sci. Technol. 7*, 91-95.

[7] Arihara, K., Cassens, R.G. & Luchansky, J.B. (1993). Characterization of bacteriocins from *Enterococcus faecium* with activity against *Listeria monocytogenes*. *Int. J. Food Microbiol. 19*, 123-134.

[8] Arumugaswamy, R.K., Rusul Rahamat Ali, G. & Nadzriah Bte Abd. Hamid, S. (1994). Prevalence of *Listeria monocytogenes* in foods in Malaysia. *Int. J. Food Microbiol. 23*, 117-121.

[9] Aureli, P., Costantini, A. & Zolea, S. (1992). Antimicrobial activity of some plant essential oils against *Listeria monocytogenes*. *J. Food Prot. 55*, 344-348.

[10] Austin, J.W., Dodds, K.L., Blanchfield, B. & Farber, J.M. (1998). Growth and toxin production by *Clostridium botulinum* on inoculated fresh-cut packaged vegetables. *J. Food Prot. 61*, 324-328.

[11] Badawy, A.S., Gerba, C.P. & Kelley, L.M. (1985). Survival of rotavirus SA-11 on vegetables. *Food Microbiol. 2*, 199-205.

[12] Baldwin, E.A., Nisperos-Carriedo, M.O. & Baker, R.A. (1995). Use of edible coatings to preserve quality of lightly (and slightly) processed products. *Crit. Rev. Food Sci. Nutr. 35*, 509-524.

[13] Barbour, A.H., Rampling, A. & Hormaeche, C.E. (1996). Comparison of the infectivity of isolation of *Listeria monocytogenes* following intragastric and intravenous inoculation in mice. *Microb. Path. 20*, 247-253.

[14] Barry-Ryan, C. & O'Beirne, D. (2000). Effects of peeling method on the quality of ready-to-use carrots. *Int. J. Food Sci. Technol.* 35, 243-254.

[15] Barry-Ryan, C. & O'Beirne, D. (1998). Effects of slicing method on the quality and storage-life of modified atmosphere packaged carrot discs. *J. Food Sci. 63*, 851 - 856.

[16] Barry-Ryan, C., Pacussi, J.M. & O'Beirne, D. (2000). Quality of shredded carrots as affected by packaging film and storage temperature. *J. Food Sci.* 65,726-730.

[17] Beckers, H.J., In't Veld, P.H., Soentoro, P.S.S. & Delfgou-van Ash, E.H.M. (1989). The occurrence of *Listeria* in foods, symposium proceedings *Foodborne Listeriosis*. Wiesbaden, Hamburg, September 7, 1988, 83-97.

[18] Behrsing, J., Winkler, S., Franz, P. & Premier, R. (2000). Efficacy of chlorine for inactivation of *Escherichia coli* on vegetables. *Postharvest Biol. Technol. 19*, 187-192.

[19] Bell, C. & Kyriakides, A. (1998). Background. In C. Bell & A. Kyriakides (Eds.), *Listeria a practical approach to the organism and its control in foods* (pp. 3-29). London: Chapman and Hall.

[20] Bennik, M.H.J., Peppelenbos, H.W., Nguyen-the, C., Carlin, F., Smid, E.J. & Gorris, L.G.M. (1996). Microbiology of minimally processed, modified atmosphere packaged chicory endive. *Postharvest Biol. Technol. 9*, 209-221.

[21] Bennik, M.H.J., Smid, E.J., Rombouts, F.M. & Gorris, L.G.M. (1995). Growth of psychrotrophic foodborne pathogens in a solid surface model system under the influence of carbon dioxide and oxygen. *Food Microbiol. 12*, 509-519.

[22] Bennik, M.H.J., Vorstman,W., Smid, E.J. & Gorris, L.G.M. (1998). The influence of oxygen and carbon dioxide on the growth of prevalent *Enterobacteriaceae* and *Pseudomonas* species isolated from fresh and controlled-atmosphere-stored vegetables. *Food Microbiol. 15*, 459-469.

[23] Berrang, M.E., Brackett, R.E. & Beuchat, L.R. (1989a). Growth of *Aeromonas hydrophila* on fresh vegetables stored under a controlled atmosphere. *Appl. Environ. Microbiol. 55*, 2167-2171.

[24] Berrang, M.E., Brackett, R.E. & Beuchat, L.R. (1989b). Growth of *Listeria monocytogenes* on fresh vegetables stored under controlled atmosphere. *J. Food Prot. 52*, 702-705.

[25] Best, M., Kennedy, M.E. & Coates, F. (1990). Efficacy of a variety of disinfectants against *Listeria* spp. *Appl. Environ. Microbiol. 56*, 377-380.

[26] Beuchat, L.R. (1999). Survival of enterohemorrhagic *Escherichia coli* O157:H7 in bovine feces applied to lettuce and the effectiveness of chlorinated water as a disinfectant. *J. Food Prot. 62*, 845-849.

[27] Beuchat, L.R. (1998). Surface decontamination of fruits and vegetables eaten raw: a review, World Health Organization Food Safety Unit, WHO/FSF/FOS/98.2. Available from: www.who.int/fsf/fos982~1

[28] Beuchat, L.R. (1996). Pathogenic microorganisms associated with fresh produce. *J. Food Prot. 59*, 204-216.

[29] Beuchat, L.R. (1992). Surface disinfection of raw produce. *Dairy Food Environ. Sanitat. 12*, 6-9.

[30] Beuchat, L.R., Brackett, R.E. & Doyle, M.P. (1994). Lethality of carrot juice to *Listeria monocytogenes* as affected by pH, sodium chloride and temperature. *J. Food Prot. 57*, 470-474.

[31] Beuchat, L.R. & Brackett, R.E. (1991). Behaviour of *Listeria monocytogenes* inoculated into raw tomatoes and processed tomato products. *Appl. Environ. Microbiol. 57*, 1367-1371.

[32] Beuchat, L.R. & Brackett, R.E. (1990a). Survival and growth of *Listeria monocytogenes* on lettuce as influenced by shredding, chlorine treatment, modified atmosphere packaging and temperature. *J. Food Sci. 55*, 755-758, 870.

[33] Beuchat, L.R. & Brackett, R.E. (1990b). Inhibitory effect of raw carrots on *Listeria monocytogenes*. *Appl. Environ. Microbiol. 56*, 1734-1742.

[34] Beuchat, L.R., Brackett, R.E., Hao, D.Y.-Y. & Conner, D.E. (1986). Growth and thermal inactivation of *Listeria monocytogenes* in cabbage and cabbage juice. *Can. J. Microbiol. 32*, 791-795.

[35] Beuchat, L.R. & Ryu, J.-H. (1997). Produce handling and processing practices. *Emerg. Infect. Dis. 3*, 459-465.

[36] Bialka, K.L. & Demirci, A. (2007). Utilization of gaseous ozone for the decontamination of *Escherichia coli* O157:H7 and *Salmonella* on raspberries and strawberries. *J. Food Prot. 70*, 1093-1098.

[37] Bidawid, S., Farber, J.M. & Sattar, S.A. (2001). Survival of hepatitis A virus on modified atmosphere packaged (MAP) lettuce. *Food Microbiol. 18*, 95-102.

[38] Bolin, H.R., Stafford, A.E., King, A.D. Jr. & Huxsoll, C.C. (1977). Factors affecting the storage stability of shredded lettuce. *J. Food Sci. 42*, 1319-1321.

[39] Brackett, R.E. (1999). Incidence, contributing factors, and control of bacterial pathogens in produce. *Postharv. Biol. Technol. 15*, 305-311.

[40] Brackett, R.E. (1994). Microbiological spoilage and pathogens in minimally processed refrigerated fruits and vegetables. In R.C. Wiley (Ed.), *Minimally processed refrigerated fruits and vegetables* (pp. 269-312). New York: Chapman and Hall.

[41] Brackett, R.E. (1992). Shelf stability and safety of fresh produce as influenced by sanitation and disinfection. *J. Food Prot. 55*, 804-814.

[42] Brackett, R.E. (1987). Antimicrobial effect of chlorine on *Listeria monocytogenes*. *J. Food Prot. 50*, 999-1003.

[43] Bremer, P.J., Monk, I. & Osborne, C.M. (2001). Survival of *Listeria monocytogenes* attached to stainless steel surfaces in the presence or absence of *Flavobacterium* spp. *J. Food Prot. 64*, 1369-1376.

[44] Brocklehurst, T.F., Zaman-Wong, C.M. & Lund, B.M. (1987). A note on the microbiology of retail packs of prepared salad vegetables. *J. Appl. Bacteriol. 63*, 409-415.

[45] Brosch, R., Catimel, B., Milon, G., Buchrieser, C., Vindel, E. & Rocourt, J. (1993). Virulence heterogeneity of *Listeria monocytogenes* strains from various sources (food, human, animal) in immunocompetent mice and its association with typing characteristics. *J. Food Prot. 56*, 296-301, 312.

[46] Buncic, S. & Avery, S.M. (1998). Effects of cold storage and heat-acid shocks on growth and verotoxin 2 production of *Escherichia coli* O157:H7. *Food Microbiol. 15*, 319-328.

[47] Buncic, S., Avery, S.M., Rocourt, J. & Dimitrijevic, M. (2001). Can food-related environmental factors induce different behaviour in two key serovars, 4b and 1/2a, of *Listeria monocytogenes*? *Int. J. Food Microbiol. 65*, 210-212.

[48] Burrows, J. P., Weber, M., Buchwitz, M., Rozanov, V., Ladsttter-Weienmayer, A., Richter, A., de Beek, R., Hoogen, R., Bramstedt, K., Eichmann, K.-U., Eisinger, M. & Perner, D. (1999). The Global Ozone Monitoring Experiment (GOME): Mission Concept and First Scientific Results. *J. Atmos. Sci. 56*, 151-175.

[49] Burt, S. (2004). Essential oils: their antibacterial properties and potential application in foods – A Review. *Int. J. Food Microbiol. 94*, 223-253.

[50] Cai, Y., Ng, L.K. & Farber, J.M. (1997). Isolation and characterization of nisin-producing *Lactococcus lactis* subsp. *lactis* from bean-sprouts. *J. Appl. Microbiol. 83*, 499-507.

[51] California Food Emergency Response Team (2007). Investigation of an *Escherichia coli* O157:H7 outbreak associated with dole pre-packaged spinach, Final March, 2007. California Department of Health Services, Scaramento, CA and the US Food and Drug Admin. http://www.marlerclark.com/2006_Spinach_Report_Final_01.pdf

[52] Carlin, F., Nguyen-the, C., Abreu da Silva, A. & Cochet, C. (1996a). Effects of carbon dioxide on the fate of *Listeria monocytogenes*, of aerobic bacteria and on the development of spoilage in minimally processed fresh endive. *Int. J. Food Microbiol. 32*, 159-172.

[53] Carlin, F., Nguyen-the, C. & Morris, C.E. (1996b). The influence of the background microflora on the fate of *Listeria monocytogenes* on minimally processed fresh broad leaved endive (*Cichorium endivia* var. *latifolia*). *J. Food Prot. 59*, 698-703.

[54] Carlin, F., Nguyen-the, C. & Abreu da Silva, A. (1995). Factors affecting the growth of *Listeria monocytogenes* on minimally processed fresh endive. *J. Appl. Bacteriol. 78*, 636-646.

[55] Carlin, F. & Peck, M.W. (1996). Growth of and toxin production by non-proteolytic *Cl. botulinum* in cooked pureed vegetables at refrigeration temperatures. *Appl. Environ. Microbiol. 62*, 3069-3072.

[56] Carlin, F. & Nguyen-the, C. (1994). Fate of *Listeria monocytogenes* on four types of minimally processed green salads. *Lett. Appl., Microbiol. 18*, 222-226.

[57] CDC (1989). Epidemiological notes and reports common source outbreak of giardiasis – New Mexico. *Morb. Mortal. Wkly. Rep. 38*, 405-407.

[58] CDC (1996). Outbreak of *Escherichia coli* O157:H7 infections associated with drinking unpasteurized commercial apple juice – British Columbia, California, Colorado and Washington, October 1996. *Morb. Mortal. Wkly. Rep. 45*, 975.

[59] CDC (1999). Outbreak of *Salmonella* serotype Muenchen infections associated with unpasteurized orange juice – United States and Canada, June 1999. *Morb. Mortal. Wkly. Rep. 48*, 582-585.

[60] Chaidez, C., Soto, M., Gortares, P. & Mena, K. (2005). Occurrence of *Cryptosporidium* and *Giardia* in irrigation water and its impact on the fresh produce industry. *Int. J. Environ. Health Res. 15*, 339-345.
[61] Cliffe-Byrnes,V., McLaughlin, C.P. & O'Beirne, D. (2003). Effects of packaging film and storage temperature on quality and shelf-life of a dry coleslaw mix packaged in a modified atmosphere. *Int. J. Food Sci. Technol.* 38, 187-199.
[62] Cliver, D.O. (1997). Foodborne viruses. In M.P. Doyle, L.R. Beuchat & T.J. Montville (Eds.), *Food Microbiology: Fundamentals and Frontiers* (pp. 437-446). Washington DC: American Society for Microbiology Press.
[63] Cooley, M.B., Chao, D. and Mandrell, R.E. (2006). *Escherichia coli* O157:H7 survival and growth on lettuce is altered by the presence of epiphytic bacteria. *J. Food Prot. 69*, 2329-2335.
[64] Conner, D.E. & Kotrola, J.S. (1995). Growth and survival of *Escherichia coli* O257:H7 under acidic conditions. *Appl. Environ. Microbiol. 61*, 382-385.
[65] Conway, W.S., Leverentz, B., Saftner, R.A., Janisiewicz, W.J., Sams, C.E. & Leblanc, E. (2000). Survival and growth of *Listeria monocytogenes* on fresh-cut apple slices and its interaction with *Glomerella cingulata* and *Penicillium expansum. Plant Dis. 84*, 177-181.
[66] Conway, W.S., Janisiewicz, W.J., Watada, A.E. & Sams C.E. (1998). Survival and growth of *Listeria monocytogenes* on fresh-cut apple slices. *Phytopathol. 88*, S18.
[67] Croci, L., De Medici, D., Scalfaro, C., Fiore, A. & Toti, L. (2002). The survival of hepatitis A virus in fresh produce. *Int. J. Food Microbiol. 73*, 29-34.
[68] Daniels, J.A., Krishnamurthi, R. & Rizvi, S.S.H. (1985). A review of the effect of carbon dioxide on microbial growth and food quality. *J. Food Prot. 48*, 532-537.
[69] Day, B.P.F. (1996). High oxygen modified atmosphere packaging for fresh prepared produce. *Postharvest News Info. 7*, 1N-34N.
[70] Delaquis, P., Bach, S. & Dinu, L.-D. (2007). Behavior of *Escherichia coli* O157:H7 in leafy vegetables. *J. Food Prot. 70*, 1966-1974.
[71] Del Corral, F., Buchanan, R.L., Bencivengo, M.M. & Cooke, P.H. (1990). Quantitative comparison of selected virulence associated characteristics in food and clinical isolates of *Listeria*. *J. Food Prot. 53*, 1003-1009.
[72] Diaz, C. & Hotchkiss, J. H. (1996). Comparative growth of *E. coli* O157:H7, spoilage organisms and shelf life of shredded iceberg lettuce stored under modified atmospheres. *J. Sci. Food Agricul. 70*, 433-438.
[73] Diehl, J.F. (1995). *Safety of Irradiated Foods*. New York: Marcel Dekker, Inc.
[74] Dingman, D.W. (2000). Growth of *Escherichia coli* O157:H7 in bruised apple (*Malus domestica*) tissue as influenced by cultivar, date of harvest and source. *Appl. Environ. Microbiol. 66*, 1077-1083.
[75] Dixon, N.M. & Kell, D.B. (1989). The inhibition by CO_2 of the growth and metabolism of micro-organisms: a review. *J. Appl. Bacteriol. 67*, 109-136.
[76] Doris, L.K.Ng & Seah, H.L. (1995). Isolation and identification of *Listeria monocytogenes* from a range of foods in Singapore. *Food Control 6*, 171-173.
[77] Dorman, H.J.D. & Deans, S.G. (2000). Antimicrobial agents from plants: antibacterial activity of plant volatile oils. *J. Appl. Microbiol. 88*, 308-316.

[78] Duffy, G., Whiting, R.C. & Sheridan, J.J. (1999). The effects of a competitive microflora, pH and temperature on the growth kenetics of *Escherichia coli* O157:H7. *Food Microbiol. 16*, 299-307.

[79] Duizer, E., Bijkerk, P., Rockx, B., De Groot, A., Twisk F. & Koopmans, M. (2004). Inactivation of caliciviruses. *Appl. Environ. Microbiol. 70*, 4538-4543.

[80] El-Kest, S.E. & Marth, E.H. (1988a). *Listeria monocytogenes* and its inactivation by chlorine: A review. *Lebensm.-Wiss. u. Technol. 21*, 346-351.

[81] El-Kest, S.E. & Marth, E.H. (1988b). Temperature, pH, and strain of pathogen as factors affecting inactivation of *Listeria monocytogenes* by chlorine. *J. Food Prot. 51*, 622-625.

[82] Emer, Z., Akbas, M.Y. & Ozdemir, M. (2008). Bactericidal activity of ozone against *Escherichia coli* in whole and ground black peppers. *J. Food Prot. 71*, 914-917.

[83] Enfors, S.O., Molin, G. & Ternström, A. (1979). Effect of packaging under carbon dioxide, nitrogen or air on the microbial flora of pork stored at 4°C. *J. Appl. Bacteriol. 47*, 197-208.

[84] Ercolani, G.L. (1976). Bacteriological quality assessment of fresh marketed lettuce and fennel. *Appl. Environ. Microbiol. 31*, 847-852.

[85] Escartin, E.F., Castillo Ayala A. & Lozano, J.S. (1989). Survival and growth of *Salmonella* and *Shigella* on sliced fresh fruit. *J. Food Prot. 52*, 471-472, 483.

[86] Farber, J.M. (1991). Microbiological aspects of modified atmosphere packaging technology: A review. *J. Food Prot. 54*, 58-70.

[87] Farber, J.M. & Peterkin, P.I. (1991). *Listeria monocytogenes*: a food-borne pathogen. *Microbiol. Rev. 55*, 476-511.

[88] Farber, J.M., Wang, S.L., Cai, Y. & Zhang, S. (1998). Changes in populations of *Listeria monocytogenes* inoculated on packaged fresh-cut vegetables. *J. Food Prot. 61*, 192-195.

[89] Farkas, J. (1997). Physical methods of food preservation. In M.P. Doyle, L.R. Beuchat & T.J. Monteville (Eds.), *Food Microbiology: Fundamentals and Frontiers* (pp. 497-519). Washington DC: American Society for Microbiology Press.

[90] FDA (1998). Guide to minimize microbial food safety hazards for fresh fruits and vegetables. Washington D.C., 36p.

[91] Fett, W.F. (2006). Inhibition of *Salmonella enterica* by plant-associated pseudomonads *in vitro* and on sprouting alfalfa seed. *J. Food Prot. 69*, 719-728.

[92] Fino, V.R. & Kniel, K.E. (2008). UV light inactivation of Hepatitis A Virus, Aichi Virus and Feline Calicivirus on strawberries, green onions and lettuce. *J. Food Prot. 71*, 908-913.

[93] Francis, G.A. & O'Beirne, D. (2005). Variation among strains of *Listeria monocytogenes*: differences in survival on packaged vegetables and in response to heat and acid conditions. *Food Control 16, 687-694.*

[94] Francis, G.A. & O'Beirne, D. (2002). Effects of vegetable type and antimicrobial dipping on survival and growth of *Listeria innocua* and *E. coli*. *Int. J. Food Sci. Technol. 37*, 711-718.

[95] Francis, G.A. & O'Beirne, D. (2001a). Effects of vegetable type, package atmosphere and storage temperature on growth and survival of *Escherichia coli* O157:H7 and *Listeria monocytogenes*. *J. Ind. Microbiol. Biot. 27*, 111-116.

[96] Francis, G.A. & O'Beirne, D. (2001b). Effects of acid adaptation on the survival of *Listeria monocytogenes* on modified atmosphere packaged vegetables. *Int. J. Food Sci. Technol. 36*, 477-487.

[97] Francis, G.A. & O'Beirne, D. (1998a). Effects of storage atmosphere on *Listeria monocytogenes* and competing microflora using a surface model system. *Int. J. Food Sci. Technol. 33*, 465-476.

[98] Francis, G.A. & O'Beirne, D. (1998b). Effects of the indigenous microflora of minimally processed lettuce on the survival and growth of *L. monocytogenes*. *Int. J. Food Sci. Technol. 33*, 477-488.

[99] Francis, G.A. & O'Beirne, D. (1997). Effects of gas atmosphere, antimicrobial dip and temperature on the fate of *Listeria innocua* and *Listeria monocytogenes* on minimally processed lettuce. *Int. J. Food Sci. Technol. 32*, 141-151.

[100] Francis, G.A., Thomas, C. & O'Beirne, D. (1999). Review: The microbiological safety of minimally processed vegetables. *Int. J. Food Sci. Technol. 34*, 1-22.

[101] Frank, J.F. & Koffi, R.A. (1990). Surface-adherent growth of *Listeria monocytogenes* is associated with increased resistance to surfactant sanitizers and heat. *J. Food Prot. 53*, 550-554.

[102] FSAI (2001). Code of practice for food safety in the fresh produce supply chain in Ireland, Dublin, 74p.

[103] Gahan, C.G.M. & Hill, C. (1999). The relationship between acid stress responses and virulence in *Salmonella typhimurium* and *Listeria monocytogenes*. *Int. J. Food Microbiol. 50*, 93-100.

[104] Gahan, C.G.M., O'Driscoll, B. & Hill, C. (1996). Acid adaptation of *Listeria monocytogenes* can enhance survival in acidic foods and during milk fermentation. *Appl. Environ. Microbiol. 62*, 3128-3132.

[105] García-Gimeno, R.M., Sanchez-Pozo, M.D., Amaro-López, M.A. & Zurera-Cosano, G. (1996). Behaviour of *Aeromonas hydrophila* in vegetable salads stored under modified atmosphere at 4 and 15°C. *Food Microbiol. 13*, 369-374.

[106] Garg, N., Churey, J.J. & Splittstoesser, D.F. (1990). Effect of processing conditions on the microflora of fresh-cut vegetables. *J. Food Prot. 53*, 701-703.

[107] Gleeson, E., Francis, G.A., & O'Beirne, D. (2002). Survival and growth of *Escherichia coli* and *Listeria innocua* on fresh-cut vegetable as affected by process severity. Proc. 32nd Annual Food Science and Technology Conference, University College Cork, p.119.

[108] Gleeson, E. & O'Beirne, D. (2005). Effects of process severity on survival and growth of *Escherichia coli* and *Listeria innocua* on minimally processed vegetables. *Food Control 16*, 677-685.

[109] Gorny, J.R. & Zagory, D. (2002). Produce food safety. In K.C. Gross, M.E. Saltveit, and C.Y. Wang (Eds.), *The commercial storage of fruits, vegetables and florist and nursery stocks, US Department of Agriculture Handbook*, Washington, D.C. 35p.

[110] Gulati, B.R., Allwood, P.B., Hedberg, C.W. & Goyal, S.M. (2001). Efficacy of commonly used disinfectants for the inactivation of calicivirus on strawberry, lettuce and a food-contact surface. *J. Food Prot. 64*, 1430-1434.

[111] Gutierrez, J., Barry-Ryan, C. & Bourke, P. (2008). The antimicrobial efficacy of plant essential oil combinations and interactions with food ingredients. *Int. J. Food Microbiol. 124*, 91-97.

[112] Hammer, K.A., Carson, C.F. & Riley, T.V. (1999). Antimicrobial activity of essential oils and other plant extracts. *J. Appl. Microbiol. 86*, 985-990.

[113] Han, Y., Linton, R.H., Nielsen, S.S. & Nelson, P.E. (2001). Reduction of *Listeria monocytogenes* on green peppers (*Capsicum annuum* L.) by gaseous and aqueous chlorine dioxide and water washing and its growth at 7°C. *J. Food Prot. 64*, 1730-1738.

[114] Han, Y., Linton, R.H., Nielsen, S.S. & Nelson, P.E. (2000a). Inactivation of *Escherichia coli* O157:H7 on surface-uninjured and -injured green pepper (Capsicum annuum L.) by chlorine dioxide gas as demonstrated by confocal laser scanning microscopy. *Food Microbiol. 17*, 643-655.

[115] Han, Y., Sherman, D.M., Linton, R.H., Nielsen, S.S. & Nelson, P.E. (2000b). The effects of washing and chlorine dioxide gas on survival and attachment of *Escherichia coli* O157:H7 to green pepper surfaces. *Food Microbiol. 17*, 521-533.

[116] Harris, L.J., Daeschel, M.A., Stiles, M.E. & Klaenhammer, T.R. (1989). Antimicrobial activity of lactic acid bacteria against *Listeria monocytogenes. J. Food Prot. 52*, 384-387.

[117] Harris, L.J., Farber, J.N., Beuchat, L.R., Parish, M.E., Suslow, T.V., Garrett, E.H. & Busta, F.F. (2003). Outbreaks associated with fresh produce: incidence, growth and survival of pathogens in fresh and fresh-cut produce. *Comp. Rev. Food Sci. Food Saf. 2 (Suppl.)*, 78-141.

[118] Harvey, J. & Gilmour, A. (1993). Occurrence and characteristics of *Listeria* in foods produced in Northern Ireland. *Int. J. Food Microbiol. 19*, 193-205.

[119] Hedberg, C.W. & Osterholm, M.T. (1993). Outbreaks of foodborne and waterborne viral gastroenteritis. *Clin. Microbiol. Rev. 6*, 199-210.

[120] Herwaldt, B.L. (2000). *Cyclospora cayetanensis*: a review focusing on the outbreaks of cyclosporiasis in the 1990s. *Clin. Infect. Dis. 31*, 1040-1057.

[121] Herwaldt, B.L. & Ackers, M.L. (1997). An outbreak in 1996 of cyclosporiasis associated with imported raspberries. *New Engl. J. Med. 336*, 1548-1556.

[122] Hill, C., O'Driscoll, B., & Booth, I.R. (1995). Acid adaptation and food poisoning microorganisms. *Int. J. Food Microbiol. 28*, 245-254.

[123] Hintlian, C.B. & Hotchkiss, J.H. (1986). The safety of modified atmosphere packaging: A review. *Food Technol. 40*, 70-76.

[124] Ho, J.L., Shands, K.N., Freidland, G., Eckind, P. & Fraser, D.W. (1986). An outbreak of type 4b *Listeria monocytogenes* infection involving patients from eight Boston hospitals. *Arch. Internal Med. 146*, 520-524.

[125] Hudson, J.A., Mott, S.J. & Penney, N. (1994). Growth of *Listeria monocytogenes*, *Aeromonas hydrophila* and *Yersinia enterocolitica* on vacuum and saturated carbon dioxide controlled atmosphere packaged sliced roast beef. *J. Food Prot. 57*, 204-208.

[126] Hutin, Y.J.F., Pool, V., Cramer, E.H., Nainan, O.V., Weth, J., Williams, I.T., Goldstein, S.T., Gensheimer, K.F., Bell, B.P., Shapiro, C.N., Alter M.J. & Margolis, H.S. (1999). A multistate foodborne outbreak of Hepatitis A, National Hepatitis A Investigation Team. *New Engl. J. Med. 340*, 595-602.

[127] Iandolo, J.J., Clark, C.W., Bluhm, L. & Ordal, Z.J. (1965). Repression of *Staphylococcus aureus* in associative culture. *Appl. Microbiol. 13*, 646-649.

[128] International Commission on Microbiological Specifications for Foods. (1996). Microorganisms in Foods. In: *ICMSF Microbiological Specifications of Food Pathogens*. London: Blackie Academic and Professional.

[129] Iwamoto, M. (2007). Taco Bell and Taco Johns *Escherichia coli* in lettuce outbreaks 2006. CDC Outbreak Response and Surveillance Team Enteric Diseases Epidemiology Branch, IAFP 2007: 94th Annual Meeting, Florida.

[130] Izumi, H. (1999). Electrolyzed water as a disinfectant for fresh-cut vegetables. *J. Food Sci. 64*, 536-539.

[131] Jacxsens, L., Devlieghere, F., Falcato, P. & Debevere, J. (1999). Behaviour of *Listeria monocytogenes* and *Aeromonas* spp. on fresh-cut produce packaged under equilibrium-modified atmosphere. *J. Food Prot. 62*, 1128-1135.

[132] James, S.J. (2007). Novel physical methods for decontamination of produce. IAFP's Third European Symposium on Food Safety: Advancements in Food Safety, Rome, 18-19 October.

[133] Janisiewicz, W.J., Conway, W.S. & Leverentz, B. (1999). Biological control of postharvest decays of apple can prevent growth of *Escherichia coli* O157:H7 in apple wounds. *J. Food Prot. 62*, 1372-1375.

[134] Jerngklinchan, J. & Saitanu, K. (1993). The occurrence of *Salmonella* in bean sprouts in Thailand. *Southeast Asian J. Trop. Med. Public Health 24*, 114-118.

[135] Jones, K. & Heaton, J. (2006). Microbial contamination of fruit and vegetables: evidence and issues. *Microbiol. 7*, 28-31.

[136] Kallender, K.D., Hitchins, A.D., Lancette, G.A., Schmieg, J.A., Garcia, G.R., Solomon, H.M. & Sofos, J.N. (1991). Fate of *Listeria monocytogenes* in shredded cabbage stored at 5 and 25°C under a modified atmosphere. *J. Food Prot. 54*, 302-304.

[137] Karapinar, M. & Gonul, S.A. (1992). Effects of sodium bicarbonate, vinegar, acetic and citric acids on growth and survival of *Yersinia enterocolitica*. *Int. J. Food Microbiol. 16*, 343-347.

[138] King, A.D. Jr. & Nagel, C.W. (1975). Influence of carbon dioxide upon the metabolism of *Pseudomonas aeruginosa*. *J. Food Sci. 40*, 362-366.

[139] Klaenhammer, T.R. (1988). Bacteriocins of lactic acid bacteria. *Biochimie 70*, 337-349.

[140] Koopmans, M. & Duizer, E. (2004). Foodborne viruses: an emerging problem. *Int. J. Food Microbiol. 90*, 23-41.

[141] Konowalchuk, J. & Speirs, J.I. (1975). Survival of enteric viruses on fresh vegetables. *J. Milk Food Technol. 37*, 132-134.

[142] Korich, D.G., Mead, J.R., Madore, M.S., Sinclair, N.A. & Sterling, C.R. (1990). Effects of ozone, chlorine dioxide, chlorine and monochloroamine on *Cryptosporidium parvum* oocyst viability. *Appl. Environ. Microbiol. 56*, 1423-1428.

[143] Koseki, S., Yoshida, K., Isobe, S. & Itoh, K. (2001). Decontamination of lettuce using acidic electrolysed water. *J. Food Prot. 64*, 652-658.

[144] Krochta, J.M. & De Mulder-Johnston, C. (1997). Scientific status summary, edible and biodegradable polymer films: challenges and opportunities. *Food Technol. 51*, 61-74.

[145] Lambert, R.J.W., Skandamis, P.N., Coote, P. & Nychas, G.J.E. (2001). A study of the minimum inhibitory concentration and mode of action of oregano essential oil, thymol and carvacrol. *J. Appl. Microbiol. 91*, 453-462.

[146] Langerak, D.I. (1978). The influence of irradiation and packaging on the quality of prepacked vegetables. *Ann. Nutr. Aliment. 32*, 569-586.

[147] Leong, Y.K., Xui, O.C. and Chia, O.K. (2008). Survival of SA11 Rotavirus in fresh fruit juices of pineapple, papaya and honeydew melon. *J. Food Prot. 71*, 1035-1037.

[148] Leyer, G.L. & Johnson, E.A. (1993). Acid adaptation induces cross-protection against environmental stresses in *Salmonella typhimurium*. *Appl. Environ. Microbiol. 59*, 1842-1847.

[149] Leyer, G.L. & Johnson, E.A. (1992). Acid adaptation promotes survival of *Salmonella* spp. in cheese. *Appl. Environ. Microbiol. 58*, 2075-2080.

[150] Liao, C.-H. & Cooke, P.H. (2001). Response to trisodium phosphate treatment of *Salmonella* Chester attached to fresh-cut green pepper slices. *Can. J. Microbiol. 47*, 25-32.

[151] Liao, C.H. & Fett, W.F. (2001). Analysis of native microflora and selection of strains antagonistic to human pathogens on fresh produce. *J. Food Prot.* 64, 1110-1115.

[152] Liao, C.-H. & Sapers, G.M. (2000). Attachment and growth of *Salmonella* Chester on apple fruits and *in vivo* response of attached bacteria to sanitizer treatments. *J. Food Prot. 63*, 876-883.

[153] Liao, C.-H. & Sapers, G.M. (1999). Influence of soft rot bacteria on growth of *Listeria monocytogenes* on potato tuber slices. *J. Food Prot. 62*, 343-348.

[154] Lin, C.M., Fernando, S.Y. & Wei, C.I. (1996). Occurrence of *Listeria monocytogenes*, *Salmonella* spp., *Escherichia coli* and *E. coli* O157:H7 in vegetable salads. *Food Control 7*, 135-140.

[155] Little, C. (2007). Managing public health risks and the microbiological quality of fresh produce – UK perspective. IAFP's Third European Symposium on Food Safety: Advancements in Food Safety, Rome, 18-19 October.

[156] Lou, Y. & Yousef, A.E. (1997). Adaptation to sublethal environmental stresses protects *Listeria monocytogenes* against lethal preservation factors. *Appl. Environ. Microbiol. 63*, 1252-1255.

[157] Lund, B.M. (1992). Ecosystems in vegetable foods. *J. Appl. Bacteriol. Symposium Supplement 73*, 115S-126S.

[158] Lund, B.M. (1983). Bacterial spoilage. In C. Dennis (Ed.), *Post-harvest Pathology of Fruits and Vegetables* (pp. 219-257). London: Academic Press.

[159] Lund, B.M. & Snowdon, A.L. (2000). Fresh and processed fruits: Chapter 27. In B.M. Lund, T.C. Baird-Parker & G.W. Gould (Eds.), *The microbiological safety and quality of food* (Volume 1, pp. 738-758). Gaithersburg: Aspen.

[160] MacGowan, A.P., Bowker, K., McLauchlin, J., Bennett, P.M. & Reeves, D.S. (1994). The occurrence and seasonal changes in the isolation of *Listeria* spp. in shop bought food stuffs, human faeces, sewage and soil from urban sources. *Int. J. Food Microbiol. 21*, 325-334.

[161] Maki, D. (2006). Don't eat the spinach – controlling foodborne infectious disease. *New Engl. J. Med. 355*, 1952-1955

[162] Manvell, P. & Ackland, M. (1986). Rapid detection of microbial growth in vegetable salads at chill and abuse temperatures. *Food Microbiol. 3*, 59-65.

[163] Marchetti, R., Casadei, M.A. & Guerzoni, M.E. (1992). Microbial population dynamics in ready-to-use vegetable salads. *Ital. J. Food Sci. 4*, 97-108.

[164] Marshall, D.L. & Schmidt, R.H. (1991). Physiological evaluation of stimulated growth of *Listeria monocytogenes* by *Pseudomonas* species in milk. *Can. J. Microbiol. 37*, 594-599.

[165] Matos, A. & Garland, J. (2005). Effects of community versus single strain inoculants on the biocontrol of *Salmonella* and microbial community dynamics in alfalfa sprouts. *J. Food Prot. 68*, 40-48.

[166] Matthews, K.R. (2006). *Emerging Issues in Food Safety: Microbiology of Fresh Produce.* Washington DC: American Society for Microbiology Press.

[167] McLauchlin, J. & Gilbert, R.J. (1990). *Listeria* in foods, Report from the PHLS Committee on *Listeria* and listeriosis. *PHLS Microbiol. Dig. 7*, 54-55.

[168] Mintz, E.D., Hudson-Wragg, M., Mshar, P., Cartter, M.L. & Hadler, J.L. (1993). Foodborne giardiasis in a corporate office setting. *J. Infect. Dis. 167*, 250-253.

[169] Monge, R. & Chinchilla, M. (1996). Presence of *Cryptosporidium* oocysts in fresh vegetables. *J. Food Prot. 59*, 202-203.

[170] Moorman, M.A., Thelemann, C.A., Zhou, S., Pestka, J.J., Linz, J.E. & Ryser, E.T. (2008). Altered hydrophobicity and membrane composition in stress-adapted *Listeria innocua*. *J. Food Prot. 71*, 182-185.

[171] National Advisory Committee on Microbiological Criteria for Foods NACMCF. (1999). Microbiological safety evaluations and recommendations on fresh produce. *Food Control 10*, 117-143.

[172] Nguyen-the, C. & Carlin, F. (1994). The microbiology of minimally processed fresh fruits and vegetables. *Crit. Rev. Food Sci. 34*, 371-401.

[173] Nguyen-the, C., Halna-du-Fretay, B. & Abreu da Silva, A. (1996). The microbiology of mixed salad containing raw and cooked ingredients without dressing. *Int. J. Food Sci. Technol. 31*, 481-487.

[174] Nguyen-the, C. & Lund, B.M. (1991). The lethal effect of carrot on *Listeria* species. *J. Appl. Bacteriol. 70*, 479-488.

[175] Niemira, B.A. (2007). Relative efficacy of sodium hypochlorite wash versus irradiation to inactivate *Escherichia coli* O157:H7 internalized in leaves of romaine lettuce and baby spinach. *J. Food Prot. 70*, 2526-2532.

[176] Nychas, G.J.E. (1995). Natural antimicrobials from plants. In G.W. Gould (Ed.), *New Methods of Food Preservation*, London: Blackie Academic and Professional.

[177] O'Beirne, D. (1989). Irradiation of fruits and vegetables: applications and issues. *Prof. Hort. 3*, 12-19.

[178] O'Beirne, D. & Francis, G.A. (2003). Analysing pathogen survival in modified atmosphere packaged (MAP) produce. In R. Ahvenainen (Ed), *Novel Food Packaging Techniqies* (pp. 231-275). Cambridge, London: Woodhead Publishing Ltd.

[179] O'Driscoll, B., Gahan, C.G.M. & Hill, C. (1996). Adaptive acid tolerance response in *Listeria monocytogenes*: isolation of an acid-tolerant mutant which demonstrates increased virulence. *Appl. Environ. Microbiol. 62*, 1693-1698.

[180] Odumeru J.A., Mitchell S.J., Alves D.M., Lynch J.A., Yee A.J., Wang S.L., Styliadis S. & Farber J.M. (1997). Assessment of the microbiological quality of ready-to-use vegetables for health-care food services. *J. Food Prot. 60*, 954-960.

[181] Park, C.-M. & Beuchat, L.R. (1999). Evaluation of sanitizers for killing *Escherichia coli* O157:H7, *Salmonella* and naturally occurring microorganisms on cantaloupes, honeydew melons and asparagus. *Dairy Food Environ. Sanit. 19*, 842-847.

[182] Park, C.-M., Hung, Y.-C., Doyle, M.P., Ezeike, G.O.I. & Kim, C. (2001). Pathogen reduction and quality of lettuce treated with electrolysed oxidizing and acidified chlorinated water. *J. Food Sci. 66*, 1368-1372.

[183] Paster, N., Juven, B.J., Shaaya, E., Menasherov, M. & Nitzan, R. (1990). Inhibitory effect of oregano and thyme essential oils on moulds and foodborne bacteria. *Lett. Appl. Microbiol. 11*, 33-37.

[184] PHLS, Public Health Laboratory System (2001). *Salmonella* Newport infection in England associated with the consumption of ready to eat salad. *Eurosurveillance Weekly*, June, 26.

[185] Piagentini, A.M., Pirovani, M.E., Güemes, D.R., Di Pentima, J.H. & Tessi, M.A. (1997). Survival and growth of *Salmonella hadar* on minimally processed cabbage as influenced by storage abuse conditions. *J. Food Sci. 62*, 616-618, 631.

[186] Popa, I., Hanson, E.J., Todd, E.C.D., Schilder, A.C. & Ryser, E.T. (2007). Efficacy of chlorine dioxide gas sachets for enhancing the microbiological quality and safety of blueberries. *J. Food Prot. 70*, 2084-2088.

[187] Price, R.J. & Lee, J.S. (1970). Inhibition of *Pseudomonas* species by hydrogen peroxide producing lactobacilli. *J. Milk Food Technol. 33*, 13-18.

[188] Raccach, M. & Baker, R.C. (1979). The effect of lactic acid bacteria on some properties of mechanically deboned poultry meat. *Poultry Sci. 58*, 144-147.

[189] Rajkowski, K.T. & Thayer, D.W. (2000). Reduction of *Salmonella* spp. and strains of *Escherichia coli* O157:H7 by gamma radiation of inoculated sprouts. *J. Food Prot. 63*, 871-875.

[190] Ranson, S.L., Walker, D.A. & Clarke, I.D. (1960). Effects of carbon dioxide on mitochondrial enzymes from *Ricinus*. *Biochem. J. 76*, 216-221.

[191] Richert, K.J., Albrecht, J.A., Bullerman, L.B. & Sumner, S.S. (2000). Survival and growth of *Escherichia coli* O157:H7 on broccoli, cucumber and green pepper. *Dairy Food Environ. Sanit. 20*, 24-28.

[192] Ringlé, P., Vincent, J.P. & Catteau, M. (1991). Evolution de *Listeria* dans les produits de 4ème gamme. In C. Lahellec (Ed.), *Les Microorganismes Contaminants dans les Industries Agroalimentaires: Colonisation, Détection, Maitrise* (pp. 324-328). Paris: 7ème Colloque de la Section Microbiologie Alimentaire, 13-14 March 1991, Société Française de Microbiologie.

[193] Rocourt, J. (1994). *Listeria monocytogenes*: the state of the science. *Dairy Food Environ. Sanit. 14*, 70-82.

[194] Rodriguez, A.M.C., Alcala, E.B., Gimeno, R.M.G. & Cosano, G.Z. (2000). Growth modelling of *Listeria monocytogenes* in packaged fresh green asparagus. *Food Microbiol. 17*, 421-427.

[195] Rosenblum, L.S., Mirkin, I.R., Allen, D.T., Safford, S. & Hadler, S.C. (1990). A multifocal outbreak of hepatitis A traced to commercially distributed lettuce. *AJPH 80*, 1075-1079.

[196] Rouquette, C., de Chastellier, C., Nair, S. & Berche, P. (1998). The ClpC ATPase of *Listeria monocytogenes* is a general stress protein required for virulence and promoting early bacterial escape from the phagosome of macrophages. *Mol. Microbiol. 27*, 1235-1245.

[197] Saddik, M.F., El-Sherbeeny, M.R. & Bryan, F.L. (1985). Microbiological profiles of Egyptian raw vegetables and salads. *J. Food Prot. 48*, 883-886.

[198] Sapers, G.M., Gorny, J.R. & Yousef, A.E. (2006). *Microbiology of fruits and vegetables*. London: CRC Taylor and Francis.

[199] Sapers, G.M. & Simmons, G.F. (1998). Hydrogen peroxide disinfection of minimally processed fruits and vegetables. *Food Technol. 52*, 48-52.

[200] Sattar, S.A., Springthorpe, V.S. & Ansari, S.A. (1994). In Y.H. Hui, J.R. Gorham, and D.O. Cliver (Eds.), *Rotavirus. Foodborne disease Handbook* (Volume 2, pp. 81-111). New York: Marcel Dekker.

[201] Schlech, W.F., Lavigne, P. M., Bortolussi, R.A., Allen, A.C., Haldane, E.V., Wort, A.J., Hightower, A.W., Johnson, S.E., King, S.H., Nicholls, E.S. & Broome, C.V. (1983). Epidemic listeriosis-evidence for transmission by food. *New Eng. J. Med. 308*, 203-206.

[202] Seo, K.H. & Frank, J.F. (1999). Attachment of *Escherichia coli* O157:H7 to lettuce leaf surface and bacterial viability in response to chlorine treatment as demonstrated by using confocal scanning laser microscopy. *J. Food Prot. 62*, 3-9.

[203] Sears, D.F. & Eisenberg, R.M. (1961). A model representing a physiological role of CO_2 at the cell membrane. *J. Gen. Physiol. 44*, 869-887.

[204] Singh, N., Singh, R.K., Bhunia, A.K. & Stroshine, R.L. (2002). Efficacy of chlorine dioxide, ozone and thyme essential oil or a sequential washing in killing *Escherichia coli* O157:H7 on lettuce and baby carrots. *Lebensm.-Wiss. u. Technol. 35*, 720-729.

[205] Sizmur, K. & Walker, C.W. (1988). *Listeria* in prepacked salads. *Lancet i*, 1167.

[206] Skandamis, P.N. & Nychas, G.J.E. (2001). Effect of oregano essential oil on microbiological and physical-chemical attributes of minced meat stored in air and modified atmosphere. *J. Appl. Microbiol. 91*, 1011-1022.

[207] Smith-Palmer, A., Stewart, J. & Fyfe, L. (1998). Antimicrobial properties of plant essential oils and essences against five important foodborne pathogens. *Lett. Appl. Microbiol. 26*, 118-122.

[208] Sofos, J.N., Beuchat, L.R., Davidson, P.M. & Johnson, E.A. (1998). Naturally occurring antimicrobials in food. Report 132, Council for Agric. Sci. Technol. Task Force.

[209] Solomon, H.M., Kautter, D.A., Lilly, T. & Rhodehamel, E.J. (1990). Outgrowth of *Cl. botulinum* in shredded cabbage at room temperature under modified atmosphere. *J. Food Prot. 53*, 831-833.

[210] Stopforth, J.D., Mai, T., Kottapalli, B. & Samadpour, M. (2008). Effect of acidified sodium chlorite, chlorine, and acidic electrolyzed water on *Escherichia coli* O157:H7, *Salmonella* and *Listeria monocytogenes* inoculated onto leafy greens. *J. Food Prot. 71*, 625-628.

[211] Szabo, E.A., Scurrah, K.J. & Burrows, J.M. (2000). Survey for psychrotrophic bacterial pathogens in minimally processed lettuce. *Lett. Appl. Microbiol. 30*, 456-460.

[212] Tabouret, M., Derycke, J., Audurier, A. & Poutrel, B. (1991). Pathogenicity of *Listeria monocytogenes* isolates in immunocompromised mice in relation to listeriolysin production. *J. Med. Microbiol. 34*, 13-18.

[213] Takeuchi, K. & Frank, J.F. (2001). Quantitative determination of the role of lettuce leaf structures in protecting *Escherichia coli* O157:H7 from chlorine disinfection. *J. Food Prot. 64*, 147-151.

[214] Takeuchi, K. & Frank, J.F. (2000). Penetration of *Escherichia coli* O157:H7 into lettuce tissues as affected by inoculum size and temperature and the effect of chlorine treatment on cell viability. *J. Food Prot. 63*, 434-440.

[215] Takeuchi, K., Matute, C.M., Hassan, A.N. & Frank, J.F. (2000). Comparison of the attachment of *Escherichia coli* O157:H7, *Listeria monocytogenes*, *Salmonella* Tyhimurium and *Pseudomonas fluorescens* to lettuce leaves. *J. Food Prot. 63*, 1433-1437.

[216] Tamminga, S.K., Beumer, R.R. & Kampelmacher, E.H. (1978). The hygienic quality of vegetables grown in or imported into the Netherlands: a tentative study. *J. Hyg. Camb. 80*, 143-154.

[217] Tauxe, R.V. (1997). Emerging foodborne diseases: an evolving public health challenge. *Dairy Food Environ. Sanit. 17*, 788-795.

[218] Thomas, C. & O'Beirne, D. (2000). Evaluation of the impact of short-term temperature abuse on the microbiology and shelf-life of a model ready-to-use vegetable combination product. *Int. J. Food Microbiol. 59*, 47-57.

[219] Thurston-Enriquez, J., Watt, P., Dowd, S.E., Enriquez, R., Pepper, I.L. & Gerbe, C.P. (2002). Detection of protozoan parasites and microsporidia in irrigation waters for crop production. *J. Food Prot. 65*, 378-382.

[220] Todd, E.C.D., Harris, C.K., Knight, A.J. & Worosz, M.R. (2007). Spinach and the media: how we learned about a major outbreak. *Food Prot. Trends 27*, 314-321.

[221] Ukuku D.O. & Fett, W. (2002). Behaviour of *Listeria monocytogenes* inoculated on cantaloupe surfaces and efficacy of washing treatments to reduce transfer from rind to fresh-cut pieces. *J. Food Prot. 65*, 924-930.

[222] Ukuku D.O. & Sapers, G.M. (2001). Effect of sanitizer treatments on *Salmonella stanley* attached to the surface of cantaloupe and cell transfer to fresh-cut tissues during cutting practices. *J. Food Prot. 64*, 1286-1291.

[223] Velani, S. & Roberts, D. (1991). *Listeria monocytogenes* and other *Listeria* spp. in prepacked mixed salads and individual salad ingredients. *PHLS Microbiol. Dig. 8*, 21-22.

[224] Vescovo, M., Scolari, G., Orsi, C., Sinigaglia, M., & Torriani, S. (1997). Combined effects of *Lactobacillus casei* inoculum, modified atmosphere packaging and storage temperature in controlling *Aeromonas hydrophila* in ready-to-use vegetables. *Int. J. Food Sci. Technol 32*, 411-419.

[225] Vescovo, M., Torriani, S., Orsi, C., Macchiarolo, F. & Scolari, G. (1996). Application of antimicrobial-producing lactic acid bacteria to control pathogens in ready-to-use vegetables. *J. Appl. Bacteriol. 81*, 113-119.

[226] Webb, T.A. & Mundt, J.O. (1978). Molds on vegetables at the time of harvest. *Appl. Environ. Microbiol. 35*, 655-658.

[227] Wei, C.I., Huang, T.S., Kim, J.M., Lin, W.F., Tamplin, M.L. & Bartz, J.A. (1995). Growth and survival of *Salmonella montevideo* on tomatoes and disinfection with chlorinated water. *J. Food Prot. 58*, 829-836.

[228] Wells, J.M. & Butterfield, J.E. (1997). *Salmonella* contamination associated with bacterial soft rot of fresh fruits and vegetables in the marketplace. *Plant Dis. 81*, 867-872.

[229] Wheeler, C., Vogt, T.M., Armstrong, G.L., Vaughan, G., Weltman, A., Nainan, O.V., Dato, V., Xia, G., Waller, K., Amon, J., Lee, T.M., Highbaugh-Battle, A., Hembree, C., Evenson, S., Ruta, M.A., Williams, I.T., Fiore, A.E. & Bell, B.P. (2005). An outbreak of Hepatitis A associated with green onions. *N. Engl. J. Med. 353*, 890-897.

[230] Wilcox, F., Tobback P. & Hendrickx, M. (1994). Microbial safety assurance of minimally processed vegetables by implementation of the Hazard Analysis Critical Control Point system. *Acta. Aliment. 23*, 221-238.

[231] Wolfe, S.K. (1980). Use of CO- and CO_2-enriched atmospheres for meats, fish and produce. *Food Technol. 34*, 55-58.

[232] Yuan, B., Summer, S., Eifert, J. and Marcy, J. (2003). Response of *Salmonella* and *Escherichia coli* O157:H7 to UV energy. *J. Food Prot. 66*, 1071-1073.

[233] Zhang, S. & Farber, J.M. (1996). The effects of various disinfectants against *Listeria monocytogenes* on fresh-cut vegetables. *Food Microbiol. 13*, 311-321.

[234] Zhong, Q., Black, D.G., Davidson, P.M. & Golden, D.A. (2008). Non-thermal inactivation of *Escherichia coli* K-12 on spinach leaves using dense phase carbon dioxide. *J. Food Prot. 71*, 1015-1017.

[235] Zhuang, R.-Y., Beuchat, L.R. & Angulo, F.J. (1995). Fate of *Salmonella montevideo* on and in raw tomatoes as affected by temperature and treatment with chlorine. *Appl. Environ. Microbiol. 61*, 2127-2131.

[236] Zhuang, R., Beuchat, L.R., Chinnan, M.S., Shewfelt, R.L. & Huang, Y-W. (1996). Inactivation of *Salmonella* Montevideo on tomatoes by applying cellulose-based edible films. *J. Food Prot. 59*, 808-812.

In: Consumer Product Safety Issues
Editors: P. A. Varga, M. D. Pintér
ISBN: 978-1-60456-826-4

Chapter 5

OCCURRENCE OF POLYCYCLIC AROMATIC HYDROCARBONS IN FOODS AND CONSUMER SAFETY

Monia Perugini and Pierina Visciano
Department of Food Science, University of Teramo, Viale Crispi, Teramo, Italy

ABSTRACT

In the field of consumer safety our competences are especially involved in contaminants residues in foods and, in particular, our researches are focused to investigate the presence of polycyclic aromatic hydrocarbons (PAHs), polychlorinated biphenyls and organochlorine pesticides. In this research paper we would highlight the toxicological risk caused by PAH levels in several foods, specifically in fish and fishery products. The attention will be paid to fresh matter in which the PAH content is a consequence of natural pollution, but also to processed products such as smoked, grilled or hydrogenated foods. PAHs, of which benzo(*a*)pyrene (BaP) is the most commonly studied and investigated, are formed by the incomplete combustion of organic matter. Human exposure occurs mainly by inhalation of airborne particulates containing PAHs and ingestion of food products contaminated by environmental pollution or by food processing (drying, smoking) or food cooking (roasting, grilling and frying). The major dietary sources are cereals, vegetables, oil, fats, fish and meat. The main concern for consumer safety is linked to PAHs carcinogenic potential as reported in the "Opinion of the Scientific Committee on Food on the risks to human health of Polycyclic Aromatic Hydrocarbons in food". Several agencies have classified a limited number of PAHs as probable or possible carcinogens. The U.S. Environmental Protection Agency proposed, for the risk assessment, to separate them in two subclasses consisting of the carcinogenic and the noncarcinogenic PAHs and to apply a cancer slope factor to the subclass of carcinogenic PAHs. The development and the establishment of toxic equivalency factors (TEFs) for PAHs similar to the TEF concept used in the assessment of mixtures containing polychlorinated biphenyls and dioxins could help to characterize more precisely the carcinogenic properties of PAH mixtures. Our contribution will focus on the significance and availability of TEF application for a detailed evaluation of consumer safety. The European Regulation established a maximum limit only for BaP as a marker for the occurrence and effect of carcinogenic PAHs in food, but this approach is limited because of the difficulty of defining safe levels of such a complex mixture.

INTRODUCTION

Polycyclic aromatic hydrocarbons (PAHs) constitute a large class of organic compounds, with a similar structure, containing two or more fused aromatic rings made up of carbon and hydrogen atoms. They may be formed and released during incomplete combustion or pyrolysis of organic matter, industrial processes and other human activities. Soils, surface waters and sediments may be contaminated by atmospheric fallout or deposition from sewage and oil and gasoline spills. Their sources, natural and mostly anthropogenic, in the environment are numerous and include: stubble burning and spreading of contaminated sewage sludge on agricultural fields; exhausts from mobile sources (motor vehicles and aircrafts); industrial plants (e.g. aluminum foundries, incinerators); wood preservation, use of tar coated wood; domestic heating with open fireplaces; burning of coal for thermal and electric energy; burning of automobile tires or of creosote treated wood; tobacco smoke; oil pollution of surface waters and soils; forest fires and volcanic eruptions (Annex to ECSCF, 2002).

Over 100 PAHs have been identified. They occur as complex mixtures throughout the environment in the air, water and soil. PAHs are hydrophobic compounds and they become more hydrophobic as molecular weight increases (Juhasz and Naidu, 2000). The heavier compounds are rapidly adsorbed into particles, while the lower compounds have long atmospheric residence times and their distribution is very sensitive to temperature with a large decrease in winter (Kiss et al., 1996).

The semi-volatile property of PAHs makes them highly mobile throughout the environment via deposition and re-volatilisation between air, soil and water bodies. It is possible that a large proportion of PAHs is deposited in the oceans and/or subjected to long range transport making them a widespread environmental problem. Raw foods should usually not contain high levels of PAHs. In areas remote from urban or industrial activities, the PAH levels found in unprocessed foods reflect the background contamination, which originates from long distance airborne transportation of contaminated particles and natural emissions from volcanoes and forest fires. In vegetables and fruits, the presence of PAHs originates mainly by deposition of air pollution particulates on their surfaces (Larsson, 1986). The detected levels are dependent on the location of the growing sites and on the product. In the neighborhood of industrial areas or along highways, the contamination of vegetation can be ten-fold higher than in rural areas. In plants the PAHs contamination can be due to the atmospheric exposure, to the uptake from the soil, or to the endogenous biosynthesis. The most important route of contamination is via atmospheric exposure, with the other routes playing lesser, possibly negligible, roles. The PAH pattern in plants was found to be similar to the vapour phase PAH pattern, because of the greater abundance of the lower molecular weight compounds in both environmental compartments (Wild and Jones, 1992; Wild and Jones, 1994). PAHs in soils are strongly associated with the organic matter fraction, may adhere onto root surfaces, but the potential for root uptake and translocation is low. The lower molecular weight PAHs (with two and three rings in particular) may also readily volatilise from contaminated soil onto foliage. The concentrations of PAHs are generally greater on plant surfaces, such as in the peel of fruit and outer leaves of cabbage, than in the internal tissues. The highest PAH levels in plants grown in areas with air pollution are especially found in vegetables with a large surface area exposed, e.g., lettuce, kale and spinach (Larsson, 1986) and can be a significant source of PAHs in the human diet.

As a result of their widespread distribution in the atmosphere PAHs, by atmospheric fallout, are also strongly represented in the marine environment. PAHs released into the aquatic ecosystem tend to adsorb rapidly on suspended materials and sediments and they are available to fish and other marine organisms through the food chain, as waterborne compounds and from contaminated sediments. Of these three possible routes, uptake of waterborne PAHs across the gills is considered to be the most significant. Moreover, PAHs uptake always depends on their bioavailability as well as the physiology of the organisms (Meador et al., 1995). When PAHs containing particulates fall out into surface water, they are transported in suspension and may end up in the fresh water or marine sediments to which they are strongly bound. These sediments constitute a pollution reservoir from which PAHs may be released. Filter-feeding bivalves such as mussels and oysters may accumulate particles contaminated by these compounds because they filter large volumes of water but are not capable of efficiently metabolising all PAHs.

PAHs fate in marine organisms is considered to be species dependent. Generally metabolic capacity in edible aquatic species appears to be best developed in fish, intermediate in crustaceans and least in molluscs (James, 1989; Stegeman and Lech, 1991). Filtrating organisms such as mussels or oysters accumulate these compounds and present elimination rates much less than those observed in vertebrates (Livingstone, 1994).

PAHs in fish are subjected to an active metabolizing process, through the P-450 cytochrome oxidase, and these oxidation and conjugation reactions facilitate their excretion. During the metabolism of some PAH compounds, reactive intermediates can form that, binding to macromolecules such as DNA and RNA, produce covalently bonded adducts that are a necessary step towards the development of PAH-induced cancer (Dunn, 1991). Fish may be victims of these carcinogenic metabolites while many invertebrates, like mussels, are protected against PAHs induced cancer by their metabolic incapacity (Knutzen, 1995). Furthermore many studies (Billiard et al., 1999; Marty et al., 1997) suggest that some PAHs are capable of inducing the embryo-toxic effects of dioxin interacting with the aryl hydrocarbon receptor (AHr). The inevitable contamination from the accumulation of PAHs is a significant concern that involves both benthic and pelagic marine organisms, but also the human population.

On the toxicity of PAHs, numerous studies indicate that one-, two- and three-ring compounds are acutely toxic, while higher molecular weight PAHs are considered to be genotoxic (Mersch-Sundermann et al., 1992; Nylund et al., 1992). Recently, several studies (Villeneuve et al., 2002; Clemons et al., 1998) stated that some PAHs are capable of inducing dioxin-like responses in vitro interacting with the AHr.

Several compounds of this class of contaminants have been shown to be potent carcinogens in experimental animals (ATSDR, 1995) and some of them may be regarded as potentially genotoxic and carcinogenic to humans. PAHs differ in carcinogenic potency. The U.S. Environmental Protection Agency (US EPA, 2002) has classified benzo(*a*)anthracene, benzo(*a*)pyrene, benzo(*b*)fluoranthene, benzo(*k*)fluoranthene, chrysene, dibenzo(*a,h*)anthracene and indeno(1,2,3-*cd*)pyrene as group B2 (probable human carcinogens). Among them, benzo(*a*)pyrene (BaP) has been well toxicologically characterized and it is referred as "one of the most potent known carcinogens". It was demonstrated that these compounds caused tumours in laboratory animals when they breathed them in the air, when they ate them or when they had long periods of skin contact with them. Studies on people showed that individuals exposed by breathing or skin contact for long periods to mixtures containing PAHs and other compounds could also develop cancer (ATSDR, 1995).

Generally, the evaluation of the PAHs toxicological risk is carried out through comparison of measured concentrations of a certain compound in the environment with the levels imposed by law or following guidelines proposed by different international organizations. This approach is obviously a generic one, as it tends to protect the world population globally and it cannot in any way account for the different eating habits and the different consumption rates of local populations. It is possible to use toxic equivalent factors (TEFs) also for PAHs and the toxicity relative to each single compound is referred to benzo(*a*)pyrene, which represents the most potentially carcinogenic polycyclic aromatic hydrocarbon. The development and establishment of TEFs for PAHs could help to characterize more precisely the carcinogenic properties of PAH mixtures (Fang et al., 2004). The US EPA separated PAHs into two subclasses consisting of carcinogenic and noncarcinogenic compounds. Benzo(*a*)pyrene was used as a reference compound, a TEF of 1.0 was applied to all classified carcinogenic PAHs and a TEF of zero to the noncarcinogenic PAHs (US EPA, 1984). However, the available information indicates that the other carcinogenic PAHs have lower potency than BaP and that the EPA approach therefore leads to an overestimation of the risk. After that, further TEFs proposed by Nisbet and LaGoy (Nisbet and LaGoy, 1992) completed a new list of TEFs which seems to better reflect the actual state of knowledge on the relative potency of individual PAHs (Petry et al., 1996).

PAHs Occurrence in Food

Food ingestion is the major route of exposure to PAHs compared to inhalation. Studies conducted on human exposure to BaP (Lioy et al., 1988) revealed that the range of dietary exposures (2 to 500 ng/day) were larger than for inhalation (10 to 50 ng/day). Among foodstuffs, cereals and oils/fats contribute in some countries the major part (approximately one third), fruits and vegetables provide one quarter and meat, fish and milk make relatively minor contributions (Dennis et al., 1983; Falcó et al., 2003). Dairy products often contain lower concentrations of PAHs than their nondairy equivalents. Thus, margarine contains higher levels than butter, and cream substitutes made from vegetable oils have higher levels that cream.

Smoked food are contaminated by PAHs during the incomplete combustion of wood used for the generation of smoke or using the liquid smoke flavourings. Data reported in the literature on PAH concentrations in smoked foods are highly variable due to the differences in the procedures used for smoking. These differences depend upon the type and composition of wood, the smoke generation conditions (internal or external generator), oxygen accessibility, temperature of pyrolysis and smoking time. The traditional smoking process includes drying and smoking of fish in the same chamber, where the smoke is produced directly below the hanging fish. In modern automatic smokehouses, the smoke develops in an external generator under controlled conditions of temperature and air access, while circulation is forced and controlled by mechanical equipment (Karl and Leinemann, 1996). Temperature generally plays an important role because the amount of PAHs in smoke formed during pyrolysis increases linearly with the smoke generation temperature within the interval 400 – 1,000 °C (Simko, 2002).

The rate of PAHs deposition on smoked food depends upon certain smoke characteristics (temperature, humidity, flow rate and density), water solubility and volatility of the particular compounds and also on the food properties, above all the lipid content.

In our published study (Visciano et al., 2006), aimed to determine the PAH content both in raw and smoked fish (*Salmo salar*) coming from a modern smokehouse, which used a smoking process at low temperature (24 °C), no significant difference ($p > 0.01$) was observed from raw to smoked samples in the contents of six PAHs (acenaphthene, phenanthrene, chrysene, benzo(*b*)fluoranthene, benzo(*k*)fluoranthene, benzo(*a*)pyrene). As a consequence of the smoking process, only fluorene, anthracene and benzo(*ghi*)perylene showed a statistically significant increase ($p < 0.01$), whereas for fluoranthene and benzo(*a*)anthracene a significant decrease ($p < 0.01$) was observed. Also total PAHs showed no significant change ($p > 0.01$). It must be underlined that environmental pollution represents an important source of PAHs in fish and three or four rings-PAHs are the most representative in marine ecosystem (Sprovieri et al., 2007; Vives et al., 2004; Baumard et al., 1998).

We also found that the low molecular weight PAHs (acenaphthene, phenanthrene, anthracene and fluoranthene) showed the highest levels in samples of smoked fillets (*Thunnus albacares*, *Xiphias gladius, Salmo salar*), randomly purchased in local markets and supermarkets in Italy (unpublished data). Among species, Atlantic salmon showed the highest levels for all the detected compounds. This result was likely a consequence of the contamination of commercial salmon feed. The fish oil and fish meal used by the manufacturers of salmon feed could be derived from marine food chains anywhere in the world depending on nutrient quality, price and availability. Since many contaminants of the aquatic ecosystem, like PAHs, are lipophilic, the use of fish oil and meal in feed may act as a direct source of these compounds into the human food chain. Easton et al. found the highest levels of PAHs in salmon feed samples, rather than in farmed fish and in the Pacific Coast wild fish (Easton et al., 2002).

The effects of different smoking techniques on PAH levels were studied in farmed rainbow trout (*Oncorhynchus mykiss*) smoked by traditional flue gas smoking and by liquid smoke flavourings. The following compounds, anthracene, fluoranthene, pyrene, benzo(*a*)anthracene, chrysene, benzo(*b*)fluoranthene, benzo(*k*)fluoranthene and benzo(*ghi*) perylene were detected in all samples, but no significant difference ($p > 0.05$) was found neither between fresh and processed samples nor between the two different smoking techniques, except for chrysene and benzo(*b*)fluoranthene (Visciano et al., 2008). These results show that samples were already naturally contaminated and the smoking process did not increase PAH levels, even if a traditional smokehouse was used and a direct smoking was applied. If these mild smoking techniques could not be considered as a source of PAHs in smoked fish, the raw matter contamination represents the main problem to be checked.

The PAHs presence in fresh fish was also investigated in our study (Perugini et al., 2007a), where PAHs were determined in bivalves (*Mitylus galloprovincialis*), cephalopods (*Todarodes sagittatus*), crustaceans (*Aristeus antennatus*) and fish (*Mullus surmeletus*, *Scomber scombrus*, *Micromesistius poutassou*, *Merluccius merluccius*) caught in the Gulf of

Naples (Tyrrhenian Sea). Independent from the season, PAHs composition pattern in bivalves was dominated by the presence of PAHs with 4-rings, followed from those with 5-rings, 6-rings and 3-rings. This pattern was not confirmed in fish, where the higher molecular PAHs (benzo(*a*)pyrene, dibenzo(*a,h*)anthracene, benzo(*ghi*)perylene and indeno(1,2,3-*cd*)pyrene) were never detected. Mussels adsorb the lower molecular weight PAHs through interstitial filtered water, and the heavier molecular weight PAHs through the digestive system. Their metabolic activity is low, so an equilibrium is reached between the PAHs accumulated and those present in the marine environment. In contrast, fish take up PAHs mainly from the dissolved form and they can rapidly convert up to 99% PAHs to metabolites within 24 hours of uptake, changing the pattern and the concentrations in the several tissues (Varanasi et al., 1989; Meador et al., 1995). The pollutants presence depends on the geographical origin too and crustaceans, cephalopods and fish that showed lower PAH concentrations and no presence of BaP, were less exposed to petroleum because they came from an area not closer to the Campi Flegrei Bay. The pattern and relative PAH concentrations in fish were quite irregular, but this variability was confirmed by different authors (Albaiges et al., 1987; Amodio Cocchieri et al., 1990). Depending on the season, no significant difference ($p > 0.05$) was found for crustaceans, cephalopods and fish. The total PAH content observed in mussels in this study was slightly lower (126 ng g^{-1} fresh weight) than that obtained by Amodio Cocchieri et al. in the same area (Amodio Cocchieri et al., 2003). In the two-year period (1998-1999 and 2000-2001) they reported for *Mitylus galloprovincialis* total PAH concentrations of 334 ng g^{-1} wet weight and 241 ng g^{-1} wet weight, respectively. This decreasing of hydrocarbons levels could be due to the environmental protection actions engaged during last years in the Gulf of Naples like the dismantling of industrial steel plant of ILVA. Because the releases of chemical contaminants, including PAHs, are already under regulatory control, further reductions in the environment will depend on the use of best management practices and possibly new regulatory improvements. Special attention should be paid to the PAHs input to Campi Flegrei Bay from local sources, because these waters are extensively used for mariculture purpose.

In the same species caught in the Adriatic Sea (Perugini et al., 2007b), fluorene, phenanthrene, anthracene, fluoranthene, benzo(*a*)anthracene, benzo(*b*)fluoranthene and benzo(*k*)fluoranthene were detected at different concentrations, whereas chrysene was detected only in mussels. The composition pattern was dominated by the presence of PAHs with 3 rings (62%) followed by those with 4-rings (37%) and 5 rings (1%). Acenaphthene, benzo(*a*)pyrene, dibenzo(*a,h*)anthracene, benzo(*ghi*)perylene and indeno(1,2,3-*cd*)pyrene showed levels below the instrumental detection limit in all samples.

Given the different PAH concentrations reported in our studies, it should be difficult to evaluate the risk for human health due to contaminated fish consumption. For this purpose the use of TEFs to generate data expressed as Equivalents of benzo(*a*)pyrene (BaPEs) has proved to be a useful framework for evaluating the potential hazard for the consumers. Considering this approach we used the TEFs proposed by Nisbet and LaGoy in order to calculate the BaPEs in our analysed samples. Data of both sum of PAHs and benzo(*a*)pyrene equivalents, calculated on the basis of the results of our studies (Perugini et al., 2007a; Perugini et al., 2007b; Visciano et al., 2006; Visciano et al., 2008), are reported in Tables 1 – 2 – 3 and 4, respectively.

Table 1. PAH concentrations (ng g^{-1} wet weight) expressed both as the sum of PAHs and as BaPEs in selected fish species from Tyrrhenian Sea (Perugini et al., 2007a)

Species	Σ PAHs *	BaPEs
Mediterranean mussels	126.40	10.67
Blue-red shrimp	11.96	0.58
Red mullet	16.14	0.87
European flying squid	43.81	1.25
Atlantic mackerel	15.70	0.96
Blue whiting	18.43	0.88
European hake	6.07	0.47

* Σ PAHs: anthracene, fluoranthene, pyrene, benzo(*a*)anthracene, chrysene, benzo(*b*)fluoranthene, benzo(*k*)fluoranthene, benzo(*a*)pyrene, dibenzo(*a,h*)anthracene, benzo(*ghi*)perylene, indeno(1,2,3-*cd*)pyrene.

By the comparison of PAHs sum and BaPEs, Atlantic mackerel, fished in the Adriatic Sea and reporting the highest PAHs contamination levels, showed a low level of BaPEs, whereas Mediterranean mussels, with a medium degree of contamination, had one of the highest levels of BaPEs (Table 2). This is due to the presence, in mussels tissues, of high levels of benzo(*a*)anthracene, showing a TEF of 0.1. Opposingly, Atlantic mackerel, which seemed the more contaminated species, with the highest total PAH concentrations, showed a low carcinogenic potency, being mainly polluted with fluoranthene that presents a TEF of 0.001. Also mussels collected in the Tyrrhenian Sea reported the highest BaPEs value (Table 1), being mainly contaminated by BaP, which has a TEF of 1. On the basis of this comparison it can be assumed that mussels are the most contaminated marine organisms contributing to the carcinogenic PAHs introduction by dietary exposure.

Table 2. PAH concentrations (ng g^{-1} wet weight) expressed both as the sum of PAHs and as BaPEs in selected fish species from Adriatic Sea (Perugini et al., 2007b)

Species	Σ PAHs *	BaPEs
Mediterranean mussels	34.73	1.56
Norway lobster	15.80	0.11
Red mullet	16.52	0.63
European flying squid	14.74	0.03
Atlantic mackerel	63.33	0.14
Blue whiting	55.53	2.76
European hake	44.14	1.75

* Σ PAHs: fluorene, phenanthrene, anthracene, fluoranthene, benzo(*a*)anthracene, chrysene, benzo(*b*)fluoranthene, benzo(*k*)fluoranthene.

Regarding to smoked fish (Tables 3 and 4), it can be observed that salmon fillets showed higher BaPE values than rainbow trout fillets and this could be linked to the different natural contamination of fresh matter.

Table 3. PAH concentrations (ng g^{-1} wet weight) expressed both as the sum of PAHs and as BaPEs in fresh and cold-smoked Atlantic salmon fillets (Visciano et al., 2006)

Samples	Σ PAHs *	BaPEs
Fresh Atlantic salmon fillets	67.68	4.04
Cold-smoked Atlantic salmon fillets	87.92	4.26

* Σ PAHs: acenaphthene, fluorene, phenanthrene, anthracene, fluoranthene, benzo(*a*)anthracene, chrysene, benzo(*b*)fluoranthene, benzo(*k*)fluoranthene, benzo(*a*)pyrene, benzo(*ghi*)perylene.

We specify moreover that BaPE values here reported must be considered only a minimum estimate of the toxicological risk at the state of knowledge, since many other additional PAHs are known or suspected carcinogens, but for these compounds toxic equivalency factors have not yet been developed.

Table 4. PAH concentrations (ng g^{-1} wet weight) expressed both as the sum of PAHs and as BaPEs in farmed rainbow trout processed by traditional flue gas smoking and by liquid smoke flavourings (Visciano et al., 2008)

Samples	Σ PAHs *	BaPEs
Fresh rainbow trout fillets	52.38	1.08
Rainbow trout fillets smoked by traditional flue gas smoking	59.53	1.13
Rainbow trout smoked as whole fishes by traditional flue gas smoking	50.99	1.11
Rainbow trout fillets smoked by liquid smoke flavourings on muscle	45.52	0.97
Rainbow trout fillets smoked by liquid smoke flavourings on skin	52.24	1.08

* Σ PAHs: anthracene, fluoranthene, pyrene, benzo(*a*)anthracene, chrysene, benzo(*b*)fluoranthene, benzo(*k*)fluoranthene, benzo(*ghi*)perylene.

Dietary Intake Assessment

Because of the widespread distribution of PAHs in the environment, most types of food contain measurable levels of PAHs, generally in the parts per billion, or microgram per kilogram, range. Although foods such as smoked or barbecued meat and fish may contain relatively high levels, unless the diet consists of very frequent consumption of such foods, it is cereals and vegetables, and their fats and oils, that make the major contributions to human dietary exposure. It is clear that diet contributes substantially to nonoccupational exposure to PAHs. For nonsmokers and non-occupationally exposed, more than 70% of exposure is attributable to diet (Menzie et al., 1992; Waldman et al., 1991; Fontcuberta et al., 2006). On the toxicity of PAHs, numerous studies indicate that the noncarginogenic compounds show oxidative stress as well as effect on the immune system, endocrine regulation and development (Hylland, 2006; Gozgit et al., 2004; Rhodes et al., 2005), whereas high molecular weight PAHs are considered to be carcinogenic. Benzo(*a*)pyrene, because of its carcinogenic potency in experimental animals, represents the most potent PAH carcinogen (Falcó et al., 2003). The carcinogenic potential of different PAHs is associated with the propensity of phase I enzymes to generate reactive epoxides that may be detoxified through phase II or bind to other cellular components, such as DNA (Collier et al., 1992). Legal regulation is limited, partly because of the difficulty of setting safe levels of such complex mixtures. Recently, Commission Regulation No 208/2005/EC (European Community, 2005), in force since 2005, set limits only for BaP contents in oils, edible fat and some foodstuffs. It defined a threshold level for muscle meat of smoked fish and smoked fishery products at 5 μg kg $^{-1}$ wet weight. Although some analyses of PAH content in several diet showed that the major contribution came from cereals, vegetables, oils and fat, and that there are numerous differences between the diets of northern and southern Europe, there is a high market demand for smoked fish, due to their desirable smoky flavours. *Salmo salar* is the more usual smoked fish, and about 40% to 50% of the Norwegian production of Atlantic salmon reaches the consumers as a cold-smoked product (Birkeland et al., 2004). In recent years also other species of fish, such as trout, tuna, swordfish, cod, herring, are processed by salting and smoking procedures and appreciated by consumers. According to the Food and Agriculture Organization the total world production of smoked herring and salmon are about 38,000 and 86,000 tons, respectively (Stolyhwo et al., 2006). High fish consumption is often associated with a belief that this food is important for human health, due to the beneficial effects on cardiovascular diseases of some natural components of fish, such as the n-3 polyunsaturated fatty acids. However, fish and seafood represent only a small percentage (about 10%) of the human diet, but it has been demonstrated that this means is one of the major routes of some contaminants, like PAHs, into the human body (Binelli and Provini, 2003). Because of the widespread distribution of PAHs in the marine environment, the use of smoke curing of fish and the PAHs bioavailability following ingestion, food such as smoked fish may contain relatively high levels and contribute to the human daily intake (Weyand et al., 1991; Phillips, 1999).

The presence of PAHs in food is a matter of concern that requires continuous monitoring. For non-occupationally exposed subjects who are also nonsmokers, more than 70% of exposure to these compounds arises from the diet (Falcó et al., 2003). In different studies the dietary intake of PAHs is reported (Ibanez et al., 2005; Lodovici et al., 1995; Dennis et al., 1983). Comparative intake data were collated from six European countries: Austria, United Kingdom, Italy, The Netherlands, Sweden and Germany (ECSCF, 2002). According to these studies the mean or averaged dietary intake of BaP for an adult person was estimated in the range 0.05 to 0.29 μg kg $^{-1}$. The major contributors to the daily BaP intake were oils and fats (47%) and cereals (36%), followed by vegetables (12%). When considering BaP and total PAHs (COT, 2002), cereals were found to be the largest contributor (24% and 35% respectively), followed by vegetables (12% and 13%) and by oils and fats (6% and 3%). A significant contribution was also from beverages (28% and 8%) and from milk and dairy products group (12% and 9%). It should be noted that oils and fats show usually the highest individual PAH levels, whereas cereals, although never having high individual PAH concentrations, are a major contributor due to their relative weight to the total diet. Also smoked and grilled foods could have high PAH levels, but they make only a modest contribution since they are minor components of the usual diet. However, if such foods are usually present in the diet, their contribution to PAH intake increases. Thus, in contrast to the available European surveys, grilled/barbecued meat contributed 21% to the mean daily intake of BaP in the USA and was the food group contributing to the second intake of PAHs after the bread, cereal and grain group which contributed 29% (Kazerouni et al., 2001). Recently, Ibanez et al. estimated the dietary intake of PAHs in the Spanish adult population. In this study (Ibanez et al., 2005), cereals (40.1%), followed by meat and meat products (20.9%) and fish and shellfish (6.1%) were the food group with the highest contributions to total dietary intake of PAHs. Most studies to assess the carcinogenic potential of single PAH were carried out following dermal, subcutaneous or inhalation exposure. Only a limited number of studies dealt with oral administration. The carcinogenic risk associated with the dietary intake was assessed on the basis of animal carcinogenicity data. Benzo(*a*)pyrene, when administered by the oral route, produced tumours of the gastrointestinal tract, liver, lungs and mammary glands of mice and rats. Of the few other PAHs tested for carcinogenicity by the oral route, dibenzo(*a,h*)anthracene and benzo(*a*)anthracene produced tumours of the gastrointestinal tract, liver and lungs in mice. No increases in tumour incidences were seen in rats after oral administration of benzo(*a*)anthracene, phenanthrene, fluorene or naphthalene (ATSDR, 1995). Even if it has been demonstrated that rodents are rather insusceptible to BaP taken by mouth, it cannot be assumed that humans are similarly resistant although they might be (Lijinsky, 1991). Moreover, other PAHs are also classified as probable human carcinogens and data from various cancer tests used to rank carcinogenic potentials relative to that of BaP have also been used to assess cancer risk. While it would be improper to exaggerate the carcinogenic risk of exposure to very low concentrations of PAHs in foods, it is prudent to assume that exposures in the range often found present some risk of cancer and should be reduced or avoided.

Conclusion

Aside from carcinogenicity, PAHs have also been reported to cause hemato, cardio, renal, neuro, immuno, reproductive and developmental toxicities in humans and laboratory animals (Ramesh at al., 2004). Studies on the bioavailability of these contaminants from various entry routes are important not only from the standpoint of scientific knowledge, but also from the public health perspective.

The Commission Regulation 208/2005/EC established maximum admissible levels only for BaP. According to the Scientific Committee on Food, benzo(*a*)pyrene can be used as a marker for the occurrence and effect of carcinogenic PAHs in food, including also benzo(*a*) anthracene, benzo(*b*)fluoranthene, benzo(*j*)fluoranthene, benzo(*k*)fluoranthene, benzo(*ghi*)-perylene, chrysene, cyclopenta(*c,d*)pyrene, dibenzo(*a,h*)anthracene, dibenzo(*a,e*)pyrene, dibenzo(*a,h*)pyrene, dibenzo(*a,i*)pyrene, dibenzo(*a,l*)pyrene, indeno(1,2,3-*cd*)pyrene and 5-methylchrysene. Nevertheless, because benzo(*a)*pyrene is only one of many possible carcinogenic PAHs present in food, further analyses of the relative proportions of these PAHs would be necessary to inform a future review of the suitability of maintaining benzo(*a*)pyrene as a marker.

REFERENCES

Albaiges, J., Algaba, J., Arambarri, P., Cabrera, F., Baluja, G., Hernandez, L. M. & Castroviejo, J. (1987). Budget of organic and inorganic pollutants in the Doñana National Park (Spain). *Science of the Total Environment*, *63*, 13-28.

Amodio Cocchieri, R., Amoroso, S., Arnese, A., Cirillo, T., Montuori, P. & Triassi, M. (2003). Pollution by mercury, arsenic, lead, chromium, cadmium, and polycyclic aromatic hydrocarbons of fish and mussels from the Gulf of Naples, Italy. *Environmental Contamination and Toxicology*, *71*, 551-560.

Amodio Cocchieri, R., Arnese, A. & Minicucci, A. M. (1990). Polycyclic aromatic hydrocarbons in marine organisms from Italian Central Mediterranean coasts. *Marine Pollution Bulletin*, *21*, 15-18.

Annex to ECSCF - European Commission Scientific Committee on Food (2002). *Polycyclic aromatic hydrocarbons – Occurrence in foods, dietary intake and health effects,* SCF/CS/CNTM/PAH/29 ADD1 Final, Brussels, Belgium, A5.

ATSDR – Agency for Toxic Substances and Disease Registry (1995). Toxicological profile for polycyclic aromatic hydrocarbons. US Department of Health and Human Services, US Public Health Service, Atlanta, 23-52.

Baumard, P., Budzinski, H., Garrigues, P., Sorbe, J., Burgeot, T. & Bellocq, J. (1998). Concentrations of PAHs (polycyclic aromatic hydrocarbons) in various marine organisms in relation to those in sediments and to trophic level. *Marine Pollution Bulletin*, *36*, 951-960.

Billiard, S., Querbach, K. & Hodson, P. (1999). Toxicity of retene to early life stages of two freshwater fish species. *Environmental Toxicology and Chemistry, 18*, 2070-2077.

Binelli A. & Provini, A. (2003). POPs in edible clams from different Italian and European markets and possible human health risk. *Marine Pollution Bulletin, 46*, 879-886.

Birkeland, S., Bencze Rørå, A. M., Skåra, T. & Bjerkeng, B. (2004). Effects of cold smoking procedures and raw material characteristics on product yield and quality parameters of cold smoked Atlantic salmon (*Salmo salar* L.) fillets. *Food Research International*, *37*, 3, 273-286.

Clemons, J. H., Allan, L. M., Marvin, C. H., Wu, Z., McCarry, B. E., Bryant, D. W. & Zacharewski, T. R. (1998). Evidence of estrogen- and TCDDlike activities in crude and fractionated extracts of PM10 air particulate material using in vitro gene expression assays. *Environmental Science and Technology, 32*, 1853–1860.

Collier, T. K., Singh, S. V., Awasthi, Y. C. & Varanasi, U. (1992). Hepatic xenobiotic metabolizing enzymes in two species of benthic fish showing different prevalences of contaminant-associated liver neoplasms. *Toxicology Applied Pharmacology, 113*, 319-324.

COT – Committee on Toxicity of Chemicals in Food, Consumer Products and the Environment (2002). Polycyclic aromatic hydrocarbons in the 2000 total diet study. Report TOX/2002/26, 1-17.

Dennis, M. J., Massey, R. C., McWeeny, D. J., Knowles, M. E. & Watson, D. (1983). Analysis of polycyclic aromatic hydrocarbons in UK total diet. *Food and Chemical Toxicology*, *21* (5), 569-574.

Dunn, B. P. (1991). Carcinogen adducts as an indicator for the public health risks of consuming carcinogen-exposed fish and shellfish. *Environmental Health Perspectives, 90*, 111-116.

Easton, M. D. L., Luszniak, D. & Von der Geest, E. (2002). Preliminary examination of contaminant loadings in farmed salmon, wild salmon and commercial salmon feed. *Chemosphere*, *46*, 1053-1074.

ECSCF - European Commission Scientific Committee on Food (2002). *Opinion of the Scientific Committee on Food on the risks to human health of Polycyclic Aromatic Hydrocarbons in food.* Background Document SCF/CS/CNTM/PAH/29 Final. European Commission Scientific Committee on Food, Brussels, Belgium, 15-24.

European Community (2005). Commission Regulation (EC) No 208/2005 of 4 February 2005 amending Commission Regulation (EC) No 466/2001 as regards polycyclic aromatic hydrocarbons. *Official Journal L 034*, 08/02/2005, 1-5.

Falcó, G., Domingo, J. L., Llobet, J. M., Teixidò, A., Casas, C. & Muller, L. (2003). Polycyclic aromatic hydrocarbons in foods: human exposure through the diet in Catalonia, Spain. *Journal of Food Protection, 66* (12), 2325-2331.

Fang, G. C., Chang, K. F., Lu, C. & Bai, H. (2004). Estimation of PAHs dry deposition and BaP toxic equivalency factors (TEFs) study at Urban, Industry Park and rural sampling sites in central Taiwan, Taichung. *Chemosphere, 55,* 787–796.

Fontcuberta, M., Arqués, J. F., Martinez, M., Suarez, A., Villalbi, J. R., Centrich, F., Serrahima, E., Duran, J. & Casas C. (2006). Polycyclic aromatic hydrocarbons in food samples collected in Barcelona, Spain. *Journal of Food Protection, 69* (8), 2024-2028.

Gozgit, J. M., Nestor, K. M., Fasco, M. J., Pentecost, B. T. & Arcaro, K. F. (2004). Differential action of polycyclic aromatic hydrocarbons on endogenous estrogen-responsive genes and on a transfected estrogen-responsive reporter in MCF-7 cells. *Toxicology Applied Pharmacology, 196*, 58-67.

Hylland, K. (2006). Polycyclic aromatic hydrocarbon (PAH) ecotoxicology in marine ecosystems. *Journal of Toxicology and Environmental Health, Part A, 69*, 109-123.

Ibanez, R., Agudo, A., Berenguer, A., Jakszyn, P., Tormo, M. J., Sanchez, M. J., Quiros, J. R., Pera, G., Navarro, C., Martinez, C., Larranaga, N., Dorronsoro, M., Chirlaque, M. D., Barricarte, A., Ardanaz, E., Amiano, P. & Gonzales, C. A. (2005). Dietary intake of polycyclic aromatic hydrocarbons in a Spanish population. *Journal of Food Protection, 68* (10), 2190-2195.

James, M. O. Biotransformation and deposition of PAH in aquatic invertebrates. In: Varanasi, U. (Ed), CRC Press Inc., *Metabolism of Polycyclic Aromatic Hydrocarbons in the Aquatic Environment*, Boca Raton, Florida, USA, 1989, 69-92.

Juhasz, A. L. & Naidu, R. (2000). Bioremediation of high molecular weight polycyclic aromatic hydrocarbons: a review of the microbial degradation of benzo(*a*)pyrene. *International Biodeterioration & Biodegradation, 45*, 57-88.

Karl H. & Leinemann M. (1996). Determination of polycyclic aromatic hydrocarbons in smoked fishery products from different smoking kilns. *Zeitschrift für Lebensmitteluntersuchung und Forschung A*, *202*, 458-464.

Kazerouni, N., Sinha, R., Hsu, C. H., Greenberg, A. & Rothman, N. (2001). Analysis of 200 food items for benzo(a)pyrene and estimation of its intake in an epidemiologic study. *Food and Chemical Toxicology*, *39*, 423-436.

Kiss, G., Varga-Puchony, Z. & Hlavay, J. (1996). Determination of polycyclic aromatic hydrocarbons in precipitation using solid-phase extraction and column liquid chromatography. *Journal of Chromatography A, 725*, 261-272.

Knutzen, J. (1995). Effects on marine organisms from polycyclic aromatic hydrocarbons (PAH) and other constituents of waste water from aluminium smelter with examples from Norway. *Science of Total Environment*, *163,* 107-122.

Larsson, B. (1986). Polycyclic aromatic hydrocarbons in Swedish foods—aspects on analysis, occurrence and intake. Ph.D. Thesis, Swedish University of Agricultural Sciences.

Lijinsky, W. (1991). The formation and occurrence of polynuclear aromatic hydrocarbons associated with foods, *Mutation Research*, *259*, 251-261.

Lioy P. L., Waldman, J. M., Greenberg, A., Harkov, A. & Pietarinen C. (1988). The total Human Environmental Exposure Study (THEES) to benzo(a)pyrene: comparison of the inhalation and food pathways. *Archives of Environmental & Occupational Health, 43*, 304-312.

Livingstone, D. R. (1994). Recent developments in marine invertebrate organic xenobiotic metabolism. *Toxicology and Ecotoxicology News*, *1*, 88-95.

Lodovici, M., Dolora, S., Canalini, C., Ciappellano, S. & Testolin, G. (1995). Polycyclic aromatic hydrocarbons contamination in the Italian diet. *Food Additives and Contaminants*, *12* (5), 703-713.

Marty, G. D., Short, J. W. & Dambach, D. M. (1997). Histopathology and cytogenetic evaluation of Pacific herring larvae exposed to petroleum hydrocarbons in the laboratory or in Prince William Sound, Alaska, after the Exxon Valdez oil spill. *Canadian Journal of Zoology, 75*, 989-1007.

Meador, J. P., Stein, J. E., Reichert, W. L. & Varanasi, U. (1995). Bioaccumulation of polycyclic aromatic hydrocarbons by marine organisms. *Reviews of Environmental Contamination and Toxicology, 143*, 79-165.

Menzie, C. A., Potocki, B. B. & Sandonato, J. (1992). Exposure to carcinogenic PAHs in the environment. *Environmental Science Technology, 26*, 1278-1284.

Mersch-Sundermann, V., Mochayedi, S., & Kevekordes, S. (1992). Genotoxicity of polycyclic aromatic hydrocarbons in *Escherichia coli* PQ37. *Mutation Research*, *278*, 1–9.

Nisbet, I. C. T. & LaGoy P. K. (1992). Toxic Equivalency Factors (TEFs) for polycyclic aromatic hydrocarbons (PAHs). *Regulatory Toxicology and Pharmacology, 16,* 290-300.

Nylund L., Heikkila P., Hameila M., Pyy L., Linnainmaa K. & Sorsa M. (1992). Genotoxic effects and chemical compositions of four creosotes. *Mutation Research*, *265*, 223–236.

Perugini, M., Visciano, P., Manera, M., Turno, G., Lucisano, A. & Amorena M. (2007a). Polycyclic aromatic hydrocarbons in marine organisms from the gulf of Naples, Tyrrhenian Sea. Journal of Agricultural Food Chemistry, 55, 2049-2054.

Perugini, M., Visciano, P., Giammarino, A., Manera, M., Di Nardo, W. & Amorena M. (2007b). Polycyclic aromatic hydrocarbons in marine organisms from the Adriatic Sea, Italy. Chemosphere, 66, 1904-1910.

Petry, T., Schmid, P. & Schlatter, C. (1996). The use of toxic equivalency factors in assessing occupational and environmental health risk associated with exposure to airborne mixture of polycyclic aromatic hydrocarbons (PAHs). *Chemosphere, 32*, 639–648.

Phillips, D. H. (1999). Polycyclic aromatic hydrocarbons in the diet. *Mutation Research, 443*, 139-147.

Ramesh, A., Walker, S. A., Hood, D. B., Guillen, M. D., Schneider, K. & Weyand, E. H. (2004). Bioavailability and risk assessment of orally ingested polycyclic aromatic hydrocarbons. *International Journal of Toxicology, 23*, 301-333.

Rhodes, S., Farwell, A., Hewitt, L. M., MacKinnon, M. & Dixon, D. G. (2005). The effects of dimethylated and alkylated polycyclic aromatic hydrocarbons on the embryonic development of the Japanese medaka. *Ecotoxicology Environmental Safety, 60*, 247-258.

Simko P. (2002). Determination of polycyclic aromatic hydrocarbons in smoked meat products and smoke flavouring food additives. *Journal of Chromatography B, 770*, 3-18.

Sprovieri, M., Feo, M. L., Prevedello, L., Salvagio Manta, D., Sammartino, S., Tamburrino, S. & Marsella, E. (2007). Heavy metal, polycyclic aromatic hydrocarbons and polychlorinated biphenyls in surface sediments of the Naples harbour (southern Italy). *Chemosphere*, *67*, 998-1009.

Stegeman, J. J. & Lech, J. J. (1991). Cytochrome P-450 monooxygenase systems in aquatic species: carcinogens metabolism and biomarkers for carcinogen and pollutant exposure. *Environmental Health Perspectives*, *90*, 101-109.

Stolyhwo, A., Kolodziejska, I. & Sikorski, Z. E. (2006). Long chain polyunsaturated fatty acids in smoked Atlantic mackerel and Baltic sprats. *Food Chemistry*, *94*, 589-595.

US EPA - Environmental Protection Agency (1984). *Health effects assessment for polycyclic aromatic hydrocarbons*. EPA 549/1-86-013. Environmental Criteria and Assessment Office, Cincinnati, OH.

US EPA - Environmental Protection Agency (2002). *Polycyclic organic matter*. National Center for Environmental Assessment, Office of Research and Development, Washington, DC.

Varanasi, U., Stein, J. E. & Nishimoto, M. Biotransformation and disposition of polycyclic aromatic hydrocarbons (PAH) in fish. In Varanasi, U. (Ed), CRC Press Inc., *Metabolism of Polycyclic Aromatic Hydrocarbons in the Aquatic Environment*, Boca Raton, Florida, USA, 1989, 94-149.

Villeneuve D. L., Khim J. S., Kannan K. & Glesy J. P. (2002). Relative Potencies of Individual Polycyclic Aromatic Hydrocarbons to Induce Dioxinlike and Estrogenic Responses in Three Different Cell Lines. *Environmental Toxicology, 17*, 128-137.

Visciano, P., Perugini, M., Amorena, M. & Ianieri, A. (2006). Polycyclic Aromatic Hydrocarbons in Fresh and Cold-Smoked Atlantic Salmon Fillets. *Journal of Food Protection, 69,* 5, 1134-1138.

Visciano, P., Perugini, M., Conte, F. & Amorena, M. (2008). Polycyclic aromatic hydrocarbons in farmed rainbow trout (*Oncorhynchus mykiss*) processed by traditional flue gas smoking and by liquid smoke flavourings. *Food and Chemical Toxicology, 46*, 1409-1413.

Vives, I., Grimalt, J. O., Fernandez, P. & Rosseland, B. (2004). Polycyclic aromatic hydrocarbons in fish from remote and high mountain lakes in Europe. *Science of the Total Environment*, *324*, 67-77.

Waldman, J. M., Lioy, P. J., Greenberg, A. & Butler, J. P. (1991). Analysis of human exposure to benzo(a)pyrene via inhalation and food ingestion in the Total Human Environmental Exposure Study (THEES). *Journal Exposure Analysis Environmental Epidemiology, 1*, 193-225.

Weyand, E. H., Wu, Y., Patel, S., Taylor, B. B. & Mauro, D. M. (1991). Urinary excretion and DNA binding of coal tar components in B6C3F1 mice following ingestion. *Chemical Research in Toxicology, 4*, 466-473.

Wild, S. R. & Jones, K. C. (1992). Polynuclear aromatic hydrocarbons uptake by carrots grown in sludge amended soil. *Journal of Environmental Quality, 21,* 217-225.

Wild, S. R. & Jones, K. C. (1994). The significance of polynuclear aromatic hydrocarbons applied to agricultural soils in sewage sludges in the U.K. *Waste Management and Research, 12*, 49-59.

In: Consumer Product Safety Issues
Editors: P. A. Varga, M. D. Pintér
ISBN: 978-1-60456-826-4

Chapter 6

THE FEDERAL FOOD SAFETY SYSTEM: A PRIMER*

Geoffrey S. Becker and Donna V. Porter

SUMMARY

Numerous federal, state, and local agencies share responsibilities for regulating the safety of the U.S. food supply, which many experts say is among the safest in the world. Nevertheless, critics view this system as lacking the organization and resources to adequately combat foodborne illness — as evidenced by a series of widely publicized food safety problems, including concerns about adulterated food and food ingredient imports, and illnesses linked to various types of fresh produce and to meat and poultry products. Numerous bills addressing various aspects of food safety have been introduced into the first session the 1 10th Congress, including proposals to reorganize oversight authorities, increase funding, tighten regulation of imported foods, and establish new authorities for enforcement, recall, and notification regarding adulterated foods, among others. Portions of several proposals have become law; more sweeping changes in the federal food safety system are still under consideration in 2008.

BACKGROUND[1]

Americans spend more than $1 trillion on food each year, nearly half of it in restaurants, schools, and other places outside the home.2 Federal laws give food manufacturers, distributors, and retailers the basic responsibility for assuring that foods are wholesome, safe, and handled under sanitary conditions. A number of federal agencies, cooperating with state,

---

* This is an edited, reformatted and augmented version of CRS Report RS22600, dated June 9, 2008

1 Background on the agencies is updated information that first appeared in CRS Report 98-91, *Food Safety Agencies and Authorities: A Primer* (out of print). Primary sources for that report included various documents and materials provided by federal food safety agencies and by the U.S. Government Accountability Office

2 Roughly two-thirds of the $1 trillion is for domestically produced farm foods; imports and seafood account for the balance. Source: USDA, Economic Research Service, data accessed January 2008 at the "Food Sector" Web page at http://www.ers.usda.gov/Browse/FoodSector/].

local, and international entities, play a major role in regulating food quality and safety under these laws.

The combined efforts of the food industry and the regulatory agencies often are credited with making the U.S. food supply among the safest in the world. Nonetheless, public health officials estimate that each year 76 million people become sick, 325,000 are hospitalized, and 5,000 die from foodborne illnesses caused by contamination from any one of a number of microbial pathogens.[3] At issue is whether the current system has the resources and structural organization to protect consumers from these dangers. Also at issue is whether federal food safety laws, first enacted in the early 1900s, have kept pace with the significant changes that have occurred in the food production, processing, and marketing sectors since then.

THE AGENCIES AND THEIR ROLES

The Government Accountability Office (GAO) has identified 15 federal agencies collectively administering at least 30 laws related to food safety. The Food and Drug Administration (FDA), which is part of the U.S. Department of Health and Human Services (HHS), and the Food Safety and Inspection Service (FSIS), which is part of the U.S. Department of Agriculture (USDA), together comprise the majority of both the total funding and the total staffing of the government's food regulatory system.[4] FSIS's FY2007 budget was approximately $930 million in appropriated funds plus another estimated $135 million in industry-paid user fees. FDA's budget for foods was $457 million in FY2007, virtually all of it appropriated.[5]

Among other agencies with smaller but still significant shares of the food safety portfolio are the National Marine Fisheries Service (NMFS), which is part of the U.S. Department of Commerce (DOC), the Environmental Protection Agency (EPA), and the Centers for Disease Control and Prevention (CDC) in HHS.

Food and Drug Administration

The FDA is responsible for ensuring that all domestic and imported food products — except for most meats and poultry — are safe, nutritious, wholesome, and accurately labeled. Examples of FDA-regulated foods are produce, dairy products, seafood, and processed foods. FDA has jurisdiction over meats from animals or birds that are not under the regulatory jurisdiction of FSIS. FDA shares responsibility for the safety of eggs with FSIS. FDA has jurisdiction over establishments that sell or serve eggs or use them as an ingredient in their products. FDA is also responsible for ensuring that most seafood products do not endanger public health (FSIS is to begin inspecting farmed catfish products under a 2008 farm bill

---

[3] U.S. Department of Health and Human Services, Centers for Disease Control and Prevention, "Foodborne Illness: Frequently Asked Questions," accessed at [http://www.cdc.gov/foodsafety/]. However, this estimate appears to be based primarily on 1997 and earlier data in a report by Paul S. Mead et al., "Food-related Illness and Death in the United States," *Emerging Infectious Diseases*, vol. 5, pp. 607-625, 1999.

[4] *High Risk Series: An Update* (GAO-07-310), January 2007.

[5] Data source: various documents of the Agriculture, Rural Development, Food and Drug Administration, and Related Agencies Subcommittee of the House Committee on Appropriations.

provision). The primary statutes governing FDA's activities are the Federal Food, Drug, and Cosmetic

Act, as amended (21 U.S.C. 301 et seq.); the Public Health Service Act, as amended (42 U.S.C. 201 et seq.); and the Egg Products Inspection Act, as amended (21 U.S.C. 1031 et seq.).

FDA's food inspection force numbers more than 1,900 in field offices throughout the United States, plus nearly 900 in the Washington, DC, area. FDA regulates food manufacturers' safety practices by relying on companies' self-interest in producing safe products, and by working with the industry to improve production practices. About 57,000 food manufacturers are subject to periodic FDA inspection for regulatory compliance. According to GAO, unannounced compliance inspections of individual establishments by FDA officials occur roughly once every five years. FDA relies on notifications from within the industry or from other federal or state inspection personnel, as well as other sources, to alert it to situations calling for increased inspection.

In the Washington, DC, area, two FDA offices are the focal point for food safety- related activities. The Center for Food Safety and Applied Nutrition (CFSAN) is responsible for (1) conducting and supporting food safety research; (2) developing and overseeing enforcement of food safety and quality regulations; (3) coordinating and evaluating FDA's food surveillance and compliance programs; (4) coordinating and evaluating cooperating states' food safety activities; and (5) developing and disseminating food safety and regulatory information to consumers and industry. FDA's Center for Veterinary Medicine (CVM) is responsible for ensuring that all animal drugs, feeds (including pet foods), and veterinary devices are safe for animals, are properly labeled, and produce no human health hazards when used in food-producing animals.

The FDA also cooperates with over 400 state agencies across the nation that carry out a wide range of food safety regulatory activities. However, the state agencies are primarily responsible for actual inspection. FDA works with the states to set the safety standards for food establishments and commodities and evaluates the states' performance in upholding such standards as well as any federal standards that may apply. FDA also contracts with states to use their food safety agency personnel to carry out certain field inspections in support of FDA's own statutory responsibilities.

Food Safety and Inspection Service

FSIS regulates the safety, wholesomeness, and proper labeling of most domestic and imported meat and poultry and their products sold for human consumption. Under the Federal Meat Inspection Act of 1906, as amended (21 U.S.C. 601 et seq.), FSIS inspects all cattle, sheep, swine, goats, and equines during slaughtering and processing. Under the Poultry Products Inspection Act of 1957, as amended (21 U.S.C. 451 et seq.), FSIS is required to inspect "any domesticated bird" being processed for human consumption; however, USDA regulations implementing this law limit the definition of domesticated birds to chickens, turkeys, ducks, geese, ratites (emus, ostriches, and rheas), and guineas. FDA has jurisdiction over exotic and alternative meats not inspected by FSIS, and shares the responsibility for egg safety with FSIS. The latter is responsible for the safety of liquid, frozen, and dried egg products, domestic and imported, and for the safe use or disposition of damaged and dirty eggs under the Egg Products Inspection Act, as amended (21 U.S.C. 1031 et seq.).

FSIS staff numbers around 9,400; roughly 8,000 of them, including about 1,000 veterinarians, are in about 6,300 meat slaughtering and/or processing plants nationwide. FSIS personnel inspect all meat and poultry animals at slaughter on a continuous basis, and at least one federal inspector is on the line during all hours the plant is operating. Processing inspection does not require an FSIS inspector to remain constantly on the production line or to inspect every item. Instead, inspectors are on site daily to monitor the plant's adherence to the standards for sanitary conditions, ingredient levels, and packaging, and to conduct statistical sampling and testing of products. Because all plants are visited daily, processing inspection also is considered to be continuous.

FSIS also is responsible for certifying that foreign meat and poultry plants are operating under an inspection system equivalent to the U.S. system before they can export their product to the United States. FSIS inspectors located at U.S. ports of entry carry out a statistical sampling program to verify the safety of imported meats from cattle, sheep, swine, goats, and equines and imported poultry meat from chickens, turkeys, ducks, geese, quail, ratites, and guineas before they are released into domestic commerce. FDA is responsible for ensuring the safety of imported meat from any other species.

Twenty-seven states operate their own meat and/or poultry inspection programs. FSIS is statutorily responsible for ensuring that the states' programs are at least equal to the federal program. Plants processing meat and poultry under state inspection can market their products only within the state (although the 2008 farm bill will permit interstate shipment in some cases). If a state chooses to discontinue its own inspection program, or if FSIS determines that it does not meet the agency's equivalency standards, FSIS must assume the responsibility for inspection if the formerly state-inspected plants are to remain in operation. FSIS also has cooperative agreements with more than two dozen states under which state inspection personnel are authorized to carry out federal inspection in meat and/or poultry plants. Products from these plants may travel in interstate commerce.

Centers for Disease Control and Prevention (HHS)

CDC is responsible for (1) monitoring, identifying, and investigating foodborne disease problems to determine the contributing factors; (2) working with FDA, FSIS, NMFS, state and local public health departments, universities, and industry to develop control methods; and (3) evaluating the effect of control methods. In 1995, CDC launched "FoodNet," a collaborative project with the FDA and USDA to improve data collection on foodborne illness outbreaks. FoodNet includes active surveillance of clinical microbiology laboratories to obtain a more accurate accounting of positive test results for foodborne illness; a physician survey to determine testing and laboratory practices; population surveys to identify illnesses not reported to doctors; and research studies to obtain new and more precise information about which food items or other exposures may cause diseases. FoodNet data allows CDC to have a clearer picture of the incidence and causes of foodborne illness and to establish baseline data against which to measure the success of changes in food safety programs. The Public Health Service Act provides legislative authority for CDC's food safety related activities.

National Marine Fisheries Service (DOC)

Although the FDA is the primary agency responsible for ensuring the safety, wholesomeness and proper labeling of domestic and imported seafood products, NMFS conducts, on a fee-for-service basis, a voluntary seafood inspection and grading program that focuses on marketing and quality attributes of U.S. fish and shellfish. The primary legislative authority for NMFS's inspection program is the Agricultural Marketing Act of 1946, as amended (7 U.S.C. 1621 et seq.). NMFS has approximately 160 seafood safety and quality inspectors, and inspection services are funded with user fees.

Environmental Protection Agency

EPA has the statutory responsibility for ensuring that the chemicals used on food crops do not endanger public health. EPA's Office of Pesticide Programs is the part of the agency that (1) registers new pesticides and determines residue levels for regulatory purposes; (2) performs special reviews of pesticides of concern; (3) reviews and evaluates all the health data on pesticides; (4) reviews data on pesticides' effects on the environment and on other species; (5) analyzes the costs and benefits of pesticide use; and (6) interacts with EPA regional offices, state regulatory counterparts, other federal agencies involved in food safety, the public, and others to keep them informed of EPA regulatory actions. The Federal Insecticide, Fungicide, and Rodenticide Act, as amended (7 U.S.C. 136 et seq.), and the Federal Food, Drug, and Cosmetic Act, as amended (21 U.S.C. 301 et seq.), are the primary authorities for EPA's activities in this area.

Other Federal Agencies with Food Safety Responsibilities

Among the other agencies that play a role in food safety, USDA's Agricultural Research Service (ARS) performs food safety research in support of FSIS's inspection program. It has scientists working in animal disease bio-containment laboratories in Plum Island, NY, and Ames, IA. USDA's Animal and Plant Health Inspection Service (APHIS) indirectly protects the nation's food supply through programs to protect plant and animal resources from domestic and foreign pests and diseases, such as brucellosis and bovine spongiform encephalopathy (BSE, or "mad cow" disease). The Department of Homeland Security (DHS) is to coordinate many food security activities, including at U.S. borders.

Congressional Committees

In the Senate, food safety issues are considered by the Committees on Agriculture, Nutrition, and Forestry; Homeland Security and Governmental Affairs; and Health, Education, Labor, and Pensions. In the House, various food safety activities fall under the jurisdiction of the Committees on Agriculture; Energy and Commerce; Oversight and Government Reform; and Science. Agriculture subcommittees of the House and Senate Appropriations Committees also serve oversight and funding roles in how the major agencies carry out food safety policies.

SELECTED ISSUES

Food safety-related incidents frequently heighten congressional scrutiny of the issue, as a number of developments in 2006 and 2007 illustrated. For example, more than 200 confirmed illnesses and three deaths were linked in fall 2006 to the consumption of bagged fresh spinach grown in California that was found to carry *E. coli* O157:H7. The incident raised public concern about the safety of all fresh leafy produce and stimulated a number of industry and government initiatives to limit future contamination. In February 2007, FDA announced a nationwide recall of peanut butter due to *Salmonella* contamination, after hundreds of illnesses, dating back to August 2006, were linked to the bacterium.[6] Throughout 2007, USDA announced numerous recalls totaling many million pounds of ground beef products due to concerns about *E. coli* O157:H7 contamination.

Attention had shifted to the safety of food imports in early 2007 when adulterated pet food ingredients imported from China sickened or killed numerous dogs and cats and subsequently were found in some hog, chicken, and fish feed.[7] In June 2007, FDA announced that it was detaining imports of certain types of farm-raised seafood from China (specifically, shrimp, catfish, basa, dace, and eel) until their shippers could confirm that they are free of unapproved drug residues.

By late 2007 a series of congressional and Administration reports had been issued that cited significant gaps in the federal food safety system. More than two dozen bills were introduced into the first session of the 1 10th Congress addressing one or more aspects of the issue. Provisions affecting food safety were included in 2007 in P.L. 110-85, the FDA amendments, including a requirement that FDA establish a registry to which companies must begin to report events involving potentially adulterated foods. Food safety provisions in the 2008 farm bill (P.L. 110-234) include subjecting farmed catfish products to FSIS mandatory inspections similar to those for red meat and poultry; creating an option for state-inspected meat and poultry plants to ship products across state lines; and requiring meat and poultry establishments to notify USDA about potentially adulterated or misbranded products. During the second session, a number of the more comprehensive food safety bills could be marked up. These include bills to impose new requirements, and possibly user fees, on food importers; to establish mandatory recall and/or traceability programs for USDA and FDA-regulated foods; and to increase oversight of domestic food producers, for example. Among the more prominent proposals are separate draft bills being circulated by the chairmen of the House Energy and Commerce Committee and the Senate Health, Education, Labor, and Pensions Committee.

---

[6] For sources and updates see the FDA website: [http://www.fda.gov/opacom/7alerts.html].

[7] FDA has the same basic safety standards for human foods and animal feeds, including pet food.

In: Consumer Product Safety Issues
Editors: P. A. Varga, M. D. Pintér

ISBN: 978-1-60456-826-4

Chapter 7

CONSUMER SAFETY & HACCP-LIKE QUALITY RISK MANAGEMENT PROGRAMS ON DAIRY FARMS: THE ROLE OF VETERINARIANS

João Cannas[*] and Jos Noordhuizen[†]
VACQA-International, Santarém, Portugal

INTRODUCTION

Consumer safety has taken, over the last decades, an important position in our society, in particular in developed countries. To ensure consumer safety, the European Union as well other countries have issued laws to ensure that the whole food chain is under control. The final objective will be the full implementation of the HACCP-concept and principles at the primary sector of production, in our case the dairy production units. As we already know, the European consumers are rapidly sensibilized by problems in areas including food safety and public health, as well as animal welfare and animal health. To illustrate this, we only need to refer to the Salmonella survey in several species in the last years. European consumers have a high impact on animal production methods, and since the 1990s we observe a growing and permanent pressure on the production sector that drives the producers to start the implementation of on-farm control programmes. (Cullor, 1997; Noordhuizen, 2004a; Radostits, 2004).European consumer concern with food goes far beyond the sole nutritional aspects or product quality issues. The possibly transmittable diseases are of major concern nowadays. Examples are salmonella as already mentioned, E. coli strains, Listeriosis, among others. So the different sectors of production of food of animal origin are very active in developing and undertaking control activities which should increase the confidence of consumers. (Vieira et al., 2007).

During the last years, the basis has been built for the implementation of systems that can meet the demands. Our dairy farmers have to implement on-farm control programmes which can show to the authorities and the consumers that the dairy industry has adequate control of

[*] Correspondence to: joao.cannas@mail.telepac.pt

the health status of the herd, and in some countries the farmer has to prove that he respects the welfare of his own animals, and that the food produced from animals is safe and the quality is good. The farmers have also new challenges in making improvements on environmental protection, and hence they need an approach which can deal with all these different challenges..

A concept like Hazard Analysis Critical Control Points (HACCP), is the one quality control concepts which best serves the objectives of both food safety and the farmers. (Vieira et al., 2007). HACCP is considered to be a systematic method, preventive and science-based, whose first priority is the safety of the products through risk identification and risk management in the production process It has a proactive, rather than reactive, approach emphasizing food hazard prevention rather than the detection of harmful defects in finished food products Its main objective is to identify, in the production process, problems before they occur, establishing control measures that are critical to maximizing food safety at each stage in the production (Vieira et al., 2007). The applicability of this kind of programme on the dairy farm is in ensuring the healthiness of the food produced (including animals) and to protect cows and food from hazards during the production process before they pass beyond the physical limits of the dairy farm to reach the next step in the food chain (Vieira et al., 2007).

The HACCP system is already very common and used in different fields like restaurants, transports, food and feed production units, for example. The veterinarians have an important place in the HACCP concept on farms. One of the principle reasons is to achieve a transparent, fully controlled chain, starting in primary production and ending at the final consumer. The veterinarian should play a very important role in all steps of the food chain, because -given his educational background- he is one of the professionals with knowledge, understanding and experience in feed, animals, food and management.

Before discussing the role of veterinarians in the HACCP development and implementation on dairy farms, we will first explain the principles of the HACCP concept. Subsequently, we introduce the chapters of a HACCP-like handbook for reasons of clarification.

After knowing all the principles and the developmental steps (Table 1), we now introduce the Handbook of a HACCP–like Quality Risk Management Program (QRMP) because from its contents we can see which pathway to follow to develop and implement the various HACCP-components. The various chapters of this handbook will be followed in the subsequent text paragraphs.

† Correspondence to: noordhu1@xs4all.nl

Table 1. Developing and implementing the HACCP-principles in 12 steps_(adapted after Cullor, 1995)

1. Assemble a multidisciplinary, facility-based, HACCP-team e.g. farmer; veterinarian; nutritionist and/or farm-economist).
2. Describe the final product and the method of distribution (e.g. formulation, processing requirements).
3. Identify the intended use of the (raw) food product and the targeted purchaser.
4. Develop a flow diagram that describes the production and distribution process (a production process decomposition diagram). Work from whole-farm level to the production step levels (e.g. milk harvesting; feeding procedures).
5. Verify the flow diagram on-site on correctness with the farm-workers.
6. Implement *Principle 1:* Prepare a list of steps in the production process at which targeted hazards occur. Identify these main hazards and their associated risk factors.
7. Apply *Principle 2*: Identify the critical control points, CCP, in the production process required to reduce or eliminate the hazard. Identify the points of particular attention, POPA's, too.
8. Apply *Principle 3*: Establish standards and critical limits, or targets, for triggering the implementation of corrective/preventive measures associated with each CCP and POPA identified when control is lost.
9. Implement *Principle 4*: Establish monitoring requirements for each CCP and POPA. Use the results of the monitoring program to adjust the procedures and to maintain control of the production process. Use monitoring also for herd performance assessment.
10. Apply *Principle 5*: Determine corrective measures, to take when monitoring indicates that a value falls outside the established critical limits or beyond a given target.
11. Apply *Principle 6*: Establish effective record-keeping procedures that document that the HACCP program has been implemented, and is operational.

Apply *Principle 7*: Establish procedures to verify that the HACCP-like program is working correctly (e.g. internal reviews and external verification, periodic revalidation of the system).

THE HANDBOOK OF A HACCP-LIKE QUALITY RISK MANAGEMENT PROGRAM

This handbook comprises, after identification of the farm and the members of the HACCP-team, the following components:

1. Production process diagram and identification of the different steps in the production process;
2. Hazard analysis chart
   Refers to the respective steps, as well as the inputs, where main hazards could occur which could cause quality problems; general preventive measures already taken; and the impact of the risk to human health; all for each step. Risk assessment.
3. Hazard audit table (identification of CCP and POPA; monitoring; measures)
   Refers to each operational activity, the potential hazards, CCP's and POPA's, screening or check points, monitoring procedures, critical limits & standards, or targets, corrective actions per CCP or POPA, records
4. Registration requirements, GVP, GDFP, GMP, farmer's own declaration
   Refers to declarations of those professionals who are active within the Quality Management Program, e.g. veterinarian, inseminator, feed mill nutritionist, to comply with the rules set within the Quality Management Program, as well as to the declaration of the farmer himself stating that he will comply with the rules set out in the law and that activities are carried out according to the program specifications. The latter also applies to farm-workers.
5. Support programs with work instructions
   Refer to personal hygiene and health of coworkers, dress/boot code, handling of hazardous materials, handling of waste materials, handling hot-spots regarding cross-contamination, hygiene & disinfection of equipment and vehicles, equipment & vehicles maintenance, calibration of measuring & dosing devices. These issues are described in technical work instructions (protocols).
6. Training programs for farmer and/or coworkers
   Trainings should be recorded on a separate sheet. Regularly the training needs of the farm have to be revisited. Training is carried out on-site and are of short duration.
7. Internal Review & External Audit
   An internal review (focus on effectiveness and compliance) is to be carried out annually for each cow worker by using a checklist. An annual review of the QMP handbook, using a checklist, is the next issue; this review should ensure that all procedures are up-to-date, documented, numbered and authorized; old documents are archived. Findings, corrections and improvements are to be listed on sheets, and archived. An external audit is to be carried out annually (focus on integrity).
8. Documents, and document control:
   To be retained for 2 years. Examples: Delivery Complaint Report; Purchase Forms; Order Forms; Training Sheet; Calibration Report; Laboratory Results Report; Internal Audit Report; External Audit Report; QMP handbook review.

After this stage, we must produce what we call Production Process Diagrams (PPD).

Production Process Diagrams

Decomposition diagrams are schematic representations of the production process e.g. on a dairy farm. It is advised to draw these diagrams on-site with farmer and farm-workers, and then verify its correctness on the farm. The basic idea is to work from the larger image downwards to the smaller images with more detail. For example, from the main farming areas down to the detailed process of the milk harvesting phase. All process steps should fit 1 page DIN A_4 in order to keep it readable. These diagrams appear to be very helpful in understanding what is happening where; they support the discussion in the HACCP-team and they clarify for farm-workers where hazards, risks, tasks and responsibilities are. Some examples are given here (Fig.1, Fig.2). Take good notice about the level they have been assigned to: overall farm-level, parts of the farm, or details as process steps.

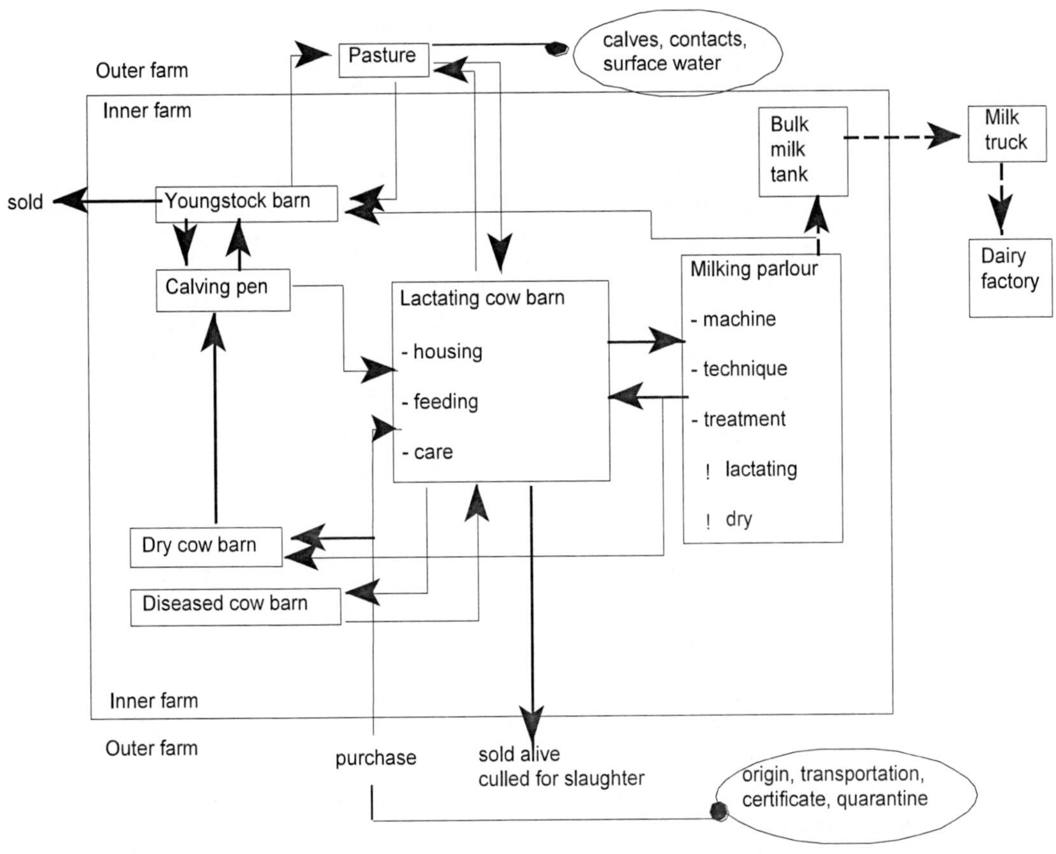

Figure 1. A general overview of a dairy production process decomposition diagram (After Lievaart et al., 2005)

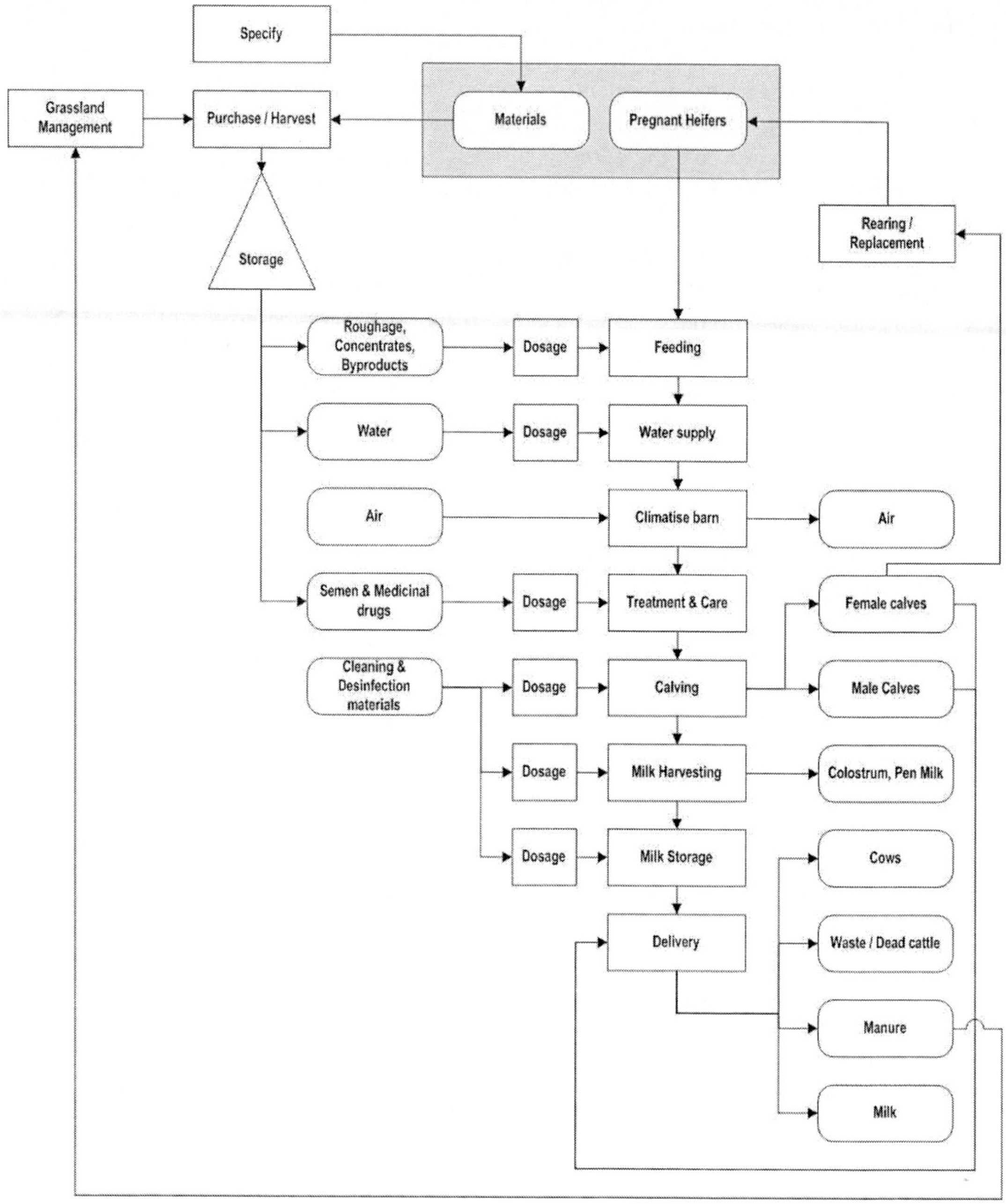

Figure 2. A detailed production diagram referring to the process steps of feeding and feeding management (after Noordhuizen et al., 2007)

Hazards & Risks Analysis Charts

The hazards and risks associated to disorders in the area of animal health, animal welfare, food safety and public health can be identified by using the *strengths-and-weaknesses assessment (SWA) sheets,* provided e.g. on the vacqa-international.com web site. If they are not available for a certain specific area, you may develop a new SWA with the help of the SWA sheet examples given, because they provide a blueprint of such an approach. Fig.3 comprises a screen image example of such a SWA-sheet.

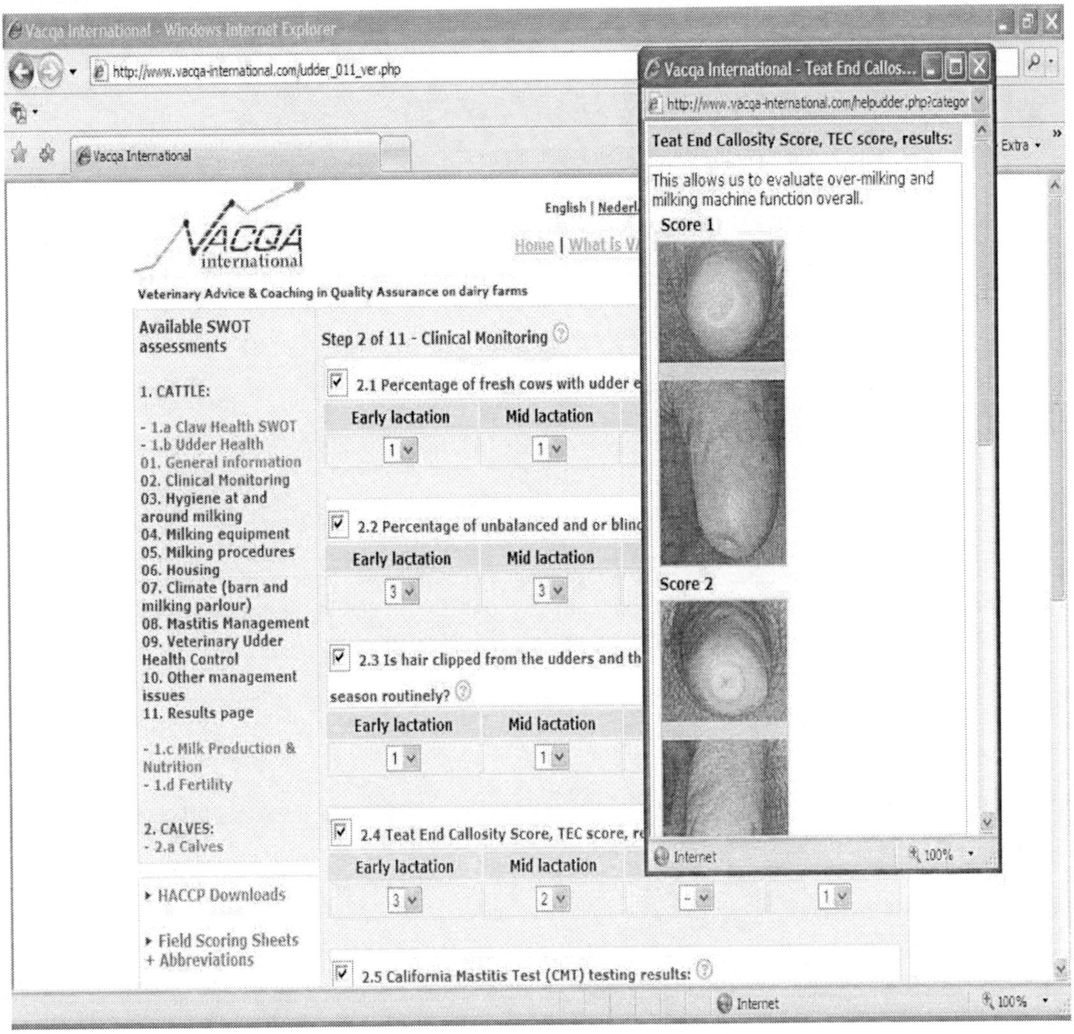

Figure 3. An example of a screen image, presenting the Vacqa-international SWA-sheets, the example of udder health.

Discussion with the farmer and with the nutritionist or farm-economist will help to prioritize the most important hazards. Weighing of risks can be done using quantitative epidemiologic information or by the on-farm HACCP-team through assigning weighing factors to the different risk conditions. The latter weighing is qualitatively conducted through estimating the probability P of occurrence and the impact I (= P*I); hence, the true risks can be identified and addressed later-on .Examples of such a risk weighting are given in Table 2 and Table 3.

Table 2. Outcome of risk estimates regarding the physical and chemical hazards of milk quality, an example

Code	C/F	Hazard	Process step M=milking T=treatment	Weighing P and I P	I	P*I	Risk yes/no	CCP/ POPA	Prevention/control measure (farmer focused)
T_1	C	Wrong drug	T	2	2	4	no	POPA	Proper diagnosis Check label Drug advisory plan See instruction leaflet
T_2	C	Residue in meat	T	1	2	2	no	POPA	Withdrawal time Drug advisory plan See instruction leaflet
T_3	C	Wrong dosage	T	2	2	4	no	POPA	Check proper dosage Use drug advisory plan
T_4	C	Drug > shelflife	T	2	1	2	no	GFP	Check shelflife See instruction leaflet
T_5	C	Cows' ID fails	T	2	3	6	yes	CCP	See instruction leaflet during withdrawal
T_6	F	Needle broken	T	1	1	1	no	GFP	Inject with care
M_1	C	Drug residues in milk	M	2	3	6	yes	CCP	Good cow ID Withdrawal time Stick to indications See instruction leaflet

T= treatment step in the process; M= milk harvesting step in process; C= chemical; F= physical; GFP= good farming practice code; CCP= critical control point; POPA= point of particular attention; P= probability; I= impact; ID= identification.

Table 3. Outcomes of risk estimates of microbiological hazards during milk harvesting, an example

Hazard	Process Step	Risk estimate P	I	P*I	Risk yes/no	CCP/POPA	Prevention/control measures	Responsible person
Milk with B.abortus	T, M	1	2	2	no	GFP	Certificate free of B.abortus	Farmer
Milk with Mycob. bovis	T, M	1	3	3	no	GFP	High hygiene level Certificate free of M. bovis	Farmer
Milk with L. monocytogenes	T, M, BT, C, F	2	4	8	yes	CCP	Hygiene at milking Infected cows separate Reduce infection spread	Farmer
Milk with S. dublin/typhimurium	T, M, F, C	1	3	3	yes	POPA	Hygiene at milking Increase health status	Farmer
Milk with C. jejuni	T, M, C	2	2	4	yes	POPA	Hygiene at milking	Farmer
Milk with Staph. aureus	T, M	3	2	6	yes	CCP	Cows with infection to be separated; reduce infection spread (see drug advisory plan & instruction leaflet)	Farmer
Milk with E.coli $O_{157}H_7$	T, M, C	2	3	6	yes	CCP	Hygiene at milking	Farmer
Milk with Y. enterocolitica	T, M, C	1	2	2	no	GFP	Hygiene at milking	Farmer

T= treatment step in process; M= milk harvesting step; F= feeding step; C= barn climate step; BT= bulk milk tank; GFP= good farming practice code; CCP= critical control point; POPA= point of particular attention

The decision-Tree Approach to Determine CCP's

To determine whether or not a control point is named formally as 'critical' we use the decision-tree scheme as given in Fig.4. True and formal critical control points, CCP, have to meet certain formal criteria (see next paragraph). Mark that if a critical point is not a true CCP, it could still be defined as a POPA when considered truly relevant for process control.

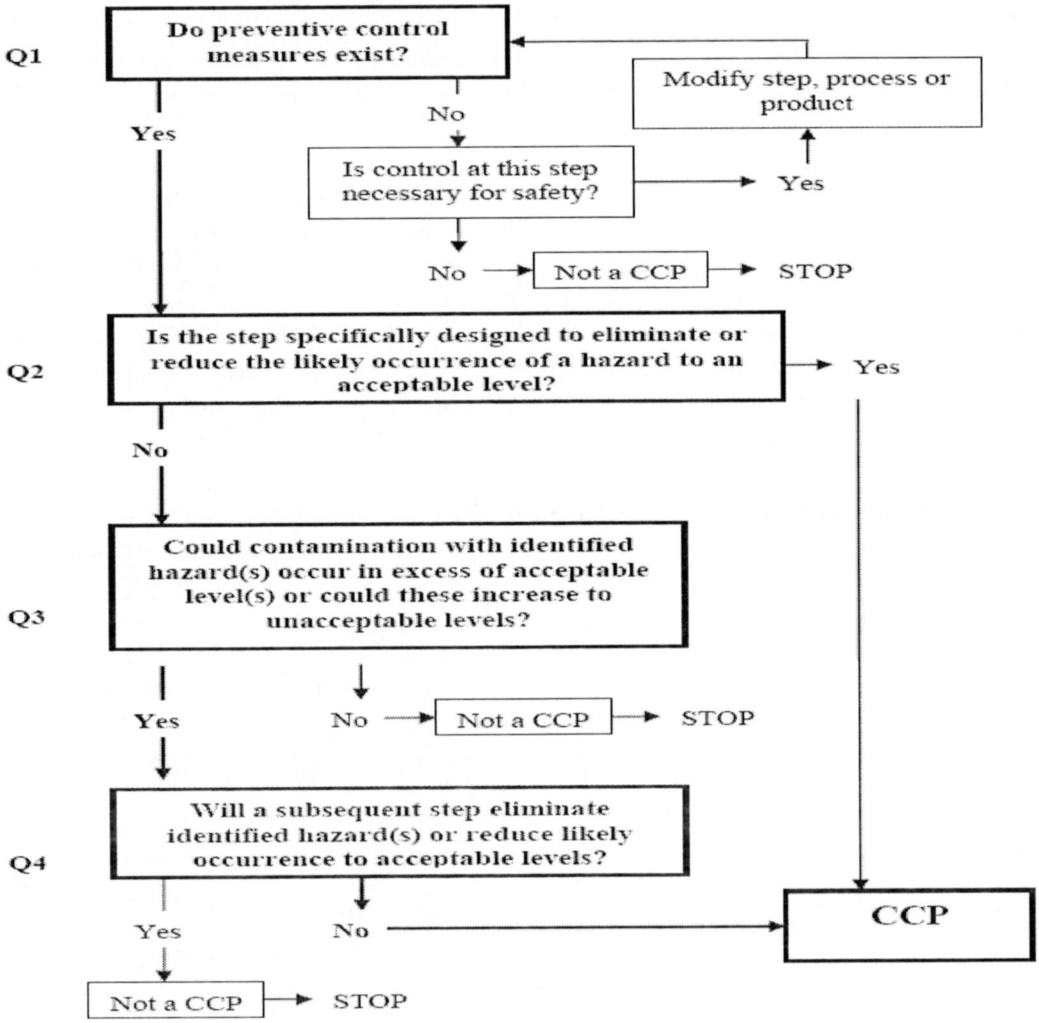

Figure 4. The decision-tree approach to determine which control points can be considered as critical control points, CCP.

Explanation to CCP's and POPA's

Critical Control Points (CCP) and Points of Particular Attention (POPA) are usually derived from the risk conditions found in earlier steps of the HACCP design.

CCP's have to meet the following formal five criteria :

- be associated with the hazard of concern
- be measurable or observable

- have norms/standards and tolerance limits
- be accompanied by corrective measures which
- must fully restore control, once it was lost.

Usually, cost—benefit calculations should be associated with the alternative corrective measures to facilitate least costs interventions and optimal integration with farm management. POPA's fail to meet at least one of the forenamed CCP-criteria, namely because they lack formal standards or because they are not fit to fully restore control. Sometimes they are called CCP-2 but we prefer the sharper distinction with true CCP. The CCP's and the POPA's are to be put into a *monitoring scheme* for the dairy farm. In this scheme the items for monitoring, as well as their frequency, the method of monitoring and the responsibility for monitoring must be listed. An issue of concern relates to the *standards* in livestock production. Physical features and processes are commonly provided with standards and tolerance limits, for example the temperature of the cleaning water for rinsing the milking machine can be set at 80 ºC +- 2 °C. However, in the case of biological processes with live animals we deal with biological variation among animals and with diagnostic tests, we deal with false-positive and false-negative test results and variable cut-off points. In such situations we cannot speak about standards but should rather introduce the term "*targets*" or reference values. E.g. the target for clinical mastitis cases can be set at < 25% per year.

Hazards audit Tables_(Monitoring CCP's and POPA's; Corrective Measures)

An example of part of a HACCP-like QRM program on a dairy farm for chemical and physical hazards in the process step of cow treatment T_i (see also the general production process diagrams) is given in Table 3.

Table 3. An example of a hazards audit Table in a HACCP-based QRM

Item no.	CCP/ POPA	Standard + Tolerance Or Target	Monitoring How	Freq.	Who	Instruction (prevention)	Corrective measures	Records
T_1	POPA	Use only proper drug	Check label	At use	Farmer	See DAP	Use proper drugs Evaluate other drugs Consult the vet	"Drug Record R"
T_2	POPA	No residues	Check drugs (R)	At delivery	Farmer	See DAP	Respect the with-drawal periods	(R)
T_3	POPA	Dosage in DAP	Check syringe	At use	Farmer	See DAP	Adjust dosage	(R)
T_5	CCP	Cow ID No tolerance	Visual	At drug use	Farmer	See DAP	Mark the cow	(R)

T= cow treatment step in the process; POPA= point of particular attention; CCP= critical control point; DAP.= drug advisory plan of the veterinarian for the herd

Good Dairy Farming Codes of Practice Hygiene, Cleaning & Disinfection- (Guidelines for daily farm practice)

In the Annex to the COAG/2003/6 – FAO (Rome, Italy) there are indicators provided for Good Agricultural Practice (GAP), especially in the area of animal production, animal health and animal welfare. At each point in the production process of the dairy farm, GAP-like codes of practice can be developed. These *codes of practice are general, practical guidelines* for the farmer and the farm-workers for optimizing farm performance, as well as for those professionals (and sometimes lay people) that are visiting the farm to comply with the rules set. In this context we speak about *Good Dairy Farming* codes of practice (GDF). Codes of practice must be developed for those areas, where certain hazards can occur, because they help in creating or improving awareness among the people (the attitude and mentality focus), and they help in highlighting certain responsibilities.

Examples of such guidelines under the heading of Good Dairy Farming codes of practice are:

1. Good Farm Hygiene codes of practice (incl. several Check Lists)
2. Good Medicine Application code of practice (see example following)
3. Good Feeding Management code of practice
4. Good Milking Hygiene code of practice
5. Good Calf Rearing code of practice (see two examples below)

In the evolution of risk assessment, we can find several risk factors which are very disease-specific, while other risk factors are much more common and present in several disorders. We will usually find the latter, more general risk factors in the GDF-guidelines, which are preliminarily meant for attitude and mentality improvement.

Although these guidelines are very general in nature, we have to strive to specify them as much as possible to the individual farm conditions and install them if needed.

Good Hygiene Code of Practice on Dairy Farms: Some Working Instructions

Dairy farming is about producing raw materials for food processing: milk and beef. Therefore, the cleaning, disinfection and maintenance of areas where such raw materials are being produced (milking parlour) and or stored (bulk milk tank) must meet with the highest demands. Hygiene-activities should follow strict rules. In these technical working instructions are addressed the different elements relevant for cleaning and their rationale. Commonly, hygiene is an issue that is not described but rather a more or less variable way of acting as prevailing in the head of a farmer. Describing the different components not only makes the farmer more aware of the actions he/she is taking, it is also easy for those people who are replacing the farmer or people working together with the farmer. Ultimately, by describing the hygiene working instructions it can be showed to third parties that the dairy farm is indeed taking hygiene seriously as one way to safeguard milk production and food safety from chemical and microbiological contamination. These working instructions on hygiene are of a blue print type, meaning that they have to be further specified for each individual farm; in some cases this means delete items, in others add items. It can also be regarded as a check for consulting afterwards. The basis for these working instructions is in scientific evidence and field experiments.

Procedure for Cleaning

There are 6 steps in the cleaning procedure to be followed:

- pre-treatment (in order to eliminate loose dirt mechanically);
- cleaning (to loosen dirt by applying certain products);
- rinsing (to eliminate loosened dirt and neutralize cleaning product residues);
- disinfection (to destroy bacteria that survived cleaning by applying disinfectants);
- rinsing (to eliminate residues of disinfectants);
- drying (to eliminate the last rinsing water).

These 6 steps are integrated into *3 working methods*, depending on the areas where more or less contact may exist with the raw food being produced. These methods are:

A. Cleaning & Drying
   For areas where no direct contact exists between surfaces to be cleaned and raw material (eg milk) being produced.
B. Cleaning, Rinsing & Drying
   For areas, materials and equipment where/on-which contact with residues of cleaning products with raw material (eg milk) must be avoided.
C. Cleaning, Rinsing, Disinfection, Rinsing & Drying
   For surfaces of equipment and materials which are in direct contact with the raw material being produced (eg milk) and which are not subjected to heat-treatment.

Hygiene Rules to Be Followed

In addition to the application of the methods described in the paragraph "Procedure for Cleaning", the farm worker(s) responsible for executing hygiene measures should follow some strict rules. Among these rules –which are basically part of *Good Farming Practice*— are for example:

- the instructions given should be strictly followed
- personal hygiene of the people involved is a prerequisite (clean hands/nails; clothes; boots)
- parts of equipment which can be taken apart, should be regularly subjected to cleaning/disinfection methods (eg, milking machine)
- in case of purchase or in situations of reconstruction, the surveillance of hygienically working remains paramount

- in case of replacing certain parts, mounting instructions must be strictly followed
- the working instructions on hygiene must be executed......
  - on the right moment
  - at the proper frequency
  - by using adequate dosage of products
  - while using clean materials
  - and without neglecting rinsing/drying after cleaning and disinfection.

Check Points in the Cleaning/Disinfection Procedure

In order to provide the farmer with the certainty that the effects of cleaning and disinfection procedures are being achieved, it is worthwhile to insert a few check points in the whole procedure. These check points too are part of Good Farming Practice.

Examples of *check points* are:

- Timing of the cleaning/disinfection: should not be conducted during moments that the raw material (e.g. milk) is being produced, so not during milking time.
- Proper dosage of cleaning/disinfection products: too low dosage will negatively affect the effect; a too high dosage is too expensive and environmentally undesired.
- Clean the working materials in order to prevent re-contamination.
- Duration of the different steps and procedures: disinfection should last at least 5 min to be effective.
- Rinsing after disinfection of surfaces in contact with e.g. milk is needed to avoid contamination of milk by product residues.
- Separation of dirty and clean parts of equipment, materials and surfaces in order to avoid re-contamination and insufficient cleaning.
- Check temperature of refrigerator, milking machine cleaning temperature of boiler, and bulk milk tank temperature daily.
- Do not store chemical products close to the bulk milk tank.
- Keep refrigerator free of ice and do not turn it off.
- Check regularly the expiration date of cleaning and disinfection products.

Special attention must be given to professional and lay visitors, because potentially they represent a risk of introducing or spreading pathogens into and on the farm. In the next paragraph we will present some details.

Visitors (Professional and Lay People)

General check points for the *first* implementation of a working instruction on hygiene regard the issues named in Table 5.

Table 5. Some relevant issues for checking the hygiene elements related to visitors.

	YES	NO	na
Does the farm have a disinfection tub for cars?	O	O	O
Is there a specific parking place for professional visitors?	O	O	O
Is the parking place on the dirt part of the farm?	O	O	O
Is the entrance to the farm clearly indicated?	O	O	O
Have barns/sites been clearly identified?	O	O	O
Is there a hygiene barrier present?	O	O	O
Is the hygiene barrier well located?	O	O	O
Is the hygiene barrier well equipped?	O	O	O
Is the hygiene barrier functional?	O	O	O
Is the on-farm working order from young to older cattle?	O	O	O
Are there clean boots and clothes per barn provided?	O	O	O
Does every barn have its own materials/instruments?	O	O	O
Are there different sites to wash hands?	O	O	O

Na= not applicable

Once these general issues have been addressed and adjusted when needed, the Hygiene Instruction for Visitors can be implemented. It is best to put that instruction on a board at the entrance and or give it to those professional visitors that come to the farm most frequently. An example is presented below.

Hygiene Instruction for Visitors

Welcome to our dairy farm!
We expect you to strictly follow the instructions on hygiene given below; thank you.

1. Cars & trucks
   Use only the parking place as indicated on the map by P.
2. Do change your boots and clothes in the hygiene barrier (see map) before further entering our farm. Report your arrival by using the phone as indicated there.
3. If you need to make contact with animals, take along disposable gloves. If needed, use the available farm equipment and not your own.
4. When having entered the farm, the farmer or his replacer will tell you the working order of the farm. At all times, follow our hygiene instructions.
5. Follow the routine working order of the farm. Use disinfection tubs where-ever they are present; change clothes/boots and wash hands whenever this is indicated.
6. Do not make contact with our cattle, pets and other animals if not really necessary.
7. When the farm visit is ended, clean the boots. All materials introduced on our farm are considered to be dirty materials (irrespective of being used or not) and should not be used on other farms. Change clothes and boots, dispose of gloves, wash your hands thoroughly.
8. Record medicinal products which have been used and or delivered in the MEDICINES' Log, and
9. Register date/name/ time in the VISITORS' Log.
10. Delivered medicinal and other products should be stored in the right place as indicated on the map or handed over to the farmer.

Thank you for your visit and thank you for complying to these rules! You have contributed to minimizing the risk of introducing pathogens into our dairy farm.
We hope to see you again.

The hygiene barrier at the entrance of a farm is one of the first levels of defense in biosecurity. Equipment, cars, and trucks from third parties (inseminator; hired labour; cooperative; claw trimmer; veterinarian etc) as well as the people themselves are potential sources from where micro-organisms can be introduced into the farm, eg through manure. In principle they should stay at the dirt side of the farm. The visitors deserve particular attention and have been addressed above. In the next Table 6 we address the hygiene in the hygiene barrier by presenting a checklist.

By entering a date in a respective box, farmers have the overview of the actions planned and taken; moreover, it shows the visitors which activities farmers have conducted to contribute to higher hygiene standards. Items can be added when wanted.

Table 6. A Check list for Cleaning the Hygiene Barrier: an example

Remind that the hygiene barrier is meant to prevent introduction of pathogens

	Daily	**Twice/week**	**Once/week**	**Monthly**
Clean towels	-			
Clean overalls	-			
Check boots on cleanness and dryness	-			
Clean washing tub	-			
Check delivery of medicines and other products	-			
Empty the dustbin, clean and dry it		-		
Empty the storage of chemical waste, clean and dry it		-		
Empty the storage of needles, clean and dry it		-		
Clean the floor, desinfect, rinse, dry it		-		
Check soap and desinfectant volumes			-	
Clean and dry doors, walls and windows			-	
Check visitors' log on completeness, add new pages			-	
Check delivery log, add new pages			-	
Empty the refrigerator, clean, rinse, desinfect, rinse, dry it				-
Empty the whole hygiene barrier, clean, rinse, desinfect, rinse, dry it				-
Check expiration dates of the cleaning products and disinfectants				-

Hang this checklist inside the hygiene barrier; list the dates of execution of actions

Hygiene Measures in the Milking Parlour and it's Environment

The most direct contact between the raw material (milk) being produced and the environment of production occurs in the milking parlour. Particular attention with regard to highest hygienic standards has therefore to be given to this site in order to prevent milk from becoming contaminated. It must be common practice that the milking machine maintenance and evaluation take place at least every 6 months by a certified technician. A written evaluation report indicating the findings, including defects, must be provided at each machine check up. Milking machine evaluation must be conducted with the machine in full action, the so-called " wet evaluation". Only then, potential failures in functioning can be detected properly. A proper milking machine function provides the best way to obtain the milk from the cows' udder, and minimizes the risk of pathogen-transmission to and micro-trauma in teats and quarters, hence limiting the risk of udder infections (mastitis and contamination of milk). Clean milking equipment is meant to keep bacteria counts in milk low. For that purpose critical components need to be checked at regular intervals; screening for dirt and calcification is part of this procedure. Droplets on surfaces may point to dirt or fatty bio-film.

Machine cleaning procedures may be checked on the following elements:

- Boiler temperature at start of cleaning procedure (at least 80 'C)
- Boiler temperature at end of cleaning procedure (at least 40 'C)
- Proper dosage of chemical cleaning products
- Volume of water needed for the cleaning process
- Distribution of cleaning fluids over milking clusters
- Proper level of the vacuum.

It is stressed here that standard cleaning products (containing alkalic products) destroys bacteria by disinfection and removes milk remainders, while acidic products clean and removes calcifications; rules set by the dealer must be followed. Rubber teat liners and long milking tubes must be replaced every 2500 milkings, while silicon teat liners need to be replaced every 5000 milkings.

Checklists are available for monitoring the hygiene measures taken in and around the milking parlour, as well as for the milking machine. The farmer should realise that this is the place where he earns his income. Therefore, highest priority must be given to the hygiene standard in and around the milking parlour. Checklists are also available for exploring increasing bacteria counts in bulk tank milk. Contributing risk factors should be evaluated on their impact and adequate measures defined. Once the causal factors are found and intervention is conducted, we should evaluate the outcome and adjust when needed.

Other Hygiene Measures

The milkers who are responsible for the routine milking have a particular task in the whole process. They are the ones who are in direct contact with both the animals, their udder and milk, and the milking machine.

They may play a role in the transmission of udder pathogens when personal hygiene is not optimal, they play a role in proper milking machine handling, and they are responsible for good udder preparation, milking itself, and sanitation procedures.

In case there rises any doubt about the proper milking method applied, it is recommended to make a time study of milking on a sample of the lactating cows. In such a time study the respective intervals in seconds or minutes between all milking-related actions per cow are recorded and compared to reference values for a group of cows.

To be sure that cleaning products meant for hygienic purposes delivered to the farm are in good order, they should be checked upon delivery. They should be checked on the following:

- expiration date, damaged packing material
- correctness of packaging bill
- having been recorded in the respective LOG or bill put in archives
- if products are to be stored in cool place, check on this
- delivered products should never be put directly on the floor.

An other issue of hygiene that deserves attention is the hygiene and cleanness of feeding alleys and exercise areas and especially those where crossing of people, animals, trucks etc occur (those spots are high risk spots for pathogen transmission).

Moreover, proper and hygienic manure storage is a relevant issue for safeguarding the farm from becoming contaminated.

A further area of concern may be the equipment and people from outside the farm providing services on the farm (e.g. feed harvesting, manure spreading). They may not only introduce or spread pathogens into/on the farm but also from farm to farm. These people and their equipment too must comply to the hygiene rules set by the farmer.

The forenamed "risk groups" can be provided with particular working instructions; such instructions can be found in detail in Noordhuizen et al. (2007).

Quality Risk Management Program-Related Support Materials: The Farm Calendar

In addition to the formal QRM-sheets presented above, QRM support materials can be very useful to the dairy farmer and his farm-workers because they focus their attention of certain points in the production process where regular checking is needed. They contribute to awareness and motivation, and to timely action. An example is the farm calendar with seasonal issues which repeatedly are addressed on the farm. Table 7 comprises such a farm calendar for one particular dairy farm.

Table 7. A Farm Calendar for health and nutrition: an example

	Jan	Feb	Mrch	Apr	May	Jun	Jul	Aug	Sept	Oct	Nov	Dec
CMT herd screening		■	■						■	■		
Milking machine & technique			■	■						■	■	
Herd claw trimming	■					■				■		
Formalin foot bathing	■		■								■	
Maiden heifers to be bred	■											

Table 7. (Continued)

	Jan	Feb	Mrch	Apr	May	Jun	Jul	Aug	Sept	Oct	Nov	Dec
Lungworm vaccination		■	■									
Screen rearing young stock										■	■	
Check barn climate		■									■	
Check on liver fluke					■							
Faeces check on nematodes 5 wks						■	■					
Check on ectoparasites										■	■	■
Feeding management at change from barn to pasture			■	■								
Start calves in pasture				■	■							
Evaluation pasturing cows						■	■					
Suppl. feeding at pasturing				■	■				■	■		
Feeding management at change from pasture to barn									■	■	■	
Roughage analysis									■	■		
Barn feedstuff balance										■	■	
Evaluation barn feeding	■	■										
Check concentrate automates												■
Grassland exploitation			■	■						■		
Grassland planning			■	■	■							
Forage harvesting planning				■	■							
Feed purchase planning							■			■		
Estimate # kgs of concentrates/cow						■						■
Estimate # kgs of concentrates/100 kg milk						■						■

Quality Risk Management Program-Related Supportive Programs: Training Programs

In certain conditions – for example on large dairy farms with several farm-workers – it can be highly indicated to organize specific training programmes for all or some farm-workers in specific domains. These trainings must be of short duration and be done on-site to acquire the largest impact.

Examples of training programmes are those on:

- claw trimming and diagnosis of claw lesions;
- diagnosis of clinical and subclinical mastitis;
- proper milking procedures;
- on-farm hygiene;
- biosecurity plans;
- calf rearing systems.

Quality Risk Management Programs: Internal Reviews and External Audits

Internal reviewing or validation is needed at least once every year to make sure that the HACCP-like QRM-program is running adequately. Based on the internal reviewing, adjustments need to be made, new elements or records added to the handbook and old elements eliminated from the handbook and put in archive. The evaluation of herd performance figures in the different farm domains and the comparison with those of previous year(s) and or other farms, may indicate the stronger and weaker points on the farm, as well as the extent to which progress has been made. For that purpose the SWA-sheets on the vacqa-international.com website can be used too. If needed, new targets are set.

It is warranted that beforehand the person responsible for the Internal Review is known. At the same time it is necessary to indicate on the Internal Review who is responsible for making certain named adjustments.

Quality Risk Management Program: Documents

Examples of documents heading under the HACCP-concept have been given above and can be found in detail in Noordhuizen et al. (2007) and on the Vacqa-international.com website.

It must be clear that there is a certain overlap between those records that are applied in Herd Health & Production Management programs (Brand et al., 1996; De Kruif et al., 2007), and those implemented in a HACCP-like program of Quality Risk Management. When both programmes are integrated properly, the purposes of both programmes too can be integrated as well as their documents. Nevertheless, when one has to prove that the HACCP-like program is currently running, then the necessary documents as presented above must be available on formal grounds.

In addition, under a HACCP-like program, additional documents must be made available.

Among these are: reports from laboratory examinations (bacteriology; serology; virology; myco-toxicology, etc);
reports from postmortem examinations;
reporting forms like event logs; incidence reports; improvement sheets; purchase logs; delivery complaint logs.

In the elapse of a HACCP-like quality risk management program over time, events like diseases will occur. Such events may be relevant for the maintenance of the program but they may also be important for operational farm management. They need, therefore, to be recorded. This can be done in an Events Log, which basically is a farm diary.

Incidence Reports are directly impacting on the stability of quality risk management program, for example because they reflect a point (CCP or POPA) where quality control was lost or where a hazard (a disease) has occurred. It is paramount to keep Incidence Reports because they may trigger adjusting the monitoring scheme or even the quality risk management program as a whole.

Discussion and Conclusions

Food animal production, e.g. on dairy farms, represents the first link in a (dairy) food chain quality assurance program. When most of the links in such a food chain have been certified for quality control measures, it is rather peculiar that the first link has received relatively little attention over the past years. Although the control of milk quality has a longstanding history, attention for the production process itself has increased only recently, not in the least due to changing consumer behavior and opinion and to outbreaks of highly contagious diseases largely addressed in the media. Quality should refer to public health and food safety in the first place. However, the forenamed are often influenced by the way that animal health and welfare are managed on the farm. Moreover, consumers have shown an increasing concern about animal health and welfare.

The classical veterinary approach to disease is either a method of prevention (and eradication for highly contagious diseases) and control, or a policy of reducing disease prevalence (namely in the case of endemic diseases). For the latter, the diagnosis and treatment of individual cases have been considered too costly and too much impacting on animal welfare. The last decades there is a growing tendency to pay more attention to prevention (e.g. by vaccination programs; de-worming schemes), and –the last years in particular- to risk identification and risk management. Herd health & production management programs, HHPM, as presented in literature (Brand et al., 1996; De Kruif et al., 2007) address all these issues, but still in a rather qualitative approach. These HHPM often are lacking structure, organization and planning; they are executed in a too much "free-style" manner. This can be a major factor of the failures encountered in HHPM, although overall farm management benefits from the implementation of HHPM and it has been shown that also economic benefits can be expected from HHPM (Dijkhuizen & Morris, 1996; Huirne et al., 2002).

The keywords of HACCP-like QRM are structure, organization, planning, communication and formalization, as was shown through the examples of Figs and Tables above (Noordhuizen et al., 2007). Hence, such QRM fill the gaps noted at HHPM. QRM and HHPM have several similarities, but the fact that in QRM activities are protocol-based, well structured in time and place, properly planned, and –most important—formalized, the QRM approach promises much more benefits and effect than current HHPM can. A merger between the two would be best, profiting from both the operational management focus (HHPM) and the more tactical farm focus (QRM). The QRM as proposed here is quite feasible on farm level. Farmers apparently adopt the QRM because it has a 'bottom-up' focus and not a top-down; it is highly farm-specific, integrating farm management and quality control in one approach. Farmers state that they are much more focused on problem areas and risk factor profiles through the formalized QRM approach. They experience that they do not tend to neglect or forget relevant issues; they feel well guided along the whole pathway to problem solution and along routine herd inspection (monitoring).

Because the demonstration to third parties (e.g. retailers; authorities; consumers) of the herd status regarding food safety & public health, and animal health & welfare, as well as the measures taken to improve that status, are formal part of the HACCP concept, the consumers can be provided with much more certainty about the quality of the products and the production methods on the dairy farm. This QRM largely contributes to the transparency of the on-farm production and the openness to society. This will contribute to the positive public image of the dairy (farming) sector.

Veterinarians can play a paramount role in this domain. First, because they have been properly educated in the various areas. Secondly, because they have experience in this domain. And thirdly, they have easy access to the dairy farms and can make frequent visits when needed. For developing and implementing HACCP-based QRM they have to acquire additional knowledge and skills (farm economics; nutrition; marketing & business administration; communication), but once this has been done, veterinarians in the field will have high added value to the dairy farmers as coach—consultant in quality control according to HACCP-based QRM.

References

Brand A., Noordhuizen, J.P.T.M., Schukken, Y.H. 1996. Herd health and production management in dairy practice. Wageningen Academic Publishers, Wageningen, The Netherlands

Cannas da Silva, J., Noordhuizen, J.P.T.M., Vagneur, M., Bexiga, R., Gelfert, C.C., Baumgartner, W. 2006. The future of veterinarians in bovine health management. In: Proceedings of the World Buiatrics Congress (Navetat & Schelcher, editors), *Nice, France, October 2006.* Also published in The Veterinary Quarterly 28 (1): 28-33

Cullor,J.S. 1995. Implementing the HACCP program on your clients' dairies. Veterinary Medicine X 290-295

Cullor, J.S. 1997. HACCP (Hazard Analysis Critical Control Points): Is It Coming to the Dairy Farm. *Journal of Dairy Science* 80:3449-3452

De Kruif, A, Mansfeld, R, Hoedemaker, M. 2007. *Tierärztliche Herdenbetreuung beim Milchrind. 2nd* Edition. Enke Verlag, Stuttgart, Germany (in German)

Dijkhuizen, A.A., Morris, R.S. 1996. Animal health economics : principles and applications. University of Sydney, Australia, postgraduate formation in veterinary science.

Huirne, R.B.M, Saatkamp, H.W., Bergevoet, R.H.M. 2002. Economic analysis of common health problems in dairy cattle. In: *Proceedings of the XIIth World Buiatrics Congress (Kaske, Scholz & Holtershinken, editors), 18-23 August 2002, Hannover, Germany*, 420-431

Lievaart, J.J., Noordhuizen, J.P.T.M., van Beek, E, van der Beek, C, van Risp, A, Schenkel, J, van Veersen, J.. 2005. The hazard analysis critical control points concept as applied to some chemical, physical and microbiological contaminants of milk on dairy farms. *The Veterinary Quarterly 27* (1): 21-29

Noordhuizen J.P.T.M. 2004. Dairy Herd Health and Production Management practice in Europe: state of the art. In: *Proceedings 23rd World Buiatrics Congress – Québec*, Canada July 11-16, 2004

Noordhuizen J.P.T.M. 2004. The design, testing and dissemination of a prototype EU-standard HACCP-compatible programme for managing quality-, food safety-, and animal health and welfare-*, risks on dairy farms*. 2004 not published

Noordhuizen, JPTM, Cannas da Silva, J., Boersema, JSC, Vieira, A. 2007

Radostits O.M. 2004. Bovine Practice: Successes of the Past and Challenges and Opportunities in the Future. *23rd World Buiatrics Congress – Québec*, Canada. July 11-16, 2004

Stup, R. 2001. Standard Operating Procedures: A Writing Guide. in *http://dairyalliance.psu.edu/pdf/ud011.pdf, consultado em 11-04-05* the Annex to the *COAG/2003/6 – FAO* (Rome, Italy) OIE 2006

Vieira, A., Boersema, J.S.C., Noordhuizen, J.P.T.M., Cannas da Silva, J., Baumgartner, W. 2007. The application of the Hazard Analysis Critical Control Points (HACCP) concept to a dairy farm in Portugal, in press.2007

The website *www.vacqa-international.com*

In: Consumer Product Safety Issues
Editors: P. A. Varga, M. D. Pintér

ISBN: 978-1-60456-826-4

Chapter 8

CONSUMER PRODUCT SAFETY IMPROVEMENT ACT OF 2008: P.L. 110-314*

Margaret Mikyung Lee

SUMMARY

Public alarm about the spate of recent product recalls throughout 2007, particularly of toys and other products used by children, has focused attention on the Consumer Product Safety Commission (the CPSC or the Commission). This scrutiny led to consideration of major amendments to the Consumer Product Safety Act (CPSA), which established and authorized the CPSC in 1972 in response to growing concerns about protecting the public from unsafe, defective consumer products. Jurisdiction over the administration and enforcement of several existing consumer safety statutes was transferred from other agencies to and consolidated under the CPSC. However, in the years since its establishment, the staff and resources of the CPSC have been considerably reduced, leading many observers to doubt its ability to fulfill its mission effectively.

Consequently, Congress considered major reform legislation to address organizational and systemic deficiencies. Legislative proposals in the 110th Congress included provisions targeting specific consumer product defects and hazards. On July 29, 2008, H.Rept. 110-787, the Conference Report for H.R. 4040, the Consumer Product Safety Improvement Act of 2008 (CPSIA), was released after several months of negotiations in the conference committee to reconcile differences between the House and Senate versions of the bill. The bill passed the House of Representatives and the Senate on July 30, 2008 (424-1) and July 31, 2008 (89-3), respectively. On August 14, 2008, President Bush signed the bill into law as P.L. 110-314. CPSC Chairman Nord and Commissioner Moore each expressed approval of the final legislation, with Chairman Nord expressing a desire for Congress to appropriate further funding to carry out the new mandates of the legislation.

This chapter provides an overview of the prior authority of the CPSC to establish consumer product safety standards and to inspect and recall unsafe consumer products, and

* This is an edited, reformatted and augmented version of CRS Report RL34684, dated September 22, 2008

discusses P.L. 110-314, the Consumer Product Safety Improvement Act of 2008, reforming the CPSC and strengthening enforcement of consumer product safety standards. For an overview and context of the current issues facing the Commission, see CRS Report RS2282 1, *Consumer Product Safety Commission: Current Issues*, by Bruce K. Mulock. For an overview of issues regarding safety of consumer products imported from China, see CRS Report RS22713, *Health and Safety Concerns Over U.S. Imports of Chinese Products: An Overview*, by Wayne M. Morrison. For an overview of the issue of phthalates in children's products, see CRS Report RL34572, *Phthalates in Plastics and Possible Human Health Effects*, by Linda-Jo Schierow and Margaret Mikyung Lee.

BACKGROUND

Public alarm about the spate of product recalls during 2007, particularly of toys and other products used by children, has focused attention on the Consumer Product Safety Commission (CPSC). P.L. 110-314, 122 Stat. 3016 (2008), the Consumer Product Safety Improvement Act of 2008 (CPSIA), was enacted as a result of Congress's consideration of major reform legislation to address organizational and systemic deficiencies, as well as specific consumer product defects and hazards. This report provides an overview of the prior authority of the CPSC to establish consumer product safety standards and to inspect, recall, and restrict importation of unsafe consumer products, and summarizes changes made by the CPSIA to reform the CPSC and strengthen enforcement of consumer product safety standards.

The Consumer Product Safety Act (CPSA, 15 U.S.C. §§2051 et seq.) established and authorized the CPSC in 1972 in response to growing concerns about protecting the public from unsafe, defective consumer products. However, in the years since its establishment, the staff and resources of the CPSC have been considerably reduced to the detriment of its ability to fulfill its mission effectively. Aside from the issue of adequacy of resources, the highly publicized recalls of children's toys in 2007 focused attention on alleged weaknesses in the CPSA and the authority of the CPSC to establish consumer product safety standards and to inspect, recall domestically, and block imports of unsafe consumer products.

The CPSC is the central, federal authority for the promotion and enforcement of consumer product safety. The system is designed to be a collaborative effort among the CPSC, the industries producing the broad range of consumer products, and the consuming public. The CPSC researches and promotes best practices for the industries, producing guidelines for manufacturers, importers, distributors and retailers. Although the CPSA authorizes the CPSC to promulgate mandatory consumer product safety standards, it mandates reliance upon voluntary standards whenever compliance with voluntary standards would eliminate or adequately reduce the risk of injury and substantial compliance with voluntary standards is likely.

Besides the CPSA, the CPSC also administers several other statutes whose authorities and functions were transferred to the CPSC upon its creation. The Federal Hazardous Substances Act (FHSA, 15 U.S.C. §§1261 et seq.) provides for warning/informational labeling of hazardous substances and for the banning of certain hazardous substances for which labeling would not provide adequate protection for the public against the potential hazards posed by the substances. The

Flammable Fabrics Act (FFA, 15 U.S.C. §§1 191 et seq.) provides for the establishment of safety standards regarding fabric flammability. It also prohibits the manufacture, sale, importation, transportation, or delivery in commerce of a product, fabric, or related material or of a product made of a fabric or related material that does not comply with the standards and deems that such practices constitute unfair methods of competition and unfair and deceptive acts or practices under the Federal Trade Commission Act. The Poison Prevention Packaging Act (PPPA, 15 U.S.C. §§1471 et seq.) authorizes the CPSC to establish special packaging standards for a household substance if such standards are required to protect children from serious injury or illness from using, handling or ingesting such substance, with exceptions for noncomplying packages for elderly/handicapped persons and packaging at the direction of a licensed medical practitioner. The Refrigerator Safety Act (RSA, 15 U.S.C. §§1211) prohibits the introduction into interstate commerce of any household refrigerator that does not conform with certain safety standards.

Prior to amendment by the CPSIA, some of these statutes provided for powers that were similar but not identical to those established under the CPSA. Therefore, the regulatory procedures and other actions which the CPSC is authorized to carry out with regard to the products regulated under these other statutes differed (and may still differ in some respects) from those authorized under the CPSA. Before the CPSIA, the CPSC could choose to regulate under the CPSA a consumer product that could be regulated sufficiently under these other statutes only if the CPSC determined that it was in the public interest to do so. The differences among the different statutory standards and procedures and enforcement authority arguably led to inconsistency in the enforcement of different product standards. For example, injunctive enforcement authority for states attorneys general is expressly provided by the FHSA and the FFA, but was not expressly provided by the CPSA before amendment by the CPSIA. The apparent ambiguity of the CPSA on this point led to amendments in the CPSIA to provide express authority.

The CPSC has the authority to establish consumer product safety standards for consumer products generally, defined as "any article or component part thereof, produced or distributed (i) for sale to a consumer for use in or around a permanent or temporary household or residence, a school, in recreation, or otherwise, or (ii) for the personal use, consumption or enjoyment of a consumer in or around a permanent or temporary household or residence, a school, in recreation, or otherwise."[1] There are express exemptions for products covered under other statutes, including tobacco and tobacco products, motor vehicles and motor vehicle equipment, pesticides, firearms/antique firearms and ammunition/supplies (except for fireworks), aircraft and components, boats and other marine vessels, drugs, medical devices, cosmetics, food, or any article which is not customarily produced or distributed for sale to, or use or consumption by, or enjoyment of, a consumer. The CPSC also has jurisdiction over amusement rides that are not permanently fixed to a site but rather are part of a travelling carnival or show, but does not have jurisdiction over rides that are permanently fixed to a particular site.[2] Furthermore, the CPSC has no jurisdiction to regulate a particular consumer product if the risk of injury associated with that product could be eliminated or sufficiently

1 CPSA §3 (codified at 15 U.S.C. §2052).

2 State agencies have jurisdiction over fixed rides that are located in a park in their jurisdiction.

reduced by actions taken under the Occupational Safety and Health Act of 1970 (Occupational Safety and Health Administration), under the Atomic Energy Act of 1954 (the Energy Research and Development Administration [now Department of Energy] and the Nuclear Regulatory Commission), or under the Clean Air Act (the Environmental Protection Agency). The CPSC has no authority to regulate any risk of injury associated with electronic product radiation emitted from an electronic product if such risk may be regulated under the Public Health Act (the Food and Drug Administration).

Manufacturers, distributors, retailers, and importers are obligated to report consumer product safety problems to the CPSC, which may order a recall or import ban. The CPSA provides the general authority of the CPSC over inspections, recalls and import bans for consumer products generally, with the exceptions noted above, unless other statutes provide other agencies with authority over specific products. Although some of the other statutes enforced by the CPSC contain provisions specifically addressing notice/recall, import bans, and other remedies similar to those contained in the CPSA, not all do.

The CPSC is authorized to inspect domestic facilities where a consumer product is manufactured and the conveyances by which it is transported and which may be relevant to the safety of such product.[3] It is unlawful for a person to fail or refuse to permit inspection as required under the CPSA.[4] For purposes of the CPSA, including inspection and testing, the CPSC may purchase any consumer product and it may require any manufacturer, distributor, or retailer of a consumer product to sell the product to the CPSC at cost.[5] The CPSC is authorized to establish and maintain a permanent product surveillance program, in cooperation with other appropriate Federal agencies, for the purpose of carrying out the CPSC's responsibilities under the CPSA and the other Acts it administers and preventing the entry of unsafe consumer products into the United States.[6] The U.S. Customs and Border Protection (CBP)[7] is authorized to obtain and deliver samples of consumer products being offered for importation to the CPSC, upon its request, for the purpose of inspecting such samples for compliance with the CPSA.[8] Similarly, under FHSA § 14 (15 U.S.C. § 1273), the CBP is authorized to obtain and deliver samples of hazardous substances being imported or offered for importation to the CPSC, upon its request, for the purpose of inspecting such samples for compliance with the FHSA.

Under CPSA § 19,[9] it has been unlawful, among other things, to manufacture, sell, distribute in commerce, or import into the United States any consumer product which does not comply with an applicable consumer product safety standard or which has been declared a banned hazardous product by a rule under the CPSA. Other consumer-product-related statutes contain similar provisions concerning prohibited or unlawful acts. The CPSIA has expanded the scope of prohibited acts.

---

3 CPSA §16 (codified at 15 U.S.C. §2065) and 16 C.F.R. § 1118.2.

4 CPSA §19(a)(3) (codified at 15 U.S.C. §2068(a)(3)).

5 CPSA §27 (codified at 15 U.S.C. §2076).

6 CPSA § 17(h) (codified at 15 U.S.C. §2066(h)).

7 The statute and regulations refer to the Secretary of the Treasury although such functions are now undertaken by Department of Homeland Security (U.S. Customs and Border Protection (CBP)) pursuant to the Homeland Security Act and 19 C.F.R. §§0.1-0.2.

8 CPSA § 17(b) (codified at 15 U.S.C. §2066(b)).

9 Codified at 15 U.S.C. §2068.

Under CPSA § 15,[10] every manufacturer (defined to include importers), distributor, or retailer of a consumer product distributed in commerce who obtains information reasonably supporting the conclusion that such product (1) fails to comply with an applicable consumer product safety rule or with a voluntary consumer product safety standard; (2) contains a defect which could create a substantial product hazard; or (3) creates an unreasonable risk of serious injury or death, shall immediately inform the CPSC, unless such manufacturer, distributor, or retailer has actual knowledge that the CPSC has been adequately informed of such defect, failure to comply, or risk. If the CPSC determines after a hearing that a product presents a substantial product hazard and that notification is required in order to adequately protect the public from such substantial product hazard, the CPSC may order the manufacturer or any distributor or retailer of the product to take any one or more of the following actions: (1) to give public notice of the defect or failure to comply; (2) to mail notice to each person who is a manufacturer, distributor, or retailer of such product; or (3) to mail notice to every person to whom the person required to give notice knows such product was delivered or sold. The CPSIA expanded the scope of these provisions to cover noncompliance with rules under all acts under CPSC jurisdiction and to provide additional means of providing public notice.

Prior to the CPSIA, if the CPSC determined after a hearing that a product presented a substantial product hazard and that action under that provision was in the public interest, it could order the manufacturer or any distributor or retailer of such product to take whichever of the following actions it elected to take: (1) to bring the product into compliance with the applicable product safety rule or repair the defect; (2) to replace the product with an equivalent product that does comply or is not defective; or (3) to refund the purchase price. The CPSIA removed the ability of the manufacturer, distributor or retailer to choose the remedial action to be taken and provides for the CPSC to determine which action should be taken.

In addition to its authority with regard to a substantial product hazard, under CPSA §12, the CPSC may bring an action in federal district court to have a product declared an imminent hazard, defined as a consumer product which presents imminent and unreasonable risks of death, serious illness, or severe personal injury, and to seize the product. If the court determines that a product constitutes an imminent hazard, it may grant any relief necessary to protect the public, including an order requiring public notification, recall, and remedies including repair, replacement, or refund of such product.

Under the FHSA § 15,[11] the CPSC may order a manufacturer, distributor, or dealer to take remedial action with respect to a banned hazardous substance similar to remedies under the CPSA, including (1) giving public notice that an article or substance is a banned hazardous substance; (2) mailing notice to each person who is a manufacturer, distributor, or dealer of such article or substance; and (3) mailing such notice to every person to whom the person giving the notice knows such article or substance was delivered or sold. The CPSC may also order the repair of such article or substance, replacement with an equivalent compliant article or substance, or refund. Similar notification and remedial actions may be ordered with respect to any toy or other article intended for use by children that is not a

10 Codified at 15 U.S.C. §2064.
11 Codified at 15 U.S.C. § 1274.

banned hazardous substance but that contains a defect which creates a substantial risk of injury to children.

The CPSC does not pay for the costs of a notice of product hazard or defect or any ordered repair, replacement, or refund; costs are born by the manufacturer, distributor, or retailer. An order issued under CPSA § 15[12] with respect to a product may require any person who is a manufacturer, distributor, or retailer of the product to reimburse any other person who is a manufacturer, distributor, or retailer of such product for such other person's expenses in connection with carrying out the order, if the CPSC determines such reimbursement to be in the public interest. Also, no charge shall be made to any person (other than a manufacturer, distributor, or retailer) who avails himself of any remedy provided under an order concerning repair, replacement, or refund, and the person subject to the order shall reimburse each person (other than a manufacturer, distributor, or retailer) who is entitled to such a remedy for any reasonable and foreseeable expenses incurred by such person in availing himself of such remedy. FHSA § 14(d)[13] contains reimbursement provisions similar to CPSA § 15.[14] Any person who is a manufacturer, distributor, or dealer of a noncompliant toy, article, or substance may be ordered to reimburse any other person who is a manufacturer, distributor, or dealer of such toy, article, or substance for such other person's expenses in connection with carrying out a remedial or notification order, if the CPSC determines such reimbursement to be in the public interest. Also, no charge shall be made to any person (other than a manufacturer, distributor, or dealer) who avails himself of any remedy provided under a remedial order and the person subject to the order shall reimburse each person (other than a manufacturer, distributor, or dealer) who is entitled to a remedy for any reasonable and foreseeable expenses incurred in seeking such remedy.

CPSC has the authority to establish import standards and policy with regard to statutes and products under its jurisdiction.[15] Under the CPSA, importers are made subject to the same responsibilities as domestic manufacturers in protecting American consumers from unreasonably hazardous products. This is explicitly stated in the definition of "manufacturer" as any person who manufactures or imports a consumer product.[16] Like the CPSA, the FHSA and the FFA assign responsibilities to importers comparable to those of domestic manufacturers and distributors.[17] Various statutory provisions authorize the CPSC to ban noncompliant imports. If the CPSC determines after a hearing that a product presents a substantial product hazard and that action under that provision is in the public interest, it may issue an order prohibiting the importation into the United States of that product.[18] An imported consumer product may be refused admission to the United States if it does not comply with an applicable consumer product safety rule; does not comply with labeling and certification requirements relating to applicable product safety standards; is an imminently hazardous product; or has a product defect which constitutes a substantial product hazard.[19] The CPSC may then inform the CBP[20] that an imported consumer product fails to comply

12 Codified at 15 U.S.C. §2064.
13 Codified at 15 U.S.C. § 1273(d).
14 Codified at 15 U.S.C. §2064.
15 Its policy on imported products, importers, and foreign manufacturers is set out at 16 C.F.R. § 1009.3.
16 CPSA §3(a)(4) (codified at 15 U.S.C. §2052(a)(4)).
17 See FHSA §15(f) (15 U.S.C. §1274(f)); FFA §9 (15 U.S.C. §1198).
18 CPSA § 15(d) (codified at 15 U.S.C. §2064(d)).
19 CPSA §17 (codified at 15 U.S.C. §2066).
20 CPSA § 17(b) (codified at 15 U.S.C. §2066(b)).

with an applicable consumer product safety rule and/or has a product defect which constitutes a substantial product hazard and may request the CBP to refuse admission to any such consumer product.[21] As discussed in this report, the CPSIA has expanded grounds for refusing admission to a product. Under FHSA § 14,[22] a misbranded hazardous substance or banned hazardous substance being imported or offered for import shall be refused importation. Section 9 of the Flammable Fabrics Act (FFA)[23] provides that imported products subject to flammability standards under the FFA shall not be released from customs custody except in accordance with § 499 of the Tariff Act of 1930[24] providing for release only after inspection by CBP for compliance with U.S. laws. The CBP also has authority for the redelivery or recall of products already released under bond but later found not to comply with flammability standards and for obtaining liquidated damages for breach of a condition of the bond arising out of a failure either to correct the product to comply or to redeliver it.

Such noncompliant or defective products must be destroyed or exported from the United States unless they can be modified by the owner or consignee in a manner that will enable them to be granted admission into the United States.[25] The CPSC and the CBP have the authority to give the owner or consignee the opportunity to make such modifications and to monitor such modifications.[26] Prior to the CPSIA, such products could also be ordered destroyed at the discretion of the CPSC. The

CPSIA made destruction mandatory, unless the CBP permits export in lieu of destruction and the product is exported within 90 days of such permission.[27] Before the CPSIA, at its discretion, the CPSC could condition importation of a consumer product on the manufacturer's/importer's compliance with the inspection and recordkeeping requirements of the CPSA. The CPSIA makes compliance with such requirements mandatory, meaning that an imported product must be refused entry into the United States if the importer does not comply with requirements related to its product.[28] The CPSC may seek an injunction or seizure of a consumer product that does not comply with a consumer product safety rule or that is being manufactured, sold, distributed, or imported in violation of a CPSC order for remedial action or prohibiting importation.[29]

With regard to imported products that are admitted and subsequently become the subject of a recall, if the CPSC is not able to exercise jurisdiction over a foreign manufacturer that has no U.S. subsidiary/presence, the CPSC may order the importer to undertake a recall and to be responsible for the expense of a mandatory recall, since "manufacturer" is defined to include importers under the CPSA.[30] The importer cannot obtain reimbursement under the CPSA, but might be able to obtain reimbursement from the foreign manufacturer as a

---

[21] 16 C.F.R. § 1115.21(d).

[22] Codified at 15 U.S.C. § 1273.

[23] Codified at 15 U.S.C. § 1198.

[24] Codified at 19 U.S.C. § 1499.

[25] CPSA §17(c and e) (codified at 15 U.S.C. §2066(c and e)).

[26] CPSA §17(c and d) (codified at 15 U.S.C. §2066(c and d)).

[27] CPSA §17(e) (codified at 15 U.S.C. §2066(e)).

[28] CPSA §§16(d) and 17(g) (codified respectively at 15 U.S.C. §§2065(d) and 2066(g)).

[29] CPSA § 22 (codified at 15 U.S.C. §2071).

[30] CPSA §15(e) (codified at 15 U.S.C. §2064(e)) for recall reimbursement and CPSA §3(a)(11) (codified at 15 U.S.C. §2052(a)(11)) for definition of "manufacturer."

contractual matter.[31] Under CPSA §17(f),[32] the owner or consignee of an imported consumer product denied entry into the United States must pay for all expenses in connection with its storage or destruction. In default of such payment, these expenses shall constitute a lien against any future importations made by such owner or consignee. Under FHSA §14(c),[33] the owner or consignee must pay for all expenses (including travel, per diem, or subsistence, and salaries of officers or employees of the United States) in connection with the destruction of a hazardous substance denied importation into the United States; the supervision of the relabeling or other action authorized to bring a hazardous substance denied importation into compliance with the FHSA so that it may be granted importation; and the storage for any hazardous substance denied importation. In default of such payment, the FHSA, like the CPSA, provides that such expenses shall constitute a lien against any future importations made by such owner or consignee.[34]

The Office of Compliance and Field Operations within the CPSC conducts compliance and administrative enforcement activities under all administered acts, provides advice and guidance on complying with all administered acts, and reviews proposed standards and rules with respect to their enforceability.[35] Among other things, it reviews consumer complaints, conducts inspections and in-depth investigations, and analyzes available data to identify those consumer products containing defects that pose a substantial risk of injury or do not comply with existing safety requirements. The Office negotiates and monitors corrective action plans for products that are defective or fail to comply with specific regulations. The Office of International Programs and Intergovernmental Affairs within the CPSC was established to enable a more coordinated and comprehensive approach to international cooperation with regard to harmonization of safety standards internationally and the ensuring of compliance with U.S. safety standards for products imported into the United States. Memoranda of understanding have been concluded with CPSC counterparts in various countries or regional groups, including the People's Republic of China, the European Union, and Canada.

The CPSC held a public meeting on September 4, 2008, to explain their plans to implement the CPSIA and will continue to hold a series of such meetings.[36]

CURRENT LEGISLATION: P.L. 110-314

There have been a number of proposals in the 1 10th Congress to address a range of consumer product safety issues. In addition to the two major CPSA reform bills that ultimately resulted in P.L. 110-314, H.R. 4040 and S. 2045/S. 2663,[37] there have been other comprehensive reform bills and bills addressing discrete issues, including safety standards for

31 See CPSA §15(e) (codified at 15 U.S.C. §2064(e)) and Riegel Textile Corp. v. Celanese Corp., 493 F. Supp. 511 (S.D.N.Y. 1980) (manufacturer has no cause of action against supplier under CPSA or FHSA for violation of those statutes, but state cause of action under state laws may remain).

32 Codified at 15 U.S.C. §2066(f).

33 Codified at 15 U.S.C. § 1273(c).

34 Id.

35 16 C.F.R. § 1000.21.

36 Video and slide presentations for the meeting on September 4, 2008, and information on future meetings about testing and certification and standards for all-terrain vehicles, lead, and phthalates, is available at [http://www.cpsc.gov/about/cpsia/cpsia.html], last visited September 22, 2008.

37 After the House passed H.R. 4040, the Senate took up S. 2663, a compromise version of S. 2045, and amended it on the Senate floor, then amended H.R. 4040 to substitute the text of S. 2663 and passed it.

cigarette lighters, All-Terrain-Vehicles or ATVs, furniture, swimming pools, portable gasoline containers, durable infant or toddler consumer products such as strollers and cribs, and other products; certification of safety- standard compliance; the enactment of lead content standards for consumer products and more stringent lead in paint standards; third-party testing for product safety and compliance; increases in civil and/or criminal penalties; increases in CPSC personnel assigned to ports-of-entry; the prohibition of sales or resales of products that are the subject of a recall; expanded jurisdiction of the CPSC to cover amusement park rides at a fixed site; greater coordination among the various agencies involved in consumer safety issues; and others. Language from some of the free-standing bills addressing specific issues was incorporated into the CPSIA.

The following sections will summarize the provisions of P.L. 110-314. The final text was the result of a lengthy conference negotiation. The Senate-passed version of H.R. 4040 [hereinafter Senate Text][38] and the House-passed version of H.R. 4040 [hereinafter House Text][39] contained many similar provisions strengthening the authority and resources of the CPSC and also establishing standards concerning lead content in children's toys. However, the Senate Text contained additional reform provisions, such as whistleblower protection and several provisions concerning consumer product safety standards for specific items such as all-terrain vehicles and garage door openers. Conversely, most of the provisions of the House Text had parallel provisions in the Senate Text; the major exception was the provision adding a prohibition on industry-sponsored travel by members or employees of the CPSC, a response to the much-criticized practice by the CPSC of accepting travel and lodging expenses from industry sponsors for trips related to CPSC business, that was widely reported after the S. 2045 markup. On July 29, 2008, H.Rept. 110-787, the Conference Report for H.R. 4040, the Consumer Product Safety Improvement Act of 2008, was released after several months of negotiations in the conference committee to reconcile differences between the House and Senate versions of the bill. The bill passed the House of Representatives and the Senate on July 30, 2008 (424-1) and July 31, 2008 (89-3), respectively. On August 14, 2008, President Bush signed the bill into law as P.L. 110-314. CPSC Chairman Nord and Commissioner Moore each expressed approval of the final legislation, with Chairman Nord expressing a desire for Congress to appropriate further funding to carry out the new mandates of the legislation.[40]

CHILDREN'S PRODUCT SAFETY

Lead Content and Measurement (§1 01)

Under CPSIA § 101, 15 U.S.C. § 1278a, in products for children aged 12 and younger, the permissible lead level will be phased in over three years at 600 parts per million (ppm) within 180 days of enactment, 300ppm after one year, and 100ppm after three years. If

---

[38] The short title for the Senate version of the bill is the CPSC Reform Act.

[39] The short title for the House version of the bill is the Consumer Product Safety Modernization Act.

[40] See CPSC Press Release 08-349, Chairman Nord's statement, dated July 31, 2008, available at [http://www.cpsc.gov/cpscpub/prerel/prhtml08/08349.html] (last visited September 19, 2008), and Commissioner Moore's statement, dated July 31, 2008, available at [http://www.cpsc.gov/pr/Moore073108cpsia.pdf] (last visited September 19, 2008).

100ppm is not technically feasible, the CPSC must set the lowest level that is technologically feasible. After promulgation of either the 1 00ppm level or the lowest level technologically feasible, the CPSC is required to periodically review and lower the limit at least every five years. There are exceptions for certain materials or products and inaccessible component parts. If the CPSC determines that it is not technologically feasible for certain electronic devices to comply with the lead standard, it must also issue requirements to minimize exposure or accessibility to lead in those devices. The permissible lead level in paint is reduced to 0.009 percent (90 ppm) from 0.06 percent (600ppm), subject to subsequent periodic review and reduction to the lowest lead level technologically feasible. A CPSC Office of General Counsel memorandum clarifies that inventory of noncompliant products may not be sold after this provision's effective date of February 10, 2009.[41] Although the CPSIA does not expressly ban such sale or distribution, the Office of General Counsel concludes that a reading of the CPSIA as a whole indicates such an interpretation.

Third-Party Testing and Certification of Children's Products (§1 02)

Section 102 of the CPSIA amends CPSA §14 (codified at 15 U.S.C. §2063) to require manufacturer certification of safety testing by accredited third-party laboratories of products designed or intended primarily for children aged 12 and younger. The CPSC must issue accreditation requirements for such laboratories and maintain a list of accredited laboratories; deadlines for the publication of such requirements differ according to the type of product being tested by the laboratory. Specific deadlines for publication of accreditation requirements after CPSIA enactment are established for the testing of lead paint (30 days of enactment),[42] cribs and pacifiers (60 days of enactment), small parts (90 days of enactment), children's metal jewelry (120 days of enactment), and baby bouncers/walkers/jumpers (210 days of enactment). Accreditation requirements related to the testing of other children's products must be published as early as practicable, but no later than 10 months after CPSIA enactment or, for safety rules established or revised one year or more after CPSIA enactment, not later than 90 days before such rules take effect.

For three years after enactment, CPSC proceedings for promulgating accreditation requirements are exempt from requirements of the Regulatory Flexibility Act and the rulemaking requirements of the Administrative Procedure Act. Upon request, proprietary labs insulated from the influence of the manufacturer/private labeler-owner may be accredited by the CPSC and permitted to test products if they provide equal or greater protection than available third-party labs. CPSC personnel are authorized to enter and inspect any accredited proprietary lab (CPSIA §215(a)). The CPSC is required to establish requirements for the periodic audit of third-party testing labs as a condition of continuing accreditation. The CPSC is authorized to revoke accreditation if necessary after an investigation finding that a lab failed to follow a requirement established by the CPSC or is unduly influenced by a manufacturer or government entity.

---

[41] Cheryl A. Falvey, CPSC General Counsel, the CPSC Office of General Counsel, Memorandum on Retroactive Application of CPSIA to Inventory (September 12, 2008).

[42] Third Party Testing for Certain Children's Products: Notice of Requirements for Accreditation of Third Party Conformity Assessment Bodies To Assess Conformity With Part 1303 of Title 16, Code of Federal Regulations, 73 Fed. Reg. 54564 (2008).

It is unclear which preemption provisions in the CPSA, as amended by the CPSIA, if any, apply to this provision concerning third-party testing and certification. There are two preemption provisions that are possible applicable. The new provision added by CPSIA § 106(h), discussed below, exempts from federal preemption any state/local toy safety standards in effect on the date of CPSIA enactment, if the state/locality applies for exemption within 90 days of CPSIA enactment. Under this provision, states/localities may also apply for exemption of future proposed state/local toy safety standards. If the preemption provisions of CPSIA § 106 do not apply, the general preemption provisions under CPSA §26, which predate the CPSIA, might apply to the testing and certification provisions, which are amendments to the CPSA. These provisions permit states/localities to apply for exemption of a proposed safety standard or regulation which is designed to protect against a risk of injury associated with a consumer product subject to a consumer product safety standard under the CPSA. These provisions do not permit exemption of existing state/local product safety standards or regulations which prescribe requirements for the performance, composition, contents, design, finish, construction, packaging, or labeling and which are designed to deal with the same risk of injury as the federal standard, unless such requirements are identical to the those of the federal standard. It is unclear whether testing and certification requirements may be considered consumer product safety standards for the purpose of these preemption provisions.

Tracking Labels for Children's Products (§103)

Effective one year after the CPSIA enactment, CPSIA §103 amends CPSA §14 (codified at 15 U.S.C. § 2063) to require a manufacturer to place tracking labels on children's products and packaging, to the extent practicable, containing information (manufacturer, production date, and production batch/run of the product) enabling the retailer and ultimate purchaser to identify recalled products and enhancing the ability of the manufacturer to track unsafe products to their precise sources. Advertisements, labels, and packaging for a consumer product will be prohibited from referring to a mandatory consumer product safety rule or a voluntary standard unless the product complies with such rule or standard.

Standards and Registration Forms for Durable Nursery Products (§104)

Under CPSIA §104 (15 U.S.C. §2056a), the Danny Keysar Child Product Safety Notification Act, the CPSC, in consultation with stakeholders and product engineers/experts, must establish mandatory safety standards for a variety of durable nursery products for use by children under five years of age. Such products include cribs, toddler beds, high chairs and booster/hook-on chairs, gates and other enclosures, bath seats, play yards, stationary activity centers, infant carriers, strollers walkers, swings, bassinets, and cradles. The CPSC is required to periodically review and revise such standards to ensure the highest level of safety feasible for such products. Any commercial users (including child care centers and hotels/motels) are prohibited from manufacturing, selling, reselling, leasing, or providing for use any cribs that do not comply with these mandatory safety standards. Manufacturers of durable children's products must provide consumers with registration forms in a required format/mode facilitating registration and to maintain databases of registrants to be used to notify such consumers in the event of a product recall or safety alert. The CPSC is required to study whether registration forms should also be mandatory for other children's products and to periodically review and assess the effectiveness of alternative recall notification technologies.

Labeling for Certain Toy and Game Advertising (§105)

CPSIA §105 amends FHSA §24 (codified at 15 U.S.C. § 1278), requiring choking hazard labelling for certain toys and games, to require similar cautionary statements on or immediately adjacent to advertising that provides a direct means of purchase, including advertising on Internet websites, in catalogues, or other advertising materials. This requirement applies to advertisements by a retailer, manufacturer, importer, distributor, or private labeler. A manufacturer, importer, distributor, or private labeler is required to inform a retailer to whom it provides a product of any cautionary statement requirement applicable to that product. A retailer is not liable for violating the advertising requirement if it requested applicable cautionary statement information from a manufacturer, importer, distributor, or private labeler who failed to provide such information or provided false information. The statement must be displayed in a clear and conspicuous manner and in the language primarily used in the advertisement, website, or catalogue. Certain format and display requirements apply. The advertising requirements take effect for internet websites 120 days after enactment and for catalogues 180 days after enactment; the CPSC may grant a grace period for catalogues. The distribution of a noncompliant advertisement is a prohibited act under the CPSA.

Adoption of a Mandatory Toy Safety Standard (§1 06)

CPSIA § 106 (15 U.S.C. §2056b) provides that ASTM International Standard F963-07, *Consumer Safety Specification for Toy Safety*, the voluntary toy safety standard promulgated by the American Society for Testing and Materials (ASTM), an independent standard- setting organization, shall be deemed a mandatory interim consumer product safety standard, pending review, in the form current on the date of enactment, with certain exceptions. The exceptions are §4.2 and Annex 4 or any provision that restates or incorporates an existing mandatory standard or ban promulgated by the CPSC. Within two years of CPSIA enactment, the CPSC must promulgate this interim standard, with revisions to further enhance toy safety, by a final rule after reviewing it. The CPSC is required to periodically review and revise the rule to ensure the highest level of toy safety.

ASTM is required to notify the CPSC if it proposes revisions to this standard. The proposed revision will be incorporated into the CPSC rule and the revised standard will then be considered a CPSC rule effective 180 days after the ASTM notification was received, unless the CPSC notifies ASTM within 90 days of receiving such notification that it has determined that the proposed revision does not improve the safety of the products covered by the standard. The existing standard, without the proposed revision, will then continue to be the CPSC rule.

CPSIA § 106(h) provides that nothing in the federal toy standard or in the CPSA statute concerning preemption shall prevent a state or local safety requirement for toys or children's products from remaining in effect if it was in effect on the day before the date of CPSIA enactment and the state or locality has filed the requirement with the CPSC within 90 days after CPSIA enactment. Upon such application, the CPSC shall consider a proposed state or local safety standard and shall grant the exemption if the state or local standard provides a significantly higher degree of protection than the federal standard and does not unduly burden interstate commerce.

Study of Disparities in Injury/Death Rates of Minority Children (§1 07)

Section 107 of the CPSIA requires the U.S. Government Accountability Office (GAO), within 90 days of the date of enactment of this legislation, to initiate a study, by itself or an independent contractor, assessing racial/ethnic disparities in the risks and incidence of preventable injuries and deaths related to suffocation, poisonings, and drownings, including those linked to the use of cribs, mattresses, swimming pools, toys, and other products intended for use by children. Minority populations in the study include Black, Hispanic, American Indian, Alaskan Native, Native Hawaiian, and Asian/Pacific Islander children in the United States. GAO shall consult with the CPSC as necessary. GAO is required to report its findings to the relevant congressional committees not later than one year after the date of enactment of this legislation, including recommendations for minimizing risks of preventable deaths and injuries among minority children, for awareness and prevention campaigns targeting minority populations, and for education initiatives to reduce current statistical disparities.

Ban on Specified Phthalates and Certain Alternatives in Certain Children's Products and Child Care Articles (§108)

Beginning 180 days after enactment of the CPSIA, §108 of the CPSIA (15 U.S.C. §2057c) permanently bans the three phthalates (chemical plasticizers used in toys and other children's products) whose toxicity is not disputed and temporarily bans three other phthalates, pending a review by a Chronic Hazard Advisory Panel (CHAP).[43] It prohibits children's toys or child care articles that contain more than 0.1% di-(2 ethylhexyl) phthalate (DEHP), dibutyl phthalate (DBP), or benzyl butyl phthalate (BBP). The sale of children's toys or child care articles containing concentrations of more than 0.1% of diisononyl phthalate (DINP), diisodecyl phthalate (DIDP), or di-n-octyl phthalate (DnOP) are prohibited on an interim basis until a review by a CHAP. After the CPSC receives the report from the CHAP, it must determine, by rule, whether to continue the interim ban; evaluate the CHAP findings and recommendations; and declare any children's product containing phthalates to be a banned hazardous product if it determines this necessary to protect children's health. This provision clarifies that it does not preempt state laws regulating the use of phthalate

---

[43] For an analysis of this issue, see CRS Report RL34572, *Phthalates in Plastics and Possible Human Health Effects*, by Linda-Jo Schierow and Margaret Mikyung Lee.

alternatives not specifically regulated in a standard under the CPSA. "Children's toy" is defined as a product designed or intended for a use by a child 12 years of age or younger, and "child care article" is defined as a product designed or intended for a child three years of age or younger to facilitate sleep, feeding, sucking, or teething.

STRENGTHENING COMMISSION ADMINISTRATION AND RESOURCES

Reauthorization Years and Funding (§201)

Section 201(a) of the CPSIA amends CPSA § 32(a) to authorize progressively increasing appropriations annually. Out of these amounts, funding shall be made available for travel, subsistence, and related expenses incurred for official duties of the Commissioners and employees in attending meetings. This travel money is to be used in lieu of accepting funds from outside sources, as further discussed below.

Table 1. Authorized Total Appropriations and Travel Component for CPSC, FY2010-FY2014

Fiscal Year	Authorized Amount	Authorized Travel Amount
FY2010	$118,200,000	$1,200,000
FY20 11	$115,640,000	$1,248,000
FY2012	$123,994,000	$1,297,000
FY20 13	$131,783,000	$1,350,000
FY2014	$136,409,000	$1,403,000

Source: Section 201(a) of the CPSIA.

In addition to requiring the CPSC to report on personnel development efforts, CPSIA §201(b) requires the CPSC to submit a report on funding allocation plans to the appropriate congressional committees, not later than 180 days after enactment of the act. This report must include the efforts of the CPSC to reach and educate second-hand retailers of consumer products, particularly with regard to recalls of durable nursery products. Such education efforts shall include the development of educational materials for distribution not later than one year after enactment of the CPSIA.

Full Commission Funding and Interim Quorum (§202)

Under CPSA §4(d) (codified at 15 U.S.C. §2053), three commissioners of the five constitute a quorum; two can constitute a quorum if necessary due to a vacancy on the Commission, but only for six months after the vacancy occurs. Title III of Pubic Law 1 02-

389[44] limited funding to three Commissioners from FY1 993 and thereafter. This limitation impeded the Commission's ability to meet the quorum necessary to take certain actions. The limitation dated back to the mid-1980s when Congress was contemplating restructuring the CPSC as a three-member commission or replacing the Commission with a single administrator, which had been the Senate's original proposed scheme in its 1972 version of the legislation that ultimately became the CPSA.[45] During consideration of the 1990 reauthorization of the CPSC, the Senate bill would have permanently reduced the CPSC to three members with a quorum of two, but ultimately the authorization of a temporary quorum of two was adopted.[46] If a vacancy lasted longer than six months, as had been the case recently, the Commission could not establish mandatory standards or engage in other rulemaking or procedures, including taking certain enforcement actions requiring decisions by the Commissioners, mandatory recalls, and corrective actions. The quorum requirements were temporarily superseded by § 2204 of P.L. 11 0-53,[47] permitting two Commissioners, if they were not affiliated with the same political party, to constitute a quorum for six months beginning on the date of enactment of the act (August 3, 2007); this authority expired on February 3, 2008.

Section 202(a) of the CPSIA provides that, if they are not affiliated with the same political party, two members shall constitute a quorum for one year beginning on the date of enactment of the act. CPSIA § 202(b) repeals the funding limitation, effective one year after enactment of the act, in order to restore the CPSC to its full five-member size and prevent the recent quorum problems.

Personnel (§202(c))

Section 202(c) of the CPSIA (15 U.S.C. §2053 note) requires the CPSC to increase the number of fulltime Commission employees to at least 500 by October 1, 2013, subject to the availability of appropriations and, out of this number, requires the addition of an unspecified number of personnel to be assigned to U.S. ports of entry or to inspect overseas production facilities.[48]

Under CPSIA §201(b), requiring a CPSC report to the appropriate congressional committees on funding allocation plans, the CPSC must include the number of full- time investigators and other full-time equivalents the CPSC intends to employ. This report must

---

44 106 Stat. 1571, 1596 (1992) (codified as amended at 15 U.S.C. §2053 note).

45 The funding limitation dates back to §101(b)(1) of P.L. 99-434, which incorporated by reference H.R. 5313, the Department of Housing and Urban Development-Independent Agencies Appropriations Act, 1987, as passed by the House, which contained the limitation. At that time, based on discussions with the authorizing committee and consumer groups, the House Committee on Appropriations concluded that the five-member structure could not be justified any longer and urged consideration of a change to a single administrator heading the agency. The committee's apparent interim solution was in effect to have a three-member Commission. H.Rept. 99-731 [for H.R. 5313] at 21 (1986). See also GAO Report HRD-87-47, *Consumer Product Safety Commission: Administrative Structure Could Benefit From Change* (1987); Robert S. Adler, *From "Model Agency" to Basket Case — Can the Consumer Product Safety Commission Be Redeemed*?, 41 Admin. L. Rev. 61, 82-92 (1989).

46 H.Rept. 101-914, at 18 (1990).

47 Implementing Recommendations of the 911 Commission Act of 2007, 121 Stat. 266, 543.

48 Actual staffing level was 393 full-time equivalent staff in FY2007; 420 was planned for FY2008; and 444 was requested for FY2009. CPSC, 2009 Performance Budget Request, at 78 (February 2008).

also include CPSC efforts to develop standards for training product safety inspectors and technical staff and CPSC efforts and policies encouraging scientific staff to seek appropriate publishing opportunities in peer-reviewed journals and other media.

Reports to Congress (§203)

Section 3003 of P.L. 104-66 (the Federal Reports Elimination and Sunset Act of 1995, codified as amended at 31 U.S.C. § 1113 note) provided that, with certain exceptions, reports required to be submitted to Congress, as listed in H. Doc. 103-7, were terminated.

Section 203 of the CPSIA (15 U.S.C. §2076 note) requires that after the date of enactment of the CPSIA, notwithstanding any rule, regulation or order to the contrary, the CPSC must comply with the requirement of CPSA §27(k) (codified at 15 U.S.C. §2076(k)) that it submit copies to Congress of budget recommendations, legislative recommendations and comments, and testimony that it submits to the President or the Office of Management and Budget. These copies are expressly exempted from the reporting limitations of P.L. 104-66.

Expedited Rulemaking Procedures (§204)

Critics alleged that the prior rulemaking procedures under CPSA §9 and other acts under the CPSC' s jurisdiction, the FHSA and the FFA, were unnecessarily onerous, requiring procedural steps beyond those required by the Administrative Procedures Act.

Section 204 of the CPSIA streamlines the rulemaking procedures under CPSA §9 and similar provisions under the FHSA and FFA by eliminating the requirement for an advanced notice of proposed rulemaking (ANPR), a step not required by the Administrative Procedure Act (APA). This ANPR was required to include an invitation for persons to submit existing standards as proposed consumer product safety standards or statements of intention to develop or modify a voluntary standard, as well as commentary. Amendments conforming to the elimination of this requirement are also made. CPSIA §204 clarifies that the elimination of the ANPR requirement does not preclude a person from submitting all or part of an existing standard as a proposed consumer product safety standard. The FHSA is also amended to eliminate the required use of additional rulemaking procedures under the Food, Drug, and Cosmetic Act.

Technical amendments to the FHSA replace references to the Secretary of Health, Education and Welfare (HEW) [now Health and Human Services (HHS)], which remained from the original authority of the Secretary of HEW and the Food and Drug Administration over the FHSA, with references to the CPSC. Technical amendments to the FFA replace references to the Secretary of Commerce and the Federal Trade Commission (FTC), which remained from their original authority over the FFA, with references to the CPSC. References to specific congressional committees in the rulemaking provisions of the FHSA and the FFA are changed to references to the "appropriate congressional committees."

Inspector General Audits and Reports (§205)

Section 205 of the CPSIA (15 U.S.C. §2076b) requires the Inspector General of the CPSC to:

- conduct reviews and audits to assess the CPSC's capital improvement efforts, including upgrades of its information technology system and the development of the new public safety database, and the adequacy of the accreditation and monitoring process for third-party testing laboratories;
- within one year of enactment of the CPSIA, conduct a review of (1) CPSC employee complaints concerning failures of other employees to properly enforce the rules and regulations of the laws enforced by the CPSC or otherwise carry out responsibilities if such failures raise issues of conflicts of interest, ethical violations, or the absence of good faith, and (2) CPSC actions to address such complaints and failures;
- submit annual reports with respect to the findings and recommendations resulting from these audits and reviews to the appropriate congressional committees beginning in FY20 10;
- transmit a report to the appropriate congressional committees within 60 days of enactment of the CPSIA on the activities of the Inspector General, any barriers preventing robust oversight of the CPSC by the Inspector General, and any additional resources and authority needed for effective oversight.

This provision further requires the CPSC, within thirty days of enactment of the CPSIA, to establish and maintain (1) a direct link from the homepage of the CPSC to the webpage of its Inspector General, and (2) a mechanism on the Inspector General's website by which individuals may anonymously report cases of waste, fraud, or abuse with respect to the CPSC.[49]

Ban on Industry-Sponsored Travel (§206)

The CPSC has been criticized for the ethical issues raised by its practice of accepting funds from industry groups to cover travel to meetings and conferences.[50] As noted above, CPSIA §201(a) amends the authorization of appropriations in the CPSA to provide funds annually from FY2010 to FY2014 for travel to attend meetings and similar functions in furtherance of the official duties of the Commissioners and employees. These funds are to be used in lieu of accepting payment or reimbursement for such expenses from any person seeking action from, doing business with, or conducting activities regulated by the CPSC or whose interests may be substantially affected by the performance (or nonperformance) of a Commissioner's or employee's official duties.

---

[49] The link is at the bottom of the CPSC home page at [http://www.cpsc.gov] and the anonymous report form is on the Inspector General website at [http://www.cpsc.gov/cgibin/ igform.aspx], both last visited September 19, 2008.

[50] 154 Cong. Rec. S 1561-2 (March 5, 2008) (remarks of Senator Klobuchar).

CPSC Chair Nord supported these amendments.[51] In the past, she had defended the practice of accepting such paid travel to seminars and conferences as enabling industry education and outreach concerning safety standards and CPSC procedures that otherwise would not have been possible under the previous CPSC budgets.

Section 206 of the CPSIA adds a new §39 to the CPSA (15 U.S.C. §2086) with parallel language prohibiting Commissioners and employees of the CPSC from accepting travel and related expenses for any meeting or similar function related to official duties from a person seeking action from, doing business with, or conducting activities regulated by the CPSC and whose interests may be substantially affected by the performance of the Commissioner's or employee's official duties.

Information Sharing with Other Government Agencies (§207)

Section 207 of the CPSIA amends CPSA §29 (codified at 15 U.S.C. §2078) by authorizing the CPSC to share information obtained under the CPSA with federal, state, local, or foreign government agencies, notwithstanding the public disclosure requirements of the CPSA, where there is a prior agreement or other written certification that such information will be maintained in confidence and used only for law enforcement or consumer protection purposes and certain other conditions apply. The CPSC may terminate such agreements if it determines that the other agency has failed to abide by the conditions of the agreement. The CPSC shall not be required to disclose information it obtained from a foreign government agency or foreign source, if such information was provided on the condition of confidentiality, or through a CPSC reporting mechanism sponsored in part by foreign government agencies. However, nothing in this provision authorizes the CPSC to withhold information from Congress or prevents the CPSC from complying with a federal court order in an action by the United States or the CPSC. Foreign government agencies include multinational organizations comprising foreign states and vested with law enforcement or investigative authority in civil, criminal, and administrative matters. The CPSC must notify each state's health department of any CPSC mandatory recall or any voluntary recall of which it has been notified.

Employee Training Exchanges (§208)

Section 208 of the CPSIA (15 U.S.C. §2053a) authorizes the CPSC to engage in employee exchanges with foreign government agencies so that officers and employees of the CPSC and the foreign government agencies may receive or provide training. There is no requirement for reimbursement or reciprocity; any reimbursement for expenses incurred by the CPSC shall be credited to the appropriations account from which such expenses were paid. An officer or employee of a foreign government agency who is employed by the CPSC as part of a training exchange shall be considered a federal employee during such employment only for the purposes of federal laws governing work injury compensation for federal employees, tort claims liability of the Federal Government, federal employee ethics

[51] 154 Cong. Rec. S 1575 (March 5, 2008) (remarks of Senator Klobuchar).

and government corruption crimes, and any other law or regulation governing the conduct of federal employees.

Repeal of CPSA §30(d) (§237)

Section 237 of the CPSIA streamlines certain regulatory proceedings under the CPSA by repealing CPSA § 30(d) (15 U.S .C. §2079). The FHSA, FFA, and PPPA existed at the time the CPSA was enacted in 1972 and separately provided for regulatory authority and proceedings. When the CPSA was enacted, it included a requirement in CPSA §30(d) that a product which could be regulated under the FHSA, FFA, or PPPA could only be regulated under the CPSA if the CPSC first issued a rule finding that it was in the public interest to regulate a product under the CPSA.

Cost-Benefit Analysis under the PPPA (§233)

A GAO report concerning the effectiveness of cost-benefit analyses by the CPSC found, *inter alia*, that the CPSC often conducted such analyses in considering a consumer safety standard, even when not legally required to do so.[52] GAO noted that although the CPSA, FHSA, and FFA required cost-benefit analyses in order to promulgate a standard, the PPPA did not. Nevertheless, CPSC had conducted such analyses on several occasions in considering special packaging standards under the PPPA. Apparently in response to such commentary, CPSIA §233 amends §3 of the PPPA, regarding establishment of special packaging standards, by clarifying that nothing in the act shall be construed to require a cost-benefit analysis of a potential safety standard under the act in order to promulgate such standard.

Enhanced Enforcement and Cooperation

Prohibition on Stockpiling (§213)

CPSA §9(g) (codified at 15 U.S.C. §2058(g)) authorizes the CPSC to prohibit manufacturers from stockpiling products prior to the effective date of a consumer product safety rule; that is, from manufacturing the product at a higher-than-normal rate between the date that a rule is promulgated and the date it takes effect, in an attempt to circumvent the rule. This authority only applied to safety standards and rules promulgated under the CPSA and not to standards or rules promulgated under other statutes under CPSC jurisdiction such as the FHSA or FFA.

Section 213 of the CPSIA amends this provision so that it authorizes the prohibition of stockpiling of products prior to the effective date of an applicable rule under any statute enforced by the CPSC.

---

[52] GAO Report GAO/HEHS-97-147, *Consumer Product Safety Commission: Better Data Needed to Help Identify and Analyze Potential Hazards,* at 21-22 (1997).

Prohibited Acts (§21 6)

Section 216 of the CPSIA increases the scope of acts prohibited under CPSA §19 (codified at 15 U.S.C. §2068) by adding new prohibited acts and by expanding the application of existing prohibited acts. The additional or amended prohibitions include

- the sale, manufacture, distribution, or importation of a product regulated by the CPSC that does not comply with an applicable consumer product safety rule under the CPSA *or any similar rule, regulations, standard or ban under any other act enforced by the CPSC* [prohibition extends to products regulated under other acts such as the FHSA];
- the sale, manufacture, distribution, or importation of a product that is the subject of a voluntary corrective action and recall of which either the CPSC has notified the public or the seller, distributor, or manufacturer knew or should have known;
- the sale, manufacture, distribution, or importation of a product that is the subject of a CPSC order for a recall or corrective action or a court order declaring an imminent hazard;
- the sale, manufacture, distribution, or importation of a banned hazardous substance, meaning a children's article or toy which is a hazardous substance or which contains a hazardous substance accessible to a child;
- failure to furnish a required compliance certification or issuance of a false compliance certification under any act enforced by the CPSC, including failure to comply with requirements for testing, certification, and tracking labels for certain children's products;
- sale, importation or distribution of a consumer product bearing a false safety compliance certification mark;
- misrepresentation to CPSC officers/employees of the scope of products subject to recall/corrective action or a material misrepresentation in a CPSC investigation;
- undue influence of a third-party laboratory with respect to testing a product for compliance with safety standards under any act enforced by the CPSC;
- the export for sale of any consumer product or substance regulated by the CPSC (except for the re-export of a product denied importation into the United States) that is the subject of a voluntary corrective action, a CPSC order for a recall or corrective action, a court order declaring an imminent hazard, or that is a banned hazardous substance (meaning a children's article or toy which is a hazardous substance or which contains a hazardous substance accessible to a child); and
- violation of a CPSC order prohibiting an export under new subsection 18(c) of the CPSA (see below).

In particular, the prohibition on the sale of products that are the subject of a mandatory or voluntary recall closes a gap in the prior law, which permitted the continued sale of inventory that is the subject of a recall. The export restrictions are also new, since previously there were no restrictions on exports of recalled products or banned hazardous substances, beyond a requirement to notify the CPSC and the receiving country that an export did not comply with a U.S. safety standard or was a banned hazardous substance. In general, these CPSIA amendments prohibit violations of the new requirements under the CPSIA. Section 216 also makes a conforming amendment to CPSA § 1 7(a)(2) to prohibit the importation of a consumer product that does not have a required certificate or label or has a false certificate.

Penalties (§21 7)

Section 217 of the CPSIA increases the civil and criminal penalties under the CPSA, the FHSA, and the FFA. The maximum civil penalty increases from \$8,000 to \$100,000 for each violation and from \$ 1.825 million to \$15,000,000 for a related series of violations.[53] Not later than December 1, 2011, and every five years thereafter, the maximum civil penalty must be adjusted for inflation. Several factors must be considered in determining civil penalties under the CPSA, the FHSA, and the FFA, including the nature, circumstances, extent and gravity of the violation, in addition to those currently enumerated in these acts. The CPSC must consider mitigation of undue adverse economic impacts on small businesses. Not later than one year after enactment, the CPSC must promulgate regulations providing its interpretation of the criteria to be considered in imposing civil penalties. The amendments to the civil penalties take effect on the earlier of the date on which the final criteria regulations are issued or the date one year after enactment.

The criminal penalties increase to a maximum of five years imprisonment, a fine pursuant to 18 U.S .C. §3571 (establishing the maximum monetary fines), or both, for knowing and willful violations of the CPSA and the FFA and for violations with intent to defraud or mislead or repeat offenses under the FHSA. Penalties may also include forfeiture of assets associated with the criminal violation of the CPSA or any other statutes enforced by the CPSC. Section 217 of the CPSIA also removes the requirement in the CPSA that directors, officers, and agents have knowledge of a notice of noncompliance in order to be subject to criminal penalties separate from those imposed on their corporation.

Enforcement by State Attorneys General (§21 8)

Prior to its amendment by the CPSIA, CPSA §24 (codified at 15 U.S.C. §2073), entitled "Private Enforcement of Product Safety Rules and of Section 15 Orders," provided for a cause of action that could be brought by "[a]ny interested person (including any individual or nonprofit, business, or other entity)" in a federal district court to enforce a consumer safety rule or corrective action order by an injunction. Despite the section heading and the absence of an express reference to state attorneys general, this provision apparently had been understood to authorize enforcement actions by state attorneys general, limited to injunctive relief. The FHSA and FFA were amended in 1990 to expressly authorize state attorneys general to enforce consumer safety rules under those statutes by obtaining injunctive relief. This amendment was justified and premised on the existence of similar authority under the

---

[53] In accordance with statutory guidelines, the amounts specified in the statute before amendment by the CPSIA, \$5,000 and \$1.25 million, were adjusted for inflation in 2004 to \$8,000 and \$1 .825 million for individual violations and a related series of violations, respectively. 69 Fed. Reg. 68884 (2004).

CPSA.[54] However, it appears that this authority was rarely, if ever, used by state attorneys general, as there apparently was no reported case precedent for such an action; it appears that state attorneys general more typically took action under state consumer protections laws, while urging the CPSC to take action under federal laws.

Section 218 of the CPSIA amends CPSA §24 (codified at 15 U.S.C. §2073) by renaming the section "Additional Enforcement of Product Safety Rules and of Section 15 Orders" and adding a new subsection expressly authorizing state attorneys general (or other authorized state officer) to bring an action in any federal district court where the defendant is found or transacts business to obtain injunctive relief against certain prohibited acts under the CPSA.[55] Such actions are subject to certain conditions, similar to the procedures in CPSA §24 and other similar consumer protection statutes.[56] A state must give 30-day written notice to the CPSC before filing a civil action, except when a state has determined that immediate action is necessary to protect its residents from a substantial product hazard as defined in CPSA § 15(a). In such cases, a state may file a suit immediately after notifying the CPSC of such determination. The CPSC may intervene in such civil actions to be heard on all matters arising from such actions and to appeal decisions in such actions. A state cannot bring a civil action where the same alleged violation is the subject of a pending criminal or civil action brought by the Federal Government, aside from suits alleging a violation of the ban on selling, manufacturing, distributing, or importing a product that is noncompliant, subject to a mandatory/voluntary corrective action, or a children's product that is a banned hazardous substance.

Any outside private counsel retained to assist in such state civil actions is prohibited from sharing with parties in other private civil actions arising out of the same facts any information that is subject to a litigation privilege and was obtained during discovery in the state attorney general's action, or from otherwise using such information in the other private civil actions.

---

[54] See H.Rept. 101-567 at 17 (1990), and H.Rept. 101-914 at 28 (1990), referring to private enforcement actions by interested parties including state attorneys general; see also *Consumer Federation of America and U.S. Public Interest Group v. Consumer Product Safety Commission,* 990 F.2d 1298, fn. 12 at 1304 (1993) (seventeen state attorneys general signed an amicus brief in which, inter alia, they "emphasize that they 'have the authority under [the CPSA private enforcement provision] and [their states'] consumer protection statutes to enforce [a youth] ban on behalf of [their] citizens'" (the "youth ban" refers to a rule that would ban all new adult-size ATVs for use by children under 16 years old).

[55] These violations include prohibitions on the sale of noncompliant products or recalled products, failure to certify or label products, false certifications or labeling, false safety certification marks, stockpiling of products before the effective date of a new standard, and violation of a cellulose insulation standard. The CPSIA does not authorize state attorneys general to enforce certain violations that involve requirements specific to federal authorities, such as not making facilities or records available for inspection by CPSC officials and failure to provide certain information or notification to the CPSC as required.

[56] In addition to the aforementioned FHSA and FFA provisions, codified at 15 U.S.C. § 1264(d) and 15 U.S.C. § 1194(a), respectively, see e.g., 5 U.S.C. § 15c (codified provision of the Clayton Antitrust Act); 15 U.S.C. § 57 12(a) (codified provision of the Telephone Disclosure and Dispute Resolution Act of 1992, providing for regulations concerning advertising for, operation of, and billing and collection procedures for, pay-per-call or "900 number" telephone services); 15 U.S.C. § 1603 (codified provision of the Telemarketing and Consumer Fraud and Abuse Prevention Act, providing for regulations defining and prohibiting deceptive, coercive, and invasive telemarketing acts or practices); and 15 U.S.C. § 6504 (codified provision of the Children's Online Privacy Protection Act, enabling parents to control what information is collected from their children online). These provisions authorize states or state attorneys general on behalf of their states and their citizens to bring civil suits to obtain violations of the relevant federal laws, in some cases, monetary damages as well as injunctive relief.

In addition to the new state civil enforcement authority in the CPSA, CPSIA §217 adds a conforming provision to the PPPA authorizing state civil enforcement of a standard or rule under that act, subject to the procedural requirements of the CPSA.

Nothing in the new CPSA authority for state civil enforcement actions or in the similar provisions in the FHSA, FFA, and PPPA shall be construed as preventing a state attorney general or other state officer from exercising his/her powers under the state laws or as prohibiting him/her from proceeding in state or federal court on the basis of an alleged violation of any civil or criminal state statute.

Whistleblower Protections (§21 9)

Notwithstanding criticism of whistleblower protection by CPSC Acting Chair Nord,[57] CPSIA §219 adds a new §40 to the CPSA (15 U.S.C. §2087) to provide whistleblower protections for private sector employees. Such protections are *not* provided for federal, state, and local government agency employees; the compromise reached in the conference agreement eliminated proposed coverage of government employees.[58] New CPSA §40 establishes a remedy for an employee of a manufacturer, private labeler, distributor, or retailer, who believes that he/she has been subjected to adverse employment actions in retaliation for (1) providing information to the employer, Federal Government, or a state attorney general relating to a violation of any laws, rules, orders, standards, or bans enforced by the CPSC; (2) testifying in or otherwise cooperating with a proceeding concerning such violation; or (3) objecting to or refusing to participate in any activity or policy that the employee reasonably believed would be a violation of laws, rules, orders, standards, or bans enforced by the CPSC. The remedy is not available to an employee who, of his/her own volition, intentionally caused a violation of any laws, rules, orders, standards, or bans enforced by the CPSC.

The employee can file a complaint with the Secretary of Labor within 180 days of the alleged retaliatory action. The Secretary must then notify the person named as responsible for the retaliation and give him/her an opportunity to respond to the allegations. The Secretary must dismiss a complaint unless the complainant shows that the whistleblowing activities were a contributing factor in the adverse employment action. If the complainant shows this, but the employer shows, by clear and convincing evidence, that it would have taken the same unfavorable action in the absence of the whistleblowing activities, then the Secretary cannot further investigate the complaint. Otherwise, within 60 days of the complaint filing, the Secretary of Labor must conduct an investigation to determine whether there is reasonable

---

57 Congress DailyPM (February 20, 2008), available at [http://nationaljournal.com/pubs/ congressdaily/].

58 Federal employees have whistleblower protection under the Whistleblower Protection Act of 1989, P.L. 101-12, 103 Stat. 16 (codified as amended at 5 U.S.C. § 1201 et seq. See also L. Paige Whitaker, CRS Report for Congress RL339 18, *The Whistleblower Protection Act: An Overview*. State and local employees may have protection under the laws of their state. The report for S. 2045, an earlier version of the Senate bill concerning consumer product safety reform, described the earlier whistleblower provision as containing an intergovernmental mandate as defined in the Unfunded Mandates Reform Act (UMRA) because it would have required state and local governments to comply with whistleblower protections authorized in the bill. However, the Congressional Budget Office estimated that the costs to governments of complying with the mandate would be small and would not exceed the threshold established in the UMRA. S.Rept. 110-265 at 22 (2008).

cause to believe the complaint has merit and issue findings in writing, with a preliminary relief order where merited. Within 30 days of notification of the findings, the defendant can object and request a hearing; otherwise, the order becomes final and not subject to judicial review. The new provision establishes the standards for burden of proof and evidence.

Within 120 days of the hearing, the Secretary of Labor must issue a final relief order or denial. If the Secretary determines that retaliation has occurred, the Secretary shall order the person responsible for the retaliation to take affirmative steps to redress the retaliation; to reinstate the complainant to his or her former position with compensation, including back pay, and other terms, conditions, and privileges of his/her employment; and to provide compensatory damages. At the request of the complainant, the Secretary can also assess a sum of the aggregate amount of all costs and fees reasonably incurred by the complainant for bringing the complaint.

If the Secretary of Labor finds that the complaint was frivolous or in bad faith, the employer may be awarded attorneys' fees to be paid by the complainant. If the Secretary has not issued a final decision within 210 days of the complaint filing or within 90 days of a written determination, the complainant can bring an action for *de novo* review in a federal district court with jurisdiction, without regard to the amount in controversy. The same burdens of proof for the administrative hearing apply to the judicial review. The court has jurisdiction to grant all relief necessary to make the employee whole, including injunctive relief and compensatory damages (including reinstatement with the same seniority status, back pay with interest, and special damages such as reasonable attorney's fees, expert witness fees, and litigation costs).

Except where a person has already sued in federal court as described in the above paragraph, any person adversely affected by a final administrative order can appeal to the federal appellate court having jurisdiction in the area where the violation allegedly occurred or in which the complainant resided when the violation allegedly occurred. Such appeal does not stay the relief order unless so ordered by the court. If such direct appeal is not made, the administrative order cannot be judicially reviewed in other proceedings.

The Secretary of Labor can file a civil action in a federal district court for the District of Columbia or where the violation occurred to enforce an order against a person who has failed to comply. The court can grant all appropriate relief, including, but not limited to, injunctive relief and compensatory damages. Also, a person for whom an order of relief was granted may bring an action in a federal district court to require compliance, without regard to the amount in controversy or diversity of citizenship of the parties. The court may award court costs and fees as appropriate.

Any nondiscretionary duty imposed by this section is enforceable in a mandamus proceeding under 28 U.S.C. § 1361.

Federal Law Preemption (§231)

Certain provisions of the CPSA, FHSA, FFA, and PPPA establish the extent to which those acts preempt, limit, or otherwise affect any other federal, state, or local law or affect any cause of action under state or local law. Section 231(a) of the CPSIA (15 U.S .C. §2051 note) clarifies that these provisions may not be expanded, contracted in scope, limited, modified, interpreted, or extended in application in any rule, regulation, preamble, statement of policy, etc., of the CPSC. The CPSC may not construe the statutory preemption provisions as preempting any cause of action under state or local common law or state statutes regarding damage claims.

The purpose of these provisions apparently is to prevent the CPSC from issuing directives or statements purporting to preempt state common-law tort causes of action in the preambles to regulations that it promulgates, such as it did with the "Standard for the Flammability (Open Flame) of Mattress Sets."[59] Pursuant to Executive Order 12988 of February 5, 1996,[60] the Federal Register notice issuing the final rule for the mattress standard explained the Commission's understanding of the preemptive effect of the Flammable Fabrics Act. Citing congressional intent evidenced in the legislative history and statutory text, the "Commission intends and expects that the new mattress flammability standard will preempt inconsistent state standards and requirements, whether in the form of positive enactments or court created requirements." Such preemption directives apparently have become more common in recent years, causing concern among opponents[61] and observations by some legal scholars that these directives have become a "backdoor" method of expanding federal objectives.[62] In the absence of express congressional, statutory direction with regard to preemption, agencies may interpret the preemptive intent through such directives and regulations.[63] Commentators have noted that the federal courts are divided on the issue of the level of deference the courts should give to such agency preemption directives.[64]

Section 231(b) of the CPSIA clarifies that nothing in the CPSIA or the FHSA shall be interpreted as preempting or otherwise affecting state warning requirements established under state laws in effect prior to August 31, 2003.

As discussed on page 11 of this chapter, the CPSIA includes certain preemption provisions regarding state toy standards. The CPSIA does not provide for federal preemption of state third-party testing and certification requirements.

---

[59] 71 Fed. Reg. 13472, 13496-7 (2006), promulgating 16 CFR Part 1633, with a preemption directive at § N of the notice for the final rule.

[60] Section 3(b)(1)(A) of the Executive Order, 61 Fed. Reg. 4729, 4732 (2006), requires agencies to make every reasonable effort to ensure that a regulation, as appropriate, "specifies in clear language the preemptive effect, if any, to be given to the regulation."

[61] Catherine M. Sharkey, Symposium: Is the Rule of Law Waning in America? Twelfth Annual Clifford *Symposium on Tort Law and Social Policy: Article: Preemption by Preamble: Federal Agencies and the Federalization of Tort Law,* 56 DePaul L. Rev. 227, 233 (2007) (citing critics of preemption preambles inside and outside the Consumer Product Safety Commission); Christine H. Kim, The Case for Preemption of Prescription Drug Failure-to-Warn Claims, 62 Food Drug L.J. 399 (2007) (noting criticism of the Bush Administration for using preemption to accomplish "silent tort reform").

[62] *Id.*

[63] Sharkey, *supra* note 61, at 245.

64 Sharkey, *supra* note 61, at 242-7; Kim, supra note 61, at 419-421 (discussing cases in the context of Food and Drug Administration preambles with preemption directives); Thomas C. Galligan, Jr., *U.S. Supreme Court Tort Reform: Limiting State Power to Articulate and Develop Tort Law - Defamation, Preemption, and Punitive Damages,* 74 U. Cin. L. Rev. 1189, 1223-43 (2006) (discussing leading cases re preemption of state tort claims); Howard L. Dorfman, Vivian M. Quinn & Elizabeth A. Brophy, *Presumption of Innocence: FDA's Authority to Regulate the Specifics of Prescription Drug Labeling and the Preemption Debate*, 61 Food Drug L.J. 585, 597-60 1 (2006) (discussing deference to agency interpretation of rules and regulations; the article generally discusses preemption issues re pharmaceuticals).

ENHANCED INSPECTION, PUBLIC NOTICE, AND RECALL

Public Disclosure of Information (§211)

The CPSIA reduces the previous protections for disclosure of product information by amending CPSA §6 (codified at 15 U.S.C. §2055). This section provides for certain safeguards for the public disclosure of information on products that are identified as specific products of named manufacturers. Proprietary/trade secret information may not be disclosed, and information protected from disclosure by the Freedom of Information Act (FOIA) may not be disclosed. The manufacturer must be notified and given the opportunity to review information to be disclosed with regard to confidentiality and accuracy within a minimum period of time prior to disclosure, unless the CPSC finds that public health and safety require a lesser period of notice and publishes this finding. If the CPSC disagrees with the manufacturer and decides to disclose allegedly confidential or inaccurate information over the objections of the manufacturer, the manufacturer may sue in federal district court to enjoin disclosure. Certain types of information disclosure are exempt from these safeguards, including information regarding an imminently hazardous product, a violation of the CPSA, a rulemaking proceeding, an adjudicatory proceeding, or other proceeding under the CPSA. Consumer advocates and retailer critics of this provision prior to amendments made by the CPSIA asserted that these safeguards unnecessarily hindered the disclosure of safety and recall information, while industry advocates sought stronger protections with regard to substantiation and disclosure of information on product categories with problems that are not common to all manufacturers.[65]

Section 211 of the CPSIA reduces prior disclosure protections in several ways:

- A manufacturer/private labeler must respond within 15 days after the CPSC notifies it of the opportunity to mark as confidential information that could permit the public to identify it as the manufacturer/labeler of a product. Although the CPSA previously did not specify a time within which the manufacturer/labeler must respond, CPSC regulations concerning disclosures under the Freedom of Information Act (16 C.F.R. § 1015.18) require a response in five working days to a notice concerning information previously submitted to the CPSC by the manufacturer/labeler. These regulations also require that a response must accompany information submitted after a CPSC notice of the opportunity to request confidentiality (the submission may indicate a final confidentiality response within 10 working days of the new submission).
- The amendment reduces from 30 to 15 days before disclosure the time within which the CPSC must notify a manufacturer/labeler of an intended disclosure and provide the opportunity to comment on accuracy, and from 10 to five days the time within which the CPSC must notify a manufacturer/labeler that it will still disclose information claimed to be inaccurate based on a CPSC determination that the disclosure is accurate and fair.

---

[65] See, e.g., GAO Report GAO/HEHS-97-147, *Consumer Product Safety Commission: Better Data Needed to Help Identify and Analyze Potential Hazards* 28-32 (1997), for a discussion of the disclosure rules and policies and the views of industry and consumer advocate groups.

- The CPSC is no longer required to publish in the Federal Register a finding that public health and safety require a lesser period of notice to the manufacturer, as it was before the CPSIA; it may publish this in any manner.
- Exemption of certain information from the disclosure protections is expanded to include violations of any rule or law enforced by the CPSC, not just the CPSA, and information for which the CPSC publishes a finding that public health and safety requires disclosure with a lesser period of notice and comment on accuracy than normally required.
- The CPSC is authorized to file a request for expedited consideration in a civil action in the federal district court for the District of Columbia to enjoin the disclosure of allegedly inaccurate information. However, such expedited consideration is not available in an action to enjoin disclosure on the grounds that it is protected or confidential under other laws. The Conference Report noted the Conferees' view that such expedited consideration should not delay action on other important matters before the court, such as Class A or B felonies.[66]

Establishment of a Public Consumer Product Safety Database (§21 2)

Section 212 of the CPSIA adds a new section 6A to the CPSA (15 U.S.C. §2055a), establishing a publicly available, searchable, internet-accessible database on the safety of consumer products within two years of enactment. This provision resolves the issues some database opponents had with the database provision of the Senate-passed version of H.R. 4040. CPSC Acting Chair Nord previously had criticized the mandatory database under the Senate-passed version of H.R. 4040 (to be established without a study concerning feasibility, effectiveness, or other concerns) because it would have permitted publication of information and complaints received from consumers before the CPSC had the opportunity to vet the information for validity or accuracy and also because its implementation would have required 25 percent of the CPSC budget.[67] However, proponents of the database pointed out that the National Highway Transportation Safety Administration already has such a database including automobile complaints reported by consumers that is publicly accessible through its website.

The CPSC currently maintains the National Electronic Injury Surveillance System (NEISS),[68] which is accessible to the public through the CPSC website.[69] According to the CPSC website, "NEIS S injury data are gathered from the emergency departments of 100 hospitals selected as a probability sample of all 5,300+ U.S. hospitals with emergency departments. The system's foundation rests on emergency department surveillance data, but the system also has the flexibility to gather additional data at either the surveillance or the

66 H.Rept. 110-787 at 70.

67 Congress DailyPM (for February 20, 2008, and for March 11, 2008), available at [http://nationalj ournal.com/pubs/congressdaily/].

68 CPSA §5(a) (1) requires the maintenance of an Injury Information Clearinghouse "to collect, investigate, analyze and disseminate injury data and information relating to the causes and prevention of death, injury and illness associated with consumer products...."

69 See [http://www.cpsc.gov/library/neiss.html].

investigation level."[70] The database mandated by the CPSIA is more comprehensive, drawing information from a variety of sources, including consumers. GAO has concluded that the current data systems, including NEISS, upon which the CPSC depends in prioritizing its regulatory and enforcement activities, are inadequate.[71]

Under new CPSA §6A, the CPSC must transmit to Congress, within 180 days of the enactment of the CPSIA, a detailed plan for establishing and maintaining this database, including integration of the database into the CPSC's overall information technology improvement plans (the CPSC must expedite such plans). The plan shall include a detailed implementation schedule and plans for a public awareness campaign. Not later than 18 months after the plan is submitted, the CPSC must establish the database. The database shall include

- reports of harm (injuries, illness, death, or risks of injuries, illness, or death) relating to the use of consumer products, and other substances regulated by the CPSC, that are received by the CPSC from consumers, government agencies, health care professionals (such as physicians, hospitals and coroners), child service providers, and public safety entities (such as police and firefighters);
- information derived from a notice for a mandatory recall of a substantial product hazard or a notice for a voluntary corrective action;
- comments that a manufacturer/private labeler requests be included in the database to respond to information concerning its products; and
- any additional information the CPSC determines to be in the public interest.

The new section establishes requirements for information to be included in reports submitted to the CPSC and for the mode of submission. The information on the database must be organized and categorized so that the information is sortable and retrievable by the date of submission, the name of the consumer product, the model name, the manufacturer's/private labeler's name, and other information fields that are in the public interest. The CPSC shall provide a clear and conspicuous notice that the CPSC does not guarantee the accuracy, completeness, or adequacy of the database contents. The name and address of a person submitting a report of harm for a product may not be disclosed by the CPSC, except that such information may be provided to the manufacturer/private labeler of the product with the express written consent of such person for the purpose of verifying the report.

The safeguard restrictions of CPSA §6(a and b), briefly discussed in the previous section of this report, do not apply to the database disclosure of reports received from consumers, health-care providers, public safety entities, and government agencies. However, such safeguards apply to database information received via a report of a safety risk from a manufacturer, distributor or retailer under CPSA §15(b)or any other mandatory or voluntary reporting program established between the CPSC and a manufacturer, retailer, or private labeler.

Reports must be available on the database within 15 business days of receipt. Any information determined to be duplicative or inaccurate shall not be included when the report

---

[70] From the CPSC website description of NEISS, available at [http://www.cpsc.gov/cpscpub/ pubs/3002.html].

[71] GAO Report GAO/HEHS-97-147, Consumer Product Safety Commission: Better Data Needed to Help Identify and Analyze Potential Hazards 13-19 (1997).

is added to the database or, if the report is already on the database, be removed or corrected within 7 business days of such determination. Within five business days of receipt, the CPSC is required to submit a report of harm to the manufacturer, who then has 10 business days to respond, to request the inclusion of its comments on the database, and to vet the report for and designate confidential or protected information. The CPSC must redact in the database any information it determines to be protected (trade secret or FOIA). If the CPSC determines that the designated information is not protected, it shall so notify the manufacturer or private labeler, who may bring an action seeking removal of such information from the database in the federal district court where it resides or has its principal place of business or in the District of Columbia.

The CPSC is required to submit an annual report to the appropriate congressional committees on the operation of the database, including the cost and the number of reports and comments received, posted, and corrected or removed. Within two years of the establishment of the database, the GAO shall submit a report to the appropriate congressional committees containing an analysis of the general usefulness of the database, including an assessment of whether a broad range of the public uses the database and finds it useful, and recommendations for measures to increase use of the database by consumers and to ensure use by a broad range of the public. The Conference Report notes that, as part of general authorizations for FY20 10 to FY20 14, the Conferees authorized $25,000,000 to establish and maintain this database and to upgrade and integrate the CPSC information technology systems.[72]

Substantial Product Hazard Reporting Requirement (§214(a)(2))

Section 15 of the CPSA (15 U.S.C. §2064) requires manufacturers, distributors, and retailers to inform the CPSC when they learn that a product they distributed fails to comply with a consumer safety rule/standard, poses a substantial product hazard, or creates an unreasonable risk of injury or death. Prior to amendment by the CPSIA, the language of the provision did not include notification of noncompliance with rules or standards promulgated by the CPSC under other acts in its jurisdiction.

Section 214(a)(2)(A) of the CPSIA expands the scope of CPSA §15(b) to require manufacturers, distributors, and retailers of any product or substance regulated by the CPSC, except for motor vehicle equipment, to notify the CPSC about products that do not comply with any rule, regulations, standard, or ban promulgated by the CPSC under any act.

Section 2 14(a)(2)(C) of the CPSIA further amends CPSA § 15(b) to provide that a notification to the CPSC that a product fails to comply with a rule promulgated under the FHSA, FFA, or PPPA may not be used as the basis for criminal prosecution under the FHSA except for offenses requiring a showing of intent to defraud or mislead. The Conference Report clarifies that the Conferees included this provision "to avoid an unjust result under a possible construction of section 5 that provides for strict liability for criminal enforcement without regard to any applicable requirement of knowledge, intent, or willfulness in such situations. . . . The Conferees do not intend for [this] to be used to shelter bad actors . . . but rather to ensure that there are no unintended impediments to the flow of information to the

---

[72] H.Rept. 110-787 at 68 (2008).

Commission."[73] The House report, in discussing a similar provision in the House-passed version of H.R. 4040, clarifies that this applies when such notification constitutes the *sole* basis for criminal liability without a requirement of knowledge, willfulness, or intent.[74] The CPSA criminal offenses require knowledge and willfulness. The FFA offenses generally require willfulness, although the FFA offense of failure to notify the CPSC of the export of a non-compliant product does not. Other acts administered and enforced by the CPSC do not provide for criminal penalties.

Enhanced Public Notice of Substantial Product Hazards

Section 214(a)(3) of the CPSIA improves public notification by amending CPSA § 15(c) (codified at 15 U.S.C. §2064(c)) to authorize the CPSC to order a manufacturer, distributor, or retailer of a product presenting a substantial product hazard to give public notice of such hazard via its website, notice to third-party internet sellers of the product, announcements in languages other than English, and announcements on radio and television where the CPSC determines that a substantial number of consumers may not be reached by other types of notice. This section further clarifies that the CPSC may require any notices under CPSA §15(c)(1) to be distributed in a language other than English, if it determines that doing so is necessary to adequately protect the public.

Enhanced Authority for Corrective Action Plans and Recalls (§21 4)

Section 214 of the CPSIA enhances the authority of the CPSC to order corrective action plans and recalls in several ways. Section 214(a)(1) of the CPSIA expands the definition of "substantial product hazard" under CPSA §15(a)(1) (codified at 15 U.S.C. §2064(a)(1)) to include failure to comply with a rule, regulation, standard, or ban under any act enforced by the CPSC, as well as with an applicable consumer product safety rule under the CPSA.

Section 214(a)(3 and 4) of the CPSIA amends CPSA §15(c and f) (codified at 15 U.S.C. §2064(c and f)) to enhance recall authority by

- expanding the scope of the hazard notification authority to include products against which an imminent hazard action has been filed in federal court as well as products determined by the CPSC to be a substantial hazard;
- authorizing the CPSC to order a manufacturer, distributor, or retailer to cease product distribution; notify other persons involved in transporting, storing, handling, or distributing the product to cease distribution; and to notify appropriate state and local health officials;
- requiring the CPSC to rescind any order concerning an allegedly imminently hazardous product if a federal district court determines that the product is not an imminently hazardous product; and

[73] H.Rept. 110-787 at 71.

[74] H.Rept. 110-501 at 39 (2007).

- clarifying that the requirement for a hearing prior to the issuance of an order to cease distribution and notify the public of a substantial product hazard does not apply to an order concerning a product against which the CPSC has filed an imminent hazard action.

Section 214(b) amends CPSA § 15(d) (codified at 15 U.S.C. §2064(d)) to strengthen CPSC authorities relating to corrective action plans by

- authorizing the CPSC to order public notice and corrective actions it determines are in the public interest and removes the ability of the manufacturer, distributor or retailer to choose which corrective action it may take;
- expanding and clarifying the scope of CPSC authority to include orders to conform with requirements of applicable rules, regulations, standards, or bans, not just applicable consumer product safety rules;
- requiring a CPSC corrective action order to include a requirement that the person to whom the order applies must submit a plan for such action to the CPSC for affirmative approval in writing, replacing the current passive requirement that the plan be satisfactory to the CPSC;
- authorizing the CPSC to order an amendment of a corrective action plan if it finds that the approved plan is ineffective or inappropriate and requiring it to consider whether a repair or replacement changes the intended functionality of the product; and
- authorizing the CPSC to revoke its approval of an action plan if it finds that a person has failed to comply substantially with its obligations under the action plan and prohibiting such person from distributing the product after receiving notice of such revocation.

Requirements for Recall Notice Content (§214(c))

CPSIA §214(c) adds a new subsection 15(i) to the CPSA (15 U.S.C. §2064(i)) that requires the CPSC, within 180 days of enactment of the CPSIA, to establish guidelines for the information to be included in any recall/corrective action notice or imminent hazard notice that would aid consumers in identifying/understanding the specific product recalled, the nature of the associated hazard, and any available remedies. The provision further details the type of information required to be contained in a recall notice of a substantial product hazard. This information includes the model or stock keeping unit (S KU) number, common product names, photograph, description of action being taken, the number of product units for which such action is being taken, description of the substantial product hazard, identification of the manufacturers and distributors of the product, the dates the product was manufactured and sold, details of any deaths or injuries associated with the product, remedies available to the consumer, and any other information the CPSC deems necessary.

Identification of Supply Chain (§215(b))

Section 215(b) of the CPSIA amends CPSA §16 (codified at 15 U.S.C. §2065), regarding inspection and recordkeeping, by adding a new subsection (c) to require importers, retailers or distributors of a consumer product or other product or substance regulated by the CPSC to identify the manufacturer upon the request of a CPSC officer or employee. Conversely, a manufacturer is similarly required to identify each retailer or distributor whom the manufacturer directly supplied with a consumer product and each subcontractor involved in the manufacture of such product or from whom the manufacturer obtained a component of such product.

Financial Responsibility (§224)

Section 224 of the CPSIA adds a new section 41 to the CPSA (15 U.S.C. §2088), requiring the CPSC, in consultation with U.S. Customs and Border Protection (CBP) and other relevant federal agencies, to identify any consumer product, or other product or substance regulated by any statute enforced by the CPSC, for which the cost of destruction would normally exceed the bond amounts under the customs laws and to recommend a bond sufficient to cover the costs of destroying the product.

The new section further provides for a GAO study to determine the feasibility of mandating an escrow, proof of insurance, or other security to cover the costs of destruction of a domestically produced product or substance regulated under any act enforced by the CPSC or the costs of an effective recall of a domestic or imported product or substance regulated by the CPSC. The GAO must submit a report on the results of this study to the appropriate congressional committees within 180 days of enactment of the CPSIA, including an assessment of whether such requirements could be implemented and recommendations for implementation.

Annual Reporting Requirement (§209)

Section 209 of the CPSIA amends the annual reporting requirement under CPSA §27(j) (codified at 15 U.S.C. §2076(j)) to include information on the number and summary of recall orders issued under CPSA § § 12 and 15; a summary of voluntary actions taken by manufacturers in consultation with the CPSC and with public notice by the CPSC; and an assessment of such orders and actions. Additionally, within one year of the date of enactment of the CPSIA, the annual report shall include progress reports and incident updates with respect to corrective action plans ordered under CPSA § 15(d); injury and death statistics for substantial product hazards under CPSA § 15(c); and the number and type of communications from consumers to the CPSC for each product for which the CPSC orders corrective action.

Inspection of Certified Proprietary Laboratories (§215(a))

Section 215(a) of the CPSIA amends CPSA §2065(a) (codified at 15 U.S.C. §2065(a)) to authorize CPSC officers and employees to enter and inspect certified proprietary laboratories.

Safety of Imported and Exported Products

Export of Recalled and Nonconforming Products (§221)

Section 221 of the CPSIA amends CPSA §18 (codified at 15 U.S.C. §2067) to authorize the CPSC to ban the exportation from the United States of any consumer product that does not comply with U.S. consumer product safety rules, unless the importing country permits importation of such product into that country. The CPSC must notify the importing country of the impending shipment. If the importing country has not notified the CPSC of its import permission within thirty days of the CPSC notice, the CPSC may take appropriate action to dispose of the product. These restrictions do not apply to the exportation of a product refused admission into the United States and permitted by U.S. customs authorities to be exported in lieu of destruction. Related conforming amendments are also made to the FFA.

Prior to amendment, CPSA § 18 permitted the export of products that do not comply with U.S. consumer safety rules to other countries, requiring only that the CPSC had to notify the appropriate agencies in the foreign countries that such products were being exported to those countries.

Development of Methodology to Identify Unsafe Imports (§222)

Section 222(a and b) of the CPSIA (15 U.S.C. §2066 note) requires the CPSC, within two years of enactment of the act, to develop a risk assessment methodology to identify consumer product shipments that are intended for import into the United States and are likely to include products that violate CPSA §17(a) (15 U.S.C. §2066(a)) and other import laws enforced by the CPSC. In developing this methodology, the CPSC is required to

- use, as far as practicable, the International Trade Data System (ITDS) established under the Tariff Act of 1930 to evaluate and assess information about shipments of consumer products intended for import into the United States;
- incorporate this methodology into its information methodology modernization plan; and
- examine how to share information maintained by the CPSC, including the public database and substantial product hazard list, for the purpose of identifying shipments of noncompliant products.

Section 222(d) of the CPSIA requires the CPSC, not later than 180 days of completion of the risk assessment methodology, to submit a report to the appropriate congressional committees including

- a plan for implementing the methodology;
- an assessment of whether the CPSC requires additional statutory authority to implement the methodology;
- the level of appropriations necessary to implement the methodology;
- changes made or to be made to the CPSC memorandum of understanding with the CBP;
- the status of CPSC access to the Automated Targeting System and the development of the Automated Targeting system rule; and
- the status of the effectiveness of the International Trade Data System in enhancing cooperation between the CPSC and CBP to identify non-compliant shipments.

Cooperation with U.S. Customs and Border Protection (§222(c))

Section 222(c) of the CPSIA (15 U.S.C. §2066 note) requires the CPSC, within one year of enactment of this act, to develop a plan for sharing information and coordinating with CBP to improve enforcement and consumer protection. This plan must consider

- the number of CPSC personnel that should be stationed at U.S. ports of entry to identify shipments of consumer products that violate import safety laws enforced by the CPSC;
- the nature and extent of cooperation between CPSC and CBP personnel in identifying such noncompliant shipments;
- the number of CPSC personnel that should be stationed at the National Targeting Center of CBP, including the nature and extent of cooperation with the CBP, the responsibilities of the CPSC personnel, and usefulness of information at the Center in identifying noncompliant shipments;
- the development of rules for the Automated Targeting System and expedited access of the CPSC to the System; and
- the information and resources necessary for the development, updating, and effective implementation of the risk assessment methodology.

Substantial Product Hazard List and Destruction of Unsafe Imports (§223)

Section 223 of the CPSIA adds a new subsection (j) to CPSA §15 (codified at 15 U.S.C. §2064(j)) requiring the CPSC to specify by rule, for any consumer product or class of products, characteristics that constitute a substantial product hazard, if such characteristics are readily observable and covered by voluntary standards that have been effective in reducing the risk of injury and experience substantial compliance. A person adversely affected by such a rule may petition for judicial review under the CPSA not later than 60 days after promulgation of the rule.

CPSIA §223 amends a couple provisions to *require* rather than *permit* certain actions with regard to imports. CPSA § 17(e) (codified at 15 U.S.C. §2066(e)), formerly permitting CBP to destroy products refused importation into the United States in lieu of exportation, is amended to require the destruction of such products unless the CBP permits export in lieu of destruction and such products are exported within ninety days of export approval. CPSA § 17(g) (codified at 15 U.S.C. §2066(g)), formerly permitting the CPSC, at its discretion, to condition importation on a manufacturer's compliance with inspection and recordkeeping requirements, is amended to require manufacturers of imports to comply with all inspection and recordkeeping requirements or the products will be refused admission. The CPSC must advise the customs authorities of who is not in compliance. A related new subsection (d) of CPSA §16 (codified at 15 U.S.C. §2065(d)) requires the CPSC, by rule, to condition manufacturing, selling, distributing, or importing any consumer product or other product on the manufacturer's compliance with the inspection and recordkeeping requirements of the CPSA and the related rules.

Study of CPSC Authority Related to Imported Products (§225)

Section 225 of the CPSIA requires GAO, within one year of the enactment of this act, to conduct a study of the authorities of the CPSA to assess their effectiveness in preventing the importation of unsafe consumer products and to submit a report of its findings to the appropriate congressional committees. This report must include recommendations with respect to plans to prevent such importation; inspection of foreign manufacturing plants by the CPSC; and a requirement that foreign manufacturers consent to the jurisdiction of U.S. courts for enforcement actions by the CPSC.

MISCELLANEOUS PROVISIONS

Adoption of a Mandatory All-Terrain Vehicles (ATVs) Safety Standard (§232)

CPSIA §232 adds a new CPSA §42 (15 U.S.C. §2089), requiring the CPSC to publish in the Federal Register, as a mandatory consumer product safety standard, the American National Standard for Four Wheel All-Terrain Vehicles Equipment Configuration and Performance Requirements developed by the Specialty Vehicle Institute of America (American National Standard ANSI/SVIA-1-2007). It is unlawful for a manufacturer or distributor to import or distribute a non-compliant ATV in commerce in the United States or, until a three- wheel ATV standard is issued, a new three-wheeled ATV. Upon revision of these standards by the standard-setting organizations, the CPSC must incorporate by a rule revisions that are related to safe performance and any additional changes necessary to reduce an unreasonable risk of injury. The CPSC must also consider strengthening additional ATV safety standards.

Formaldehyde Study (§234)

GAO is required to conduct a study on the use of formaldehyde in the manufacturing of textiles and apparel not later than two years after the enactment of the CPSIA.

Expedited Judicial Review

CPSIA §236 provides for expedited judicial review of a consumer product safety rule relating to the identification of substantial hazards under CPSA § 15(j); all-terrain vehicles under CPSA §42; durable infant and toddler products under CPSIA § 104; and mandatory toy safety standards under CPSIA § 104. A person adversely affected by such a rule may petition in the U.S. Court of Appeals for the District of Columbia Circuit for expedited judicial review of the rule within 60 days of its promulgation. The judgment of the court affirming or setting aside the rule is final, subject to review by the U.S. Supreme Court. A rule to which expedited judicial review applies is not subject to judicial review in proceedings relating to imported products under CPSA § 17 or in civil or criminal proceedings for enforcement.

Definitions

Among other technical and conforming changes, CPSIA §235 adds definitions for "appropriate congressional committees," "children's product," and "third-party-logistics provider." "Appropriate congressional committees" are defined as the Committee on Energy and Commerce of the House of Representatives and the Committee on Commerce, Science, and Transportation of the Senate.

"Children's product" is defined as a consumer product designed or intended primarily for children 12 years of age or younger. Several factors are to be considered in determining whether a product is primarily intended for a child 12 years of age or younger, including a manufacturer statement or label about the intended use of the product; whether the packaging, display, promotion or advertising of the product represents it as appropriate for use by children 12 years of age or younger; whether the product is commonly recognized by consumers as being intended for use by children 12 years of age or younger; and the Age Determination Guidelines issued by the CPSC in 2002.[75]

"Third-party logistics provider" is defined as a person who solely receives, holds or otherwise transports a consumer product in the ordinary course of business but who does not take title to the product. CPSIA §235 also adds third-party logistics providers to the list of carriers that are not deemed to be manufacturers, distributors, or retailers of a consumer product under the CPSA solely by reason of receiving or transporting a consumer product in the ordinary course of their business and thus are exempt from the requirements of the CPSA.

---

[75] Timothy P. Smith, ed., CPSC, Age Determination Guidelines: Relating Children's Ages To Toy Characteristics and Play Behavior (2002).

Pool and Spa Safety Act Technical Corrections

CPSIA §238 amends the Virginia Graeme Baker Pool and Spa Safety Act[76] by adding a definition of "state" to the act and by providing for the adoption by the CPSC of revisions made to the pool and spa safety standard by the American Society of Mechanical Engineers, an independent standard-setting organization (the earlier act had mandated adoption of the voluntary standard).

Provisions Deleted in the Conference Agreement

The Senate Text contained several provisions concerning consumer safety product standards or studies concerning specific consumer products that did not have counterparts in the House Text and ultimately were not included in the final text of the conference agreement,[77] including (section numbers from Senate Text)

- garage door openers (§31);
- carbon monoxide poisoning from portable gas generators and charcoal briquettes (§32);
- cigarette lighters (§33); and
- equestrian helmets (§41).

Although the House Text did not include such provisions because its sole focus was reform of the CPSC and its authority and enforcement powers, language in H.Rept. 110-501 (2007) directed the CPSC to take action concerning single-product issues, including

- the promulgation by the CPSC of a final rule on cigarette lighters;
- a public awareness campaign concerning smoke alarms and smoke detection systems;
- a consideration of a safety standard concerning the warning labels for the lead content of ceramic food containers or serving ware;
- a consideration of CPSC authority to regulate pet toys that could be used by children and possible rules concerning the lead content and use of lead paint in such pet toys;
- a consideration of the tipping hazards of home appliances and furniture and possible safety standards; and
- a study of injuries and deaths related to toy guns and possible rules for making of toy guns and possible rules for marking of toys guns distinguishing them from actual firearms.[78]

[76] P.L. 110-140, title XIV, 121 Stat. 1492, 1794 (2007).

[77] P.L. 110-278, the Children's Gasoline Burn Prevention Act, 122 Stat. 2602 (2008), adopted safety standards for child-resistant closures on portable gasoline containers similar to Senate Text §28.

[78] H.R. 5471, to require the Consumer Product Safety Commission to prescribe rules requiring distinctive markings on toy and look-alike firearms (cosponsored by Towns and Blackburn).

The House Committee on Energy and Commerce noted in its report that it became aware of the potential dangers posed by asbestos in toys late in the legislative process and would take up these issues in subsequent hearings and legislation.

The Conference Report echoed all the concerns noted above, directing the CPSC to consider these issues and take action regarding safety standards for such products and hazards.[79] Additionally, the Conference Report recognized nanotechnology as a new, emerging technology used in the manufacture of consumer products and expressed its expectation that the CPSC would review the technology and the safety of its application in consumer products.[80]

---

79 H.Rept. 110-787 at 67, 75-77 (2008).
80 *Id.* at 68.

In: Consumer Product Safety Issues
Editors: P. A. Varga, M. D. Pintér

ISBN: 978-1-60456-826-4

Chapter 9

CONSUMER PRODUCT SAFETY IMPROVEMENT ACT OF 2008: H.R. 4040*

Margaret Mikyung Lee

SUMMARY

Public alarm about the spate of recent product recalls, particularly of toys and other products used by children, has focused attention on the Consumer Product Safety Commission (the CPSC or the Commission). The Consumer Product Safety Act (CPSA) established and authorized the CPSC in 1972 in response to growing concerns about protecting the public from unsafe, defective consumer products. Jurisdiction over the administration and enforcement of several existing consumer safety statutes was transferred from other agencies to and consolidated under the CPSC. However, in the years since its establishment, the staff and resources of the CPSC have been considerably reduced, leading many observers to doubt its ability to fulfill its mission effectively.

Consequently, Congress is considering major reform legislation to address organizational and systemic deficiencies. Legislative proposals include provisions targeting specific consumer product defects and hazards. On December 19, 2007, the House of Representatives passed H.R. 4040, the Consumer Product Safety Modernization Act, which was subsequently placed on the Senate calendar. On December 5, 2007, the Senate Committee on Commerce, Science, and Transportation reported S. 2045, the CPSC Reform Act of 2007, which was placed on the Senate calendar. On February 15, 2008, Senate leaders announced a bipartisan agreement for compromise legislation that was introduced and placed on the calendar as S. 2663. On March 6, 2008, the Senate passed H.R. 4040, after amending it to substitute the text of S. 2663 as amended on the floor.

Conferees from each chamber were designated to reconcile differences between the House and Senate versions of the bill. On July 28, 2008, the conferees announced that a final agreement had been reached, and the conference report was released on July 29, 2008. As of the date of this report, it remains uncertain whether the bill will be brought up for a floor vote

* This is an edited, reformatted and augmented version of CRS Report RL34399, dated July 29, 2008

in both chambers before the August recess, although it appears likely. This report summarizes information drawn from secondary sources and will be updated to reflect actual conference language.

This chapter provides an overview of the current authority of the CPSC to establish consumer product safety standards and to inspect and recall unsafe consumer products, and discusses the summary of the conference agreement as well as provisions of the Senate-passed and House-passed versions of H.R. 4040 reforming the CPSC and strengthening enforcement of consumer product safety standards. For an overview and context of the current issues facing the Commission, see CRS Report RS22821, *Consumer Product Safety Commission: Current Issues*, by Bruce K. Mulock. For an overview of issues regarding safety of consumer products imported from China, see CRS Report RS22713, *Health and Safety Concerns Over U.S. Imports of Chinese Products: An Overview*, by Wayne M. Morrison. For an overview of the issue of phthalates in children's products, see CRS Report RL34572, *Phthalates in Plastics and Possible Human Health Effects*, by Linda-Jo Schierow and Margaret Mikyung Lee.

LATEST LEGISLATIVE DEVELOPMENTS

On July 29, 2008, the conference report for H.R. 4040, the Consumer Product Safety Improvement Act of 2008, was released. Previously on July 28, 2008, the conferees announced that an agreement on a final text had been reached resolving the differences between the Senate and House versions of H.R. 4040, including a resolution on several disputed items remaining after earlier approval of agreed-upon items. On July 17, 2008, at a public conference meeting, the conferees voted to approve a package of agreements on 9 items by unanimous consent pending final adoption of the complete conference report; earlier, on June 25, 2008, the conferees voted to approve a package of agreements on 21 items. In a formal vote, the House conferees rejected a provision of the Senate amendment for an import ban on certain toys.

The conferees include conference co-chairs Senator Daniel Inouye and Representative John Dingell, as well as Senators Mark Pryor, Barbara Boxer, Amy Klobuchar, Ted Stevens, Kay Bailey Hutchison, John Sununu, and Representatives Henry Waxman, Bobby Rush, Diana DeGette, Jan Schakowsky, Joe Barton, Ed Whitfield and Cliff Stearns. As of the date of this report, it remains uncertain whether the bill will be brought up for a floor vote in both chambers before the August recess.

The conference committee and Senator Pryor issued press releases containing summaries of some of the agreed-upon provisions. As of the date of this report, we have not had an opportunity to fully examine the actual text of the bill or the Joint Explanation of the Committee of Conference in the recently released conference report.[1] The summaries below of various provisions in the conference agreement are based on details in press releases by the

[1] The text of the H.R. 4040 as agreed by the conference committee is available at [http://energycommerce.-house.gov/CPSC/ConferenceReport.072908.HR4040.FINAL_00 2_xml.pdf] (last visited July 29, 2008) and the full conference report, including the Joint Explanation of the Committee of Conference, is available at [http://www.rules.house.gov/ 110/text/110_hr4040cr.pdf] (last visited on July 29, 2008).

conference, Senator Pryor, and by the media;[2] the main body of this report will be updated later to reflect the text of the conference report.

Strengthen Commission Resources and Effectiveness

Reauthorization Years and Authorization of Appropriations

The Consumer Product Safety Commission (the CPSC or the Commission) would reportedly be authorized for five years beginning in 2010 ($118 million in FY20 10, $115.6 million in FY201 1, $124 million in FY2012, $132 million in FY2013, $136 million in FY2014). A specific travel allowance of $25 million would be provided; this, together with the ban on industry-sponsored travel, is in response to the reported practice of accepting funds from industry groups to cover travel to meetings and conferences.

Full Commission Requirement and Interim Quorum

Reportedly, the current funding limitation to three CPSC members would be eliminated in order to restore the CPSC to its full five-member size and prevent recent problems achieving the quorum necessary for CPSC actions such as mandatory recalls. A temporary two- member quorum would be permitted for one year beginning on the date of enactment.

Expedited Rulemaking

The bill would streamline the rulemaking procedures of the CPSC under the various consumer safety laws it administers by eliminating the requirement under those laws for an Advanced Notice of Proposed Rulemaking (ANPR), a step not required by the Administrative Procedure Act (APA).

Repeal of Section 30(d) of the Consumer Product Safety Act (CPSA)

Section 30(d) requires the CPSC to issue a rule finding that it is in the public interest to regulate a product under the CPSA if that product could also be regulated under certain other acts under its jurisdiction.

Formaldehyde Study

The Government Accountability Office would be required to conduct a study on the use of formaldehyde in the manufacturing of textiles and apparel.

---

2 Press release of the Senate Committee on Commerce, Science and Transportation and the House Committee on Energy and Commerce, dated July 28, 2008, *CPSC Conference Report Agreement Reached*, available at [http://commerce.senate.gov/public/index.cfm? FuseAction =PressReleases.Detail&PressRelease_id=2781a005-2c39-46a2-9e25-95da65f4bd65&Month=7&Year=2008] and at [http://energycommerce.house.gov/Press_110/110nr330.shtml] (both last visited on July 29, 2008) (hereinafter conference press release; Press Release of Senator Pryor dated July 28, 2008, *Senate, House Negotiators Clear Strong Product Safety Legislation*, available at [http://pryor.senate.gov/newsroom/details.cfm?id=301415&] (last visited July 29, 2008) (hereinafter Senator Pryor press release; and Congressional Quarterly staff, *Conferees Reach a Deal on Product Safety*, available at [http://www.CQ.com] at [http://www.cq.com/document/display.do?docid=2928030] (last visited July 29, 2008) (hereinafter CQ summary).

Other

In addition to the provisions above, the conference agreement addressed the following issues:

- Industry-Sponsored Travel Ban,
- Inspector General Audits and Reports,
- Submission of Documents to Congress, and
- Cost Benefit Analysis for the Poison Prevention Packaging Act.

Enhancing Children's Product Safety

Children's Products Containing Lead and Lead Paint Rule

In products for children aged 12 years and younger, the permissible lead level would be phased in over three years at 600 parts per million (ppm) within 180 days of enactment, 300ppm after one year, and 100ppm after three years. The CPSC would be required to periodically review and lower the limit.

Adoption of a Mandatory Toy Safety Standard

ASTM International Standard F963-07 *Consumer Safety Specification for Toy Safety*, the voluntary toy safety standard promulgated by the American Society for Testing and Materials, an independent standard-setting organization, would be made the mandatory interim rule, pending review, in the form current on the date of enactment, with certain exceptions. The exceptions are §4.2 and Annex 4 or any provision that restates or incorporates an existing mandatory standard or ban promulgated by the CPSC. The CPSC would be required to promulgate this interim standard by a final rule after reviewing it and would be required to ensure the highest level of toy safety in its rule.

Third-Party Testing

Third-party safety certification of products for children aged twelve and younger would be required. Upon CPSC accreditation, proprietary labs insulated from owner/manufacturer influence could be permitted to test products if they provide equal or greater protection than available third-party labs. CPSC personnel would be authorized to enter and inspect any accredited proprietary lab and to revoke accreditation if necessary. The new federal requirements would not preempt state authority and laws concerning third-party testing and certification.

Ban on Specified Phthalates and Certain Alternatives in Certain Children's Products and Child Care Articles

According to the conference press release, the bill would permanently ban the three phthalates whose toxicity is not disputed and would temporarily ban three other phthalates pending a review by a Chronic Hazard Advisory Panel (CHAP). It would prohibit children's toys or child care articles that contain more than 0.1% di-(2 ethylhexyl) phthalate (DEHP), dibutyl phthalate (DBP), or benzyl butyl phthalate (BBP). The sale of children's toys or child care articles containing concentrations of more than 0.1% of diisononyl phthalate (DINP), diisodecyl phthalate (DIDP), or di-n-octyl phthalate (DnOP) would be prohibited on an interim basis until a review by a CHAP. After the CPSC receives the report from the CHAP, it would determine, by rule, whether to continue the interim ban.

Adoption of a Mandatory All-Terrain Vehicles (ATVs) Safety Standard

According to the conference press release and Senator Pryor's press release, the CPSC would be required to publish in the *Federal Register*, as a mandatory consumer product safety standard, the American National Standard for Four Wheel All-Terrain Vehicles Equipment Configuration and Performance Requirements developed by the Specialty Vehicle Institute of America (American National Standard ANSI/S VIA- 1-2007). It would be unlawful for a manufacturer or distributor to import or distribute a non-complaint ATV in commerce in the United States. This reportedly would effectively ban all three-wheel ATVs, which apparently have a greater tendency to roll over than four-wheel ATVs.[3] Upon revision of these standards by the standard-setting organizations, the CPSC would be required to incorporate by a rule such revisions that are related to safe performance and any additional changes necessary to reduce an unreasonable risk of injury. The Commission would also be required to consider strengthening additional ATV safety standards.

Tracking Labels for Children's Products

Manufacturers would be required to label children's products with information enabling consumers and retailers to identify recalled products and enhancing the ability track products determined to be unsafe to their sources.

Other

In addition to the provisions above, the conference agreement addressed the following issues:

- Durable Nursery Product Standards,
- Catalog and Internet Advertising, and
- Study of Injuries and Deaths in Minority Children.

Enhanced Enforcement and Cooperation

Increased Civil Penalty

The maximum civil penalty would be increased from $5,000 to $100,000 for each individual violation and from $1.25 million to $15 million for a related series of violations.

Increased Criminal Penalty

The maximum criminal penalty would be increased to five years imprisonment for a knowing and willful violation of consumer product safety laws and could include forfeiture of assets associated with the criminal violation. The conference agreement would remove the

---

[3] *See* CQ summary, *supra* note 2.

requirement that directors, officers, and agents have knowledge of a notice of noncompliance to be subject to criminal penalties separate from those imposed on their corporation.

Enforcement by State Attorneys General

State Attorneys General would have authority to enforce federal consumer product safety laws and to respond rapidly to remove unsafe products from the marketplace.

Federal Law Preemption

The bill would clarify that the requirements enacted by the bill and existing requirements under the federal consumer product safety laws shall not be interpreted as preempting or otherwise affecting state warning requirements under laws enacted prior to August 31, 2003. Certain preemption provisions were included with regard to state toy standards. As noted above, no federal preemption was provided with regard to state third-party testing and certification requirements.

Whistleblower Protections

Whistleblower protections would be provided for private sector employees, including employees of manufacturers, private labelers, retailers, and distributors. Such protections would *not* be provided for government agency employees.

Prohibited Acts

The scope of prohibited acts would be expanded to make it unlawful for retailers to sell recalled products and certain other acts prohibited in the House and Senate versions of H.R. 4040.

Other

In addition to the provisions above, the conference agreement addressed the following issues:

- Enhanced Recall Authority,
- Corrective Action Plans,
- Requirements for Recall Notices,
- Public Disclosure of Information, and
- Annual Reporting Requirement.

Enhanced Import/Export Safety

Import Safety Management and Interagency Cooperation

The CPSC would be required to develop a method to identify consumer product shipments that are intended for import into the United States and are likely to include products that do not comply with U.S. consumer product safety rules.

Destruction of Noncompliant Imports

The CPSC would have greater authority to prevent the entry of unsafe consumer products into the United States by destroying noncompliant imports. The bill would enhance cooperation and information sharing among federal agencies, including the U.S. Customs and Border Protection, to improve enforcement and consumer protection.

Export of Recalled and Nonconforming Products

The CPSC would have the authority to prohibit the exportation from the United States of a product that does not comply with U.S. consumer product safety rules unless the importing country has notified the CPSC of its permission for importation of such product into that country.

Other

In addition to the provisions above, the conference agreement addressed the following issues:

- Study of effectiveness of authorities related to imported product safety,
- Substantial Product Hazard List, and
- Financial Responsibility.

Background and Current Law

Public alarm about the spate of recent product recalls, particularly of toys and other products used by children has focused attention on the Consumer Product Safety Commission (CPSC). Congress is considering major reform legislation to address organizational and systemic deficiencies. These legislative proposals also include provisions targeting specific consumer product defects and hazards. This report will provide an overview of the current authority of the CPSC to establish consumer product safety standards and to inspect, recall, and restrict importation of unsafe consumer products and summarize the major legislative proposals for reforming the CPSC and strengthening enforcement of consumer product safety standards.

The Consumer Product Safety Act (CPSA, 15 U.S.C. §§2051 et seq.) established and authorized the CPSC in 1972 in response to growing concerns about protecting the public from unsafe, defective consumer products. However, in the years since its establishment, the staff and resources of the CPSC have been considerably reduced to the detriment of its ability to fulfill its mission effectively. Aside from the issue of adequacy of resources, the recent, highly-publicized recalls of children's toys have focused attention on alleged weaknesses in the CPSA and the authority of the CPSC to establish consumer product safety standards and to inspect, recall domestically, and block imports of unsafe consumer products.

The CPSC is the central, federal authority for the promotion and enforcement of consumer product safety. The system is designed to be a collaborative effort among the CPSC, the industries producing the broad range of consumer products, and the consuming public. The CPSC researches and promotes best practices for the industries, producing guidelines for manufacturers, importers, distributors and retailers. Although the CPSA authorizes the CPSC to promulgate mandatory consumer product safety standards, it mandates reliance upon voluntary standards whenever compliance with voluntary standards would eliminate or adequately reduce the risk of injury and substantial compliance with voluntary standards is likely.

Besides the CPSA, the CPSC also administers several other statutes whose authorities and functions were transferred to the CPSC upon its creation. The Federal Hazardous Substances Act (FHSA, 15 U.S.C. §§1261 et seq.) provides for warning/informational labeling of hazardous substances and for the banning of certain hazardous substances for which labeling would not provide adequate protection for the public against the potential hazards posed by the substances. The Flammable Fabrics Act (FFA, 15 U.S.C. §§1 191 et seq.) provides for the establishment of safety standards regarding fabric flammability and that the manufacture, sale, importation, transportation, or delivery in commerce of a product, fabric, or related material or of a product made of a fabric or related material that does not comply with the standards are unlawful and constitute an unfair method of competition and an unfair and deceptive act or practice under the Federal Trade Commission Act. The Poison Prevention Packaging Act (PPPA, 15 U.S.C. §§1471 et seq.) authorizes the CPSC to establish special packaging standards for a household substance if such standards are required to protect children from serious injury or illness from using, handling or ingesting such substance, with exceptions for noncomplying packages for elderly/handicapped persons and packaging at the direction of a licensed medical practitioner. The Refrigerator Safety Act (RSA, 15 U.S.C. § § 1211) prohibits the introduction into interstate commerce of any household refrigerator that does not conform with certain safety standards.

Some of these statutes provide for powers that are similar but not identical to those established under the CPSA. Therefore, the regulatory procedures and other actions which the CPSC is authorized to carry out with regard to the products regulated under these other statutes may differ from those authorized under the CPSA. The CPSC may choose to regulate under the CPSA a consumer product that may be regulated sufficiently under these other statutes only if the CPSC determines that it is in the public interest to do so. The differences among the different statutory standards and procedures and enforcement authority arguably lead to inconsistency in the enforcement of different product standards. For example, injunctive enforcement authority for states attorneys general is expressly provided by the FHSA and the FFA, but not by the CPSA or other statutes administered by the CPSC; the apparent ambiguity of the CPSA on this point has led to clarifying provisions in the major Senate and House bills.

The CPSC has the authority to establish consumer product safety standards for consumer products generally, defined as "any article or component part thereof, produced or distributed (i) for sale to a consumer for use in or around a permanent or temporary household or residence, a school, in recreation, or otherwise, or (ii) for the personal use, consumption or enjoyment of a consumer in or around a permanent or temporary household or residence, a school, in recreation, or otherwise."[4] There are express exemptions for products covered under other statutes, including tobacco and tobacco products, motor vehicles and motor vehicle equipment, pesticides, firearms/antique firearms and ammunition/supplies (except for fireworks), aircraft and components, boats and other marine vessels, drugs, medical devices, cosmetics, food, or any article which is not customarily produced or distributed for sale to, or

4 CPSA §3 (codified at 15 U.S.C. §2052).

use or consumption by, or enjoyment of, a consumer. The CPSC also has jurisdiction over amusement rides that are not permanently fixed to a site but rather are part of a travelling carnival or show, but does not have jurisdiction over rides that are permanently fixed to a particular site. Furthermore, the CPSC has no jurisdiction to regulate a particular consumer product if the risk of injury associated with that product could be eliminated or sufficiently reduced by actions taken under the Occupational Safety and Health Act of 1970 (Occupational Safety and Health Administration), under the Atomic Energy Act of 1954 (the Energy Research and Development Administration [now Department of Energy] and the Nuclear Regulatory Commission), or under the Clean Air Act (the Environmental Protection Agency). The CPSC has no authority to regulate any risk of injury associated with electronic product radiation emitted from an electronic product if such risk may be regulated under the Public Health Act (the Food and Drug Administration).

Manufacturers, distributors, retailers, and importers are obligated to report consumer product safety problems to the CPSC, which may order a recall or import ban. The CPSA provides the general authority of the CPSC over inspections, recalls and import bans for consumer products generally, with the exceptions noted above, unless other statutes provide other agencies with authority over specific products. Although some of the other statutes enforced by the CPSC contain provisions specifically addressing notice/recall, import bans, and other remedies similar to those contained in the CPSA, not all do.

The CPSC is authorized to inspect domestic facilities where a consumer product is manufactured and the conveyances by which it is transported and which may be relevant to the safety of such product.[5] It is unlawful for a person to fail or refuse to permit inspection as required under the CPSA.[6] For purposes of the CPSA, including inspection and testing, the CPSC may purchase any consumer product and it may require any manufacturer, distributor, or retailer of a consumer product to sell the product to the CPSC at cost.[7] The CPSC is authorized to establish and maintain a permanent product surveillance program, in cooperation with other appropriate Federal agencies, for the purpose of carrying out the CPSC's responsibilities under the CPSA and the other Acts it administers and preventing the entry of unsafe consumer products into the United States.[8] The U.S. Customs and Border Protection (CBP)[9] is authorized to obtain and deliver samples of consumer products being offered for importation to the CPSC, upon its request, for the purpose of inspecting such samples for compliance with the CPSA.[10] Similarly, under FHSA § 14 (15 U.S.C. § 1273), the CBP is authorized to obtain and deliver samples of hazardous substances being imported or offered for importation to the CPSC, upon its request, for the purpose of inspecting such samples for compliance with the FHSA.

Under CPSA § 19,[11] it is unlawful, among other things, to manufacture, sell, distribute in commerce, or import into the United States any consumer product which does not comply with an applicable consumer product safety standard or which has been declared a banned

5 CPSA §16 (codified at 15 U.S.C. §2065) and 16 C.F.R. § 1118.2.
6 CPSA §19(a)(3) (codified at 15 U.S.C. §2068(a)(3)).
7 CPSA §27 (codified at 15 U.S.C. §2076).
8 CPSA § 17(h) (codified at 15 U.S.C. §2066(h)).
9 The statute and regulations refer to the Secretary of the Treasury although such functions are now undertaken by Department of Homeland Security (U.S. Customs and Border Protection (CBP)) pursuant to the Homeland Security Act and 19 C.F.R. § §0.1-0.2.
10 CPSA § 17(b) (codified at 15 U.S.C. §2066(b)).
11 Codified at 15 U.S.C. §2068.

hazardous product by a rule under the CPSA. Other consumer-product-related statutes contain similar provisions concerning prohibited or unlawful acts.

Under CPSA § 15,[12] every manufacturer (defined as including an importer), distributor, or retailer of a consumer product distributed in commerce who obtains information reasonably supporting the conclusion that such product (1) fails to comply with an applicable consumer product safety rule or with a voluntary consumer product safety standard; (2) contains a defect which could create a substantial product hazard; or (3) creates an unreasonable risk of serious injury or death, shall immediately inform the CPSC, unless such manufacturer, distributor, or retailer has actual knowledge that the CPSC has been adequately informed of such defect, failure to comply, or risk. If the CPSC determines after a hearing that a product presents a substantial product hazard and that notification is required in order to adequately protect the public from such substantial product hazard, the CPSC may order the manufacturer or any distributor or retailer of the product to take any one or more of the following actions: (1) to give public notice of the defect or failure to comply; (2) to mail notice to each person who is a manufacturer, distributor, or retailer of such product; or (3) to mail notice to every person to whom the person required to give notice knows such product was delivered or sold.

If the CPSC determines after a hearing that a product presents a substantial product hazard and that action under that provision is in the public interest, it may also order the manufacturer or any distributor or retailer of such product to take whichever of the following actions it elects to take: (1) to bring the product into compliance with the applicable product safety rule or repair the defect; (2) to replace the product with an equivalent product that does comply or is not defective; or (3) to refund the purchase price.

In addition to its authority with regard to a substantial product hazard, under CPSA §12, the CPSC may bring an action in federal district court to have a product declared an imminent hazard, defined as a consumer product which presents imminent and unreasonable risks of death, serious illness, or severe personal injury, and to seize the product. If the court determines that a product does constitute an imminent hazard, it may grant any relief necessary to protect the public, including an order requiring public notification, recall, and remedies including repair, replacement, or refund of such product.

Under the FHSA § 15,[13] the CPSC may order a manufacturer, distributor, or dealer to take remedial action with respect to a banned hazardous substance similar to remedies under the CPSA, including (1) giving public notice that an article or substance is a banned hazardous substance; (2) mailing notice to each person who is a manufacturer, distributor, or dealer of such article or substance; and (3) mailing such notice to every person to whom the person giving the notice knows such article or substance was delivered or sold. The CPSC may also order the repair of such article or substance, replacement with an equivalent compliant article or substance, or refund. Similar notification and remedial actions may be ordered with respect to any toy or other article intended for use by children that is not a banned hazardous substance but that contains a defect which creates a substantial risk of injury to children.

---

[12] Codified at 15 U.S.C. §2064.

[13] Codified at 15 U.S.C. § 1274.

The CPSC does not pay for the costs of a notice of product hazard or defect or any ordered repair, replacement, or refund; costs are born by the manufacturer, distributor, or retailer. An order issued under CPSA § 15[14] with respect to a product may require any person who is a manufacturer, distributor, or retailer of the product to reimburse any other person who is a manufacturer, distributor, or retailer of such product for such other person's expenses in connection with carrying out the order, if the CPSC determines such reimbursement to be in the public interest. Also, no charge shall be made to any person (other than a manufacturer, distributor, or retailer) who avails himself of any remedy provided under an order concerning repair, replacement, or refund, and the person subject to the order shall reimburse each person (other than a manufacturer, distributor, or retailer) who is entitled to such a remedy for any reasonable and foreseeable expenses incurred by such person in availing himself of such remedy. FHSA § 14(d)[15] contains reimbursement provisions similar to CPSA § 15.[16] Any person who is a manufacturer, distributor, or dealer of a noncompliant toy, article, or substance may be ordered to reimburse any other person who is a manufacturer, distributor, or dealer of such toy, article, or substance for such other person's expenses in connection with carrying out a remedial or notification order, if the CPSC determines such reimbursement to be in the public interest. Also, no charge shall be made to any person (other than a manufacturer, distributor, or dealer) who avails himself of any remedy provided under a remedial order and the person subject to the order shall reimburse each person (other than a manufacturer, distributor, or dealer) who is entitled to a remedy for any reasonable and foreseeable expenses incurred in seeking such remedy.

CPSC has the authority to establish import standards and policy with regard to statutes and products under its jurisdiction.[17] Under the CPSA, importers are made subject to the same responsibilities as domestic manufacturers in protecting American consumers from unreasonably hazardous products. This is explicitly stated in the definition of "manufacturer" as any person who manufactures or imports a consumer product.[18] Like the CPSA, the FHSA and the FFA assign responsibilities to importers comparable to those of domestic manufacturers and distributors.[19] Various statutory provisions authorize the CPSC to ban noncompliant imports. If the CPSC determines after a hearing that a product presents a substantial product hazard and that action under that provision is in the public interest, it may issue an order prohibiting the importation into the United States of that product.[20] An imported consumer product may be refused admission to the United States if it does not comply with an applicable consumer product safety rule; does not comply with labeling and certification requirements relating to applicable product safety standards; is an imminently hazardous product; or has a product defect which constitutes a substantial product hazard.[21] The CPSC may inform the U.S. Customs and Border Protection (CBP)[22] that an imported

---

[14] Codified at 15 U.S.C. §2064.

[15] Codified at 15 U.S.C. § 1273(d).

[16] Codified at 15 U.S.C. §2064.

[17] Its policy on imported products, importers, and foreign manufacturers is set out at 16 C.F.R. § 1009.3.

[18] CPSA §3(a)(4) (codified at 15 U.S.C. §2052(a)(4)).

[19] See FHSA §15(f) (15 U.S.C. §1274(f)); FFA §9 (15 U.S.C. §1198).

[20] CPSA § 15(d) (codified at 15 U.S.C. §2064(d)).

[21] CPSA §17 (codified at 15 U.S.C. §2066).

[22] CPSA § 17(b) (codified at 15 U.S.C. §2066(b)).

consumer product fails to comply with an applicable consumer product safety rule and/or has a product defect which constitutes a substantial product hazard and may request the CBP to refuse admission to any such consumer product.[23] Under FHSA § 14,[24] a misbranded hazardous substance or banned hazardous substance being imported or offered for import shall be refused importation. Section 9 of the Flammable Fabrics Act (FFA)[25] provides that imported products subject to flammability standards under the FFA shall not be released from customs custody except in accordance with § 499 of the Tariff Act of 1930[26] providing for release only after inspection by CBP for compliance with U.S. laws. The CBP also has authority for the redelivery or recall of products already released under bond but later found to not comply with flammability standards and for obtaining liquidated damages for breach of a condition of the bond arising out of a failure either to correct the product to comply or to redeliver it.

Such noncompliant or defective products must be exported from the United States or may be ordered destroyed unless they can be modified by the owner or consignee in a manner that will enable them to be granted admission into the United States; the CPSC and the CBP have the authority to give the owner or consignee the opportunity to make such modifications and to monitor such modifications. The CPSC may condition importation of a consumer product on the manufacturer's/importer's compliance with the inspection and recordkeeping requirements of the CPSA. The CPSC may seek an injunction or seizure of a consumer product that does not comply with a consumer product safety rule or that is being manufactured, sold, distributed, or imported in violation of a CPSC order for remedial action or prohibiting importation.[27]

With regard to imported products that are admitted and subsequently become the subject of a recall, if the CPSC is not able to exercise jurisdiction over a foreign manufacturer that has no U.S. subsidiary/presence, the importer would be considered responsible for recall costs in the United States. That importer may then be able to obtain reimbursement from the foreign manufacturer as a contractual matter. Under CPSA §17(f),[28] the owner or consignee must pay for all expenses in connection with storage or destruction of an imported consumer product denied entry into the United States. In default of such payment, these expenses shall constitute a lien against any future importations made by such owner or consignee. Under FHSA § 14(c),[29] the owner or consignee must pay for all expenses (including travel, per diem, or subsistence, and salaries of officers or employees of the United States) in connection with the destruction of a hazardous substance denied importation into the United States; the supervision of the relabeling or other action authorized to bring a hazardous substance denied importation into compliance with the FHSA so that it may be granted importation; and the storage for any hazardous substance denied importation. In default of such payment, such expenses shall constitute a lien against any future importations made by such owner or consignee.

---

[23] 16 C.F.R. § 1115.21(d).

[24] Codified at 15 U.S.C. § 1273.

[25] Codified at 15 U.S.C. §1198.

[26] Codified at 19 U.S.C. § 1499.

[27] CPSA § 22 (codified at 15 U.S.C. §207 1).

[28] Codified at 15 U.S.C. §2066(f).

[29] Codified at 15 U.S.C. § 1273(c).

The Office of Compliance and Field Operations within the CPSC conducts compliance and administrative enforcement activities under all administered acts, provides advice and guidance on complying with all administered acts and reviews proposed standards and rules with respect to their enforceability.[30] Among other things, it reviews consumer complaints, conducts inspections and in-depth investigations, and analyzes available data to identify those consumer products containing defects that pose a substantial risk of injury or do not comply with existing safety requirements. The Office negotiates and monitors corrective action plans for products that are defective or fail to comply with specific regulations.

The Office of International Programs and Intergovernmental Affairs within the CPSC was established to enable a more coordinated and comprehensive approach to international cooperation. Memoranda of understanding have been concluded with CPSC counterparts in various countries or regional groups, including the People's Republic of China, the European Union, and Canada.

CURRENT LEGISLATION: H.R. 4040

There have been various proposals in the 1 10th Congress to address various consumer product safety issues. In addition to the two major CPSA reform bills, H.R. 4040 and S. 2045/S. 2663, there have been other comprehensive reform bills and other bills addressing discrete issues, including safety standards for cigarette lighters, All-Terrain-Vehicles or ATVs, furniture, swimming pools, portable gasoline containers, durable infant or toddler consumer products such as strollers and cribs, and other products; certification of safety-standard compliance; the enactment of lead content standards for consumer products and more stringent lead in paint standards; third-party testing for product safety and compliance; increases in civil and/or criminal penalties; increases in CPSC personnel assigned to ports-of-entry; the prohibition of sales or resales of products that are the subject of a recall; expanded jurisdiction of the CPSC to cover amusement park rides at a fixed site; greater coordination among the various agencies involved in consumer safety issues; and others. Language from many of the free-standing bills addressing specific issues has been incorporated into the comprehensive bills. This section will summarize the major provisions of the two major reform bills. The Senate took up S. 2663, a compromise version of S. 2045, and passed H.R. 4040 in lieu of S. 2663 after amending H.R. 4040 to substitute the text of S. 2663 as amended on the Senate floor. Reportedly, informal negotiations between representatives from each chamber are taking place, although conferees for a formal conference about the text have not yet been appointed.[31]

The Senate-passed version of H.R. 4040 [hereinafter Senate Text][32] and the House-passed version of H.R. 4040 [hereinafter House Text][33] contain many similar provisions strengthening the authority and resources of the CPSC and also establishing standards concerning lead content in children's toys. However, the Senate Text contains additional

---

30 16 C.F.R. § 1000.21.

31 Bureau of National Affairs, Conversations on CPSC Reform Continue Between Key House and Senate Players, 36 Product Safety and Liability Reporter 373 (April 14, 2008).

32 The short title for the Senate version of the bill is the CPSC Reform Act.

33 The short title for the House version of the bill is the Consumer Product Safety Modernization Act.

reform provisions, such as whistleblower protection and several provisions concerning consumer product safety standards for specific items such as all-terrain vehicles and garage door openers. Conversely, most of the provisions of the House Text have parallel provisions in the Senate Text; the major exception is the provision adding a prohibition on industry-sponsored travel by members or employees of the CPSC, a response to the much-criticized practice by the CPSC of accepting travel and lodging expenses from industry sponsors for trips related to CPSC business, that was widely reported after the S. 2045 markup.

Strengthening the Commission Resources and Structure

Funding

Section 3 of the Senate Text authorizes progressively increasing appropriations annually from $88,500,000 for FY2009 to $155,900,000 for FY20 15 for the Commission, and from $1,600,000 for FY2009 to $2,834,555 for FY20 15 for its Inspector General. This section also authorizes $40,000,000 for FY2009 and FY20 10 for the improvement of the Commission's research, development, and testing facilities, and also $1,000,000 for FY2009 and FY2010 for research into safety issues related to the use of nanotechnology in consumer products. Section 3 would also provide up to $1,200,000 annually from FY2009 to FY20 15 for travel, subsistence, and related expenses incurred for official duties of the Commissioners and employees in attending meetings. This travel money is to be used in lieu of accepting funds from outside sources, as discussed further below.

Section 201 of the House Text authorizes increasing appropriations annually from $80,000,000 for FY2009 to $100,000,000 for FY201 1. This section also authorizes $20,000,000 for FY2009 to FY20 11 for the repair and improvement of the Commission's research, development, and testing facilities.

Personnel

Section 4 of the Senate Text would increase the number of fulltime Commission employees to at least 500 by October 1, 2013, subject to the availability of appropriations, and requires the addition of at least 50 personnel to be assigned to U.S. ports of entry or to inspect overseas production facilities by October 1, 2010, subject to the availability of appropriations. The CPSC would also be required to develop and implement a professional career development program for professional staff to encourage staff retention and provide professional development opportunities for employees. Additionally, within 180 days of the enactment of this legislation, the CPSC would be required to develop standards for training product safety inspectors and technical staff in consultation with organizations with expertise in consumer product safety issues and to submit a report to Congress regarding such standards.

The House Text does not expressly provide for specific increases in staff, instead requiring a report to Congress concerning plans for the number of full-time inspectors and other employees under §201(b). This report would also include plans for inspection of imported products; feasibility of CPSC inspection and certification of foreign testing facilities; feasibility of mandating bonds for repeat offenders and serious hazards; and the efforts of the CPSC to educate second-hand retailers of consumer products, particularly with regard to recalls of durable nursery products.

Quorum

Three commissioners of the five constitute a quorum; two can constitute a quorum if necessary due to a vacancy on the Commission, but only for six months after the vacancy occurs. Title III of Pubic Law 102-389[34] limited funding to three Commissioners from FY1993 and thereafter. This limitation has impeded the Commission's ability to meet the quorum necessary to take certain actions. The limitation dates back to the mid-1980s when Congress was contemplating restructuring the CPSC as a three-member commission or replacing the Commission with a single administrator, which had been the Senate's original proposed scheme in its 1972 version of the legislation that ultimately became the CPSA.[35] During consideration of the 1990 reauthorization of the CPSC, the Senate bill would have permanently reduced the CPSC to three members with a quorum of two, but ultimately the current statutory language was adopted permitting a quorum of two only temporarily.[36] If a vacancy lasts longer than six months, as has been the case recently, the Commission cannot establish mandatory standards or engage in other rulemaking or procedures, including taking certain enforcement actions requiring decisions by the Commissioners, mandatory recalls, and corrective actions. The quorum requirements were temporarily superseded by §2204 of P.L. 1 10-53,[37] permitting two Commissioners, if they are not affiliated with the same political party, to constitute a quorum for six months beginning on the date of enactment of the act (August 3, 2007); this authority expired on February 3, 2008.

Section 5 of the Senate Text would repeal the funding limitation, effective on the date of enactment of the act, and urge the President to fill any vacancies expeditiously, resulting in a full panel of five Commissioners. If they are not affiliated with the same political party, two members would constitute a quorum for nine months beginning on the date of enactment of the CPSC Reform Act.

Section 202 of the House Text would repeal the funding limitation effective October 1, 2011. It would also provide that, if they are not affiliated with the same political party, two members constitute a quorum for a period beginning on the date of enactment of the act, but ending on August 3, 2008, if the President nominates a person to fill a vacancy before that date, or, if the President nominates a person to fill a vacancy after that date, ending three months after the date the President nominates a person or on February 3, 2009, whichever is earlier.

---

[34] 106 Stat. 1571, 1596 (1992) (codified as amended at 15 U.S.C. §2053 note).

[35] The funding limitation dates back to § 101 (b)(1) of P.L. 99-434, which incorporated by reference H.R. 5313, the Department of Housing and Urban Development-Independent Agencies Appropriations Act, 1987, as passed by the House, which contained the limitation. At that time, based on discussions with the authorizing committee and consumer groups, the House Committee on Appropriations concluded that the five-member structure could not be justified any longer and urged consideration of a change to a single administrator heading the agency. The committee's apparent interim solution was in effect to have a three-member Commission. H.Rept. 99-731 [for H.R. 5313] at 21 (1986). See also GAO Report HRD-87- 47, *Consumer Product Safety Commission: Administrative Structure Could Benefit From Change* (1987); Robert S. Adler, *From "Model Agency" to Basket Case — Can the Consumer Product Safety Commission Be Redeemed?*, 41 Admin. L. Rev. 61, 82-92 (1989).

[36] H.Rept. 101-914, at 18 (1990).

[37] Implementing Recommendations of the 911 Commission Act of 2007, 121 Stat. 266, 543.

Rulemaking Procedures

Critics allege that the rulemaking procedures under CPSA §8 and other acts under the CPSC's jurisdiction, the FHSA and the FFA, are unnecessarily onerous, requiring protections beyond those required by the Administrative Procedures Act.

Section 8 of the Senate Text would eliminate the requirement for an advanced notice of proposed rulemaking (ANPR), which must include an invitation for persons to submit existing standards as proposed consumer product safety standards or statements of intention to develop or modify a voluntary standard, as well as commentary. Amendments conforming to the elimination of this requirement would be made. The Senate Text would include language clarifying that the elimination of the ANPR requirement does not preclude a person from submitting all or part of an existing standard as a proposed consumer product safety standard. Similar amendments eliminating the requirement for an ANPR would be made to the FHSA and the FFA. The FHSA would also be amended to eliminate the required use of additional rulemaking procedures under the Food, Drug, and Cosmetic Act, and to replace references to the Secretary of Health, Education and Welfare (HEW) [now Health and Human Services (HHS)], which remained from the original authority of the Secretary of HEW and the Food and Drug Administration over the FHSA, with references to the CPSC. The FFA would be further amended to replace references to the Secretary of Commerce and the Federal Trade Commission (FTC), which remained from their original authority over the FFA, with references to the CPSC.

Section 204 of the House Text would make similar amendments to the CPSA, FHSA, and FFA rulemaking procedures, but would not include the language clarifying that, with regard to CPSA procedures, a person may still submit all or part of an existing standard as a proposed consumer product safety standard, notwithstanding the elimination of the mandatory invitation to do so.

Repeal of CPSA §30(d)

Section 35 of the Senate Text and § 221 of the House Text both would repeal CPSA §30(d), the requirement that the CPSC can only regulate a risk under the CPSA that can also be regulated under the FHSA, FFA, or PPPA if the CPSC makes a determination that promulgating such regulations under the CPSA is in the public interest.

Cost-Benefit Analysis under the PPPA

A GAO report concerning the effectiveness of cost-benefit analyses by the CPSC, *inter alia*, found that the CPSC often conducted such analyses in considering a consumer safety standard, even when not legally required to do so.[38] GAO noted that although the CPSA, FHSA, and FFA required cost-benefit analyses in order to promulgate a standard, the PPPA did not, yet the CPSC had conducted such analyses on several occasions in considering special packaging standards. Apparently in response to such commentary, §25 of the Senate Text would amend §3 of the Poison Prevention Packaging Act, regarding establishment of special packaging standards, by clarifying that nothing in the act shall be construed to require a cost-benefit analysis of a safety standard under the act in order to promulgate such standard. The House Text contains no similar provision.

[38] GAO Report GAO/HEHS-97-147, Consumer Product Safety Commission: Better Data Needed to Help Identify and Analyze Potential Hazards, at 21-22 (1997).

Ban on Industry-Sponsored Travel

The CPSC has been criticized for the ethical issues raised by its practice of accepting funds from industry groups to cover travel to meetings and conferences.[39] As noted above, section 3 of the Senate Text would provide up to $1,200,000 annually from FY2009 to FY2015 for travel to attend meetings and similar functions in furtherance of the official duties of the Commissioners and employees. These funds would be used in lieu of accepting payment or reimbursement for such expenses from any person seeking action from, doing business with, or conducting activities regulated by the CPSC or whose interests may be substantially affected by the performance of the Commissioner's or employee's official duties.

CPSC Chair Nord supported these amendments.[40] In the past, she had defended the practice of accepting such paid travel to seminars and conferences as enabling industry education and outreach concerning safety standards and CPSC procedures that otherwise would not have been possible under the previous CPSC budgets.

Section 222 of the House Text would add a new section with similar language to the CPSA (numbered §38 in the bill, although there already is a CPSA §38 relating to a low-speed electrical bicycle standard) which would prohibit a Commissioner or employee of the CPSC from accepting travel and related expenses for any meeting or similar function related to official duties from a person seeking action from, doing business with, or conducting activities regulated by the CPSC and whose interests may be substantially affected by the performance of the Commissioner's or employee's official duties. In addition, the section would authorize the appropriation of $1,200,000 to the CPSC for each of fiscal years 2009 to 2011 for such travel and lodging expenses necessary to the official duties of the Commissioners and employees.

Inspector General Authority and Accessibility

Section 26 of the Senate Text would require the Inspector General of the CPSC to:

- conduct reviews and audits of implementation of the CPSA by the CPSC, including assessments of the implementation of various amendments and reforms enacted by this legislation;
- submit annual reports with respect to the findings and recommendations resulting from these implementation audits and reviews to the CPSC and the relevant congressional committees for each of fiscal years 2009 to 2015;
- conduct a review of CPSC employee complaints concerning failures of other employees to properly enforce the rules and regulations of the laws enforced by the CPSC (including the negotiation of corrective action plans) and of the process by which corrective actions are negotiated by the CPSC and submit a report on findings and recommendations regarding employee complaints to the CPSC and relevant congressional committees within one year of enactment of this legislation; and
- conduct a review of whether and to what extent there have been unauthorized and unlawful information disclosures by CPSC members, officers, and employees to persons regulated by the CPSC who are not authorized to receive such disclosures

[39] 154 Cong. Rec. S 1561-2 (March 5, 2008) (remarks of Sen. Klobuchar).

and submit a report of findings and recommendations concerning unauthorized disclosures to the CPSC and relevant congressional committees within one year of enactment of this legislation.

Section 220 of the House Text would require the Inspector General to

- transmit a report Congress within 60 days of enactment of this legislation on the activities of the Inspector General and any additional resources and authority needed for effective oversight, and
- conduct a review of CPSC employee complaints concerning violations of the laws enforced by the CPSC and of the process by which corrective actions are negotiated with such employees and submit a report on findings, recommendations, and actions taken regarding employee complaints to the CPSC and to Congress within one year of enactment of this legislation.

Section 27 of the Senate Text and § 220(c)) of the House Text would require the CPSC to establish and maintain a mechanism on its website by which individuals may anonymously report cases of waste, fraud or abuse with respect to the CPSC within thirty days of enactment of this legislation. The Senate Text would require this mechanism to be on the webpage of the Inspector General with a direct link from the homepage of the CPSC while the House Text would require the mechanism to be maintained on the homepage of the CPSC itself.

Reports to Congress

Section 3003 of P.L. 104-66 (the Federal Reports Elimination and Sunset Act of 1995, codified as amended at 31 U.S.C. §1113 note) provided that, with certain exceptions, reports required to be submitted to Congress, as listed in H. Doc. 103-7, were terminated.

Section 6 of the Senate Text would require that, notwithstanding any rule, regulation or order to the contrary, the CPSC must comply with the requirement of CPSA §27(k) (codified at 15 U.S.C. §2076(k)) that it must submit copies to Congress of budget recommendations, legislative recommendations and comments, and testimony that it submits to the President or the Office of Management and Budget. These copies would be expressly exempted from the reporting limitations of P.L. 104-66.

Section 203 of the House Text would also provide that the CPSC shall comply with the requirement of CPSA §27(k) regardless of any other rule to the contrary and would exempt this requirement from the reporting limitations of P.L. 104-66.

Enhanced Enforcement and Cooperation

Prohibition on Stockpiling

CPSA §9(g) (codified at 15 U.S.C. §2058(g)) authorizes the CPSC to prohibit manufacturers from stockpiling products prior to theeffective date of a consumer product safety rule; that is, from manufacturing the product at a higher-than-normal rate between the date that a rule is promulgated and the date it takes effect, in an attempt to circumvent the

[40] 154 Cong. Rec. S 1575 (March 5, 2008) (remarks of Sen. Klobuchar).

rule. However, this authority only applies to safety standards and rules promulgated under the CPSA and not to standards or rules promulgated under other statutes under CPSC jurisdiction such as the FHSA or FFA.

Both §9 of the Senate Text and §207 of the House Text would amend this provision so that it would authorize the prohibition of stockpiling of products prior to the effective date of a an applicable rule under any statute enforced by the CPSC.

Prohibited Acts

Section 15 of the Senate Text would add several prohibited acts to CPSA §19 (codified at 15 U.S.C. §2068), including

- the sale, manufacture, distribution, or importation of a product that is the subject of a voluntary corrective action and recall, a CPSC order for a recall or corrective action, a court order declaring an imminent hazard, or a CPSC designation of a banned hazardous substance;
- sale, importation or distribution of a consumer product bearing a false safety compliance certification mark;
- misrepresenting to CPSC officers/employees the scope of products subject to recall/corrective action or making a material misrepresentation in a CPSC investigation;
- failure to furnish a required compliance certification or issuance of a false compliance certification under any act enforced by the CPSC; and
- unduly influencing a third-party laboratory with respect to testing a product for compliance with safety standards.

In particular, the prohibition on the sale of products that are the subject of a mandatory or voluntary recall would close a gap in current law, which permits the continued sale of inventory that is the subject of a recall.

This provision would also amend CPSA §18 (codified at 15 U.S.C. §2067) to authorize the CPSC to ban the export of any consumer product or substance that the CPSC determines does not comply with U.S. safety standards, yet does not violate the safety standards of the importing country, or is the subject of a voluntary corrective action, a CPSC order for a recall or corrective action, a court order declaring an imminent hazard, or a CPSC designation of a banned hazardous substance. The CPSC may permit the export of a product that has been voluntarily recalled if it meets safety standards of the importing country. Violation of an export ban would be a prohibited act under CPSA §19. Currently, CPSA §18 permits the export of products that do not comply with U.S. consumer safety rules to other countries, requiring only that the CPSC must notify the appropriate agencies in the foreign countries that such products are being exported to those countries. Related conforming amendments would be made to the FHSA and FFA.

Sections 213 and 214 of the House Text respectively would make similar amendments to restrict the export of recalled and domestically non-conforming products and prohibit the sale of recalled products. With regard to exported products, the CPSC may permit a product to be exported to a particular country if that country has notified the CPSC that it accepts importation, provided that the CPSC may take appropriate action to dispose of the product if the importing country has not notified the CPSC of acceptance within thirty days of the notice of impending exportation. The House Text does not include the other additions to prohibited acts made by the Senate Text, including those concerning false certification marks or false certifications, misrepresentations to the CPSC, and undue influence of third-party testing laboratories.

Penalties

Section 16 of the Senate Text would increase the civil and criminal penalties under the CPSA, the FHSA, and the FFA. The maximum civil penalty would be increased to $250,000 for each violation and to $20,000,000 for a related series of violations; penalties exceeding $10,000,000 could only be imposed upon a finding of aggravated circumstances. Not later than December 1, 2011, and every five years thereafter, the maximum civil penalty would be adjusted for inflation. The CPSC would promulgate regulations to establish criteria to be considered in imposing civil penalties, including repeat offenses, precedential value of earlier penalties, and other factors in addition to those currently enumerated in the CPSA. In determining the amount of civil penalties and any mitigation of such penalties, the CPSC would be required to consider mitigation of undue adverse economic impacts on small businesses.

The Senate Text would increase criminal penalties to a maximum of five years, a fine pursuant to 18 U.S .C. §3571, or both, for knowing and willful violations of the CPSA and the FFA and for violations with intent to defraud or mislead or repeat offenses under the FHSA. Penalties could also include forfeiture of assets associated with the criminal violation of the CPSA or other statutes enforced by the CPSC.

Sections 215 and 216 of the House Text respectively would increase civil penalties and include asset forfeiture in criminal penalties. Maximum civil penalties would be increased in two steps. Initially, they would be increased to $5,000,000 for each violation or related series of violations under the CPSA, the FHSA, and the FFA, beginning on the date on which final CPSC penalty regulations interpreting new criteria are issued or 360 days after the enactment of this legislation, whichever is earlier. The maximum penalty would be permanently increased to $10,000,000 for a related series of violations under the CPSA, the FHSA, and the FFA, beginning one year after the date on which the temporary increase took effect. The CPSC would promulgate regulations providing its final interpretation of new factors to be considered by the CPSC in determining the amount of civil penalties. The statutory factors would be expanded to include the nature, circumstances, extent and gravity of the violation and also any other appropriate factors.

The House Text would expand criminal penalties to include forfeiture of assets associated with the offense; it would not increase the maximum imprisonment or fines.

Information Sharing with Other Government Agencies

Section 18 of the Senate Text and §219 of the House Text would authorize the CPSC to share information obtained under the CPSA with federal, state, local, or foreign government agencies, consistent with the public disclosure requirements of the CPSA, where there is a prior agreement or other written certification that such information will be maintained in confidence and used only for law enforcement or consumer protection and certain conditions apply. The CPSC may terminate such agreements if it determines that the other agency has failed to abide by the conditions of the agreement. The CPSC shall not be required to disclose information it obtained from a foreign government agency or foreign source, if such information was provided on the condition of confidentiality, or through a mechanism sponsored in part by foreign government agencies. However, nothing shall authorize the CPSC to withhold information from Congress or prevent the CPSC from complying with a federal court order in an action by the United States or the CPSC. The House Text further would provide that the CPSC must notify each state's health department of any CPSC mandatory recall or any voluntary recall of which it has been notified.

Preemption

Section 17 of the Senate Text and §218 of the House Text would clarify that the CPSC cannot and shall not seek, in issuing any rule, regulation, preamble, statement of policy, etc., to expand, contract the scope, limit, modify, interpret, or extend the application of the provisions of the CPSA, FHSA, FFA, and PPPA that concern the extent to which those acts preempt, limit, or otherwise affect any other federal, state, or local law or affect any cause of action under state or local law.

The purpose of these provisions apparently is to prevent the CPSC from making directives or statements purporting to preempt state common-law tort causes of action in the preambles to regulations that it promulgates, such as it did with the "Standard for the Flammability (Open Flame) of Mattress Sets."[41] Pursuant to Executive Order 12988 of February 5, 1996,[42] the Federal Register notice issuing the final rule for the mattress standard explained the Commission's interpretation of the Flammable Fabrics Act, citing congressional intent evidenced in the legislative history and statutory text, that the "Commission intends and expects that the new mattress flammability standard will preempt inconsistent state standards and requirements, whether in the form of positive enactments or court created requirements." Such preemption directives apparently have become more common in recent years, causing concern among opponents[43] and observations by some legal scholars that these directives have become a "backdoor" method of expanding federal objectives.[44] In the absence of express congressional, statutory direction with regard to preemption, agencies may interpret the preemptive intent through such directives and regulations.[45] Commentators have noted that the federal courts are divided on the issue of the level of deference the courts should give to such agency preemption directives.[46]

---

[41] 71 Fed. Reg. 13472, 13496-7 (2006), promulgating 16 CFR Part 1633, with a preemption directive at § N of the notice for the final rule.

[42] 61 Fed. Reg. 4729, 4732 (2006), which, *inter alia*, at § 3(b)(1)(A) requires agencies to make every reasonable effort to ensure that a regulation, as appropriate, "specifies in clear language the preemptive effect, if any, to be given to the regulation."

[43] Catherine M. Sharkey, Symposium: Is the Rule of Law Waning in America? Twelfth Annual Clifford Symposium on Tort Law and Social Policy: Article: Preemption by Preamble: Federal Agencies and the Federalization of Tort Law, 56 DePaul L. Rev. 227, 233 (2007) (citing critics of preemption preambles inside and outside the Consumer Product Safety Commission); Christine H. Kim, The Case for Preemption of Prescription Drug Failure-to-Warn Claims, 62 Food Drug L.J. 399 (2007) (noting criticism of the Bush Administration for using preemption to accomplish "silent tort reform").

[44] *Id.*

[45] Sharkey, *supra* note 43, at 245.

[46] Sharkey, *supra* note 43, at 242-7; Kim, *supra* note 43, at 419-421 (discussing cases in the context of Food and Drug Administration preambles with preemption directives); Thomas C. Galligan, Jr., *U.S. Supreme Court Tort Reform: Limiting State Power to Articulate and Develop Tort Law - Defamation, Preemption, and Punitive Damages*, 74 U. Cin. L. Rev. 1189, 1223-43 (2006) (discussing leading cases re preemption of state

Enforcement by State Attorneys General

CPSA §24 (codified at 15 U.S.C. §2073), entitled "Private Enforcement of Product Safety Rules and of Section 15 Orders," provides for a cause of action that may be brought by "[a]ny interested person (including any individual or nonprofit, business, or other entity)" in a federal district court to enforce a consumer safety rule or corrective action order by an injunction. Despite the section heading and the absence of an express reference to state attorneys general, this provision apparently has been understood to authorize enforcement actions by state attorneys general, limited to injunctive relief. The FHSA and FFA were amended in 1990 to expressly authorize state attorneys general to enforce consumer safety rules under those statutes by obtaining injunctive relief. This amendment was justified and premised on the existence of similar authority under the CPSA.[47] However, it appears that this authority has rarely, if ever been used by state attorneys general, as there apparently is no reported case precedent for such an action; it appears that state attorneys general more typically take action under state consumer protections laws, while urging the CPSC to take action under federal laws.

Section 20 of the Senate Text would clarify the CPSA by inserting a new §26A that would expressly authorize actions to obtain injunctive relief by state attorneys general with certain conditions, such as notice to the CPSC, intervention by the CPSC, suspension of actions brought by state attorneys general where an action brought by the CPSC or other federal agency was pending, etc., similar to the procedures in CPSA §24 and other similar consumer protection statutes.[48] A state attorney general who prevails in such an action could recover reasonable costs and attorney's fees; the court would have to review any attorney's fees awarded to ensure that they are consistent with CPSA §11(f) (codified at 15 U.S.C. §2060(f)) defining computation of reasonable attorney's fees. Any outside private counsel retained to assist in such state civil actions would be prohibited from sharing with parties in other private civil actions any information that arises out of the same facts and any information subject to a litigation privilege that was obtained during discovery in the state

---

tort claims); Howard L. Dorfman, Vivian M. Quinn & Elizabeth A. Brophy, *Presumption of Innocence: FDA's Authority to Regulate the Specifics of Prescription Drug Labeling and the Preemption Debate*, 61 Food Drug L.J. 585, 597-60 1 (2006) (discussing deference to agency interpretation of rules and regulations; the article generally discusses preemption issues re pharmaceuticals).

47 See H.Rept. 101-567 at 17 (1990), and H.Rept. 101-914 at 28 (1990), referring to private enforcement actions by interested parties including state attorneys general; see also *Consumer Federation of America and U.S. Public Interest Group v. Consumer Product Safety Commission*, 990 F.2d 1298, fn. 12 at 1304 (1993) (seventeen state attorneys general signed an amicus brief in which, *inter alia*, they "emphasize that they 'have the authority under [the CPSA private enforcement provision] and [their states'] consumer protection statutes to enforce [a youth] ban on behalf of [their] citizens'" (the "youth ban" refers to a rule that would ban all new adult-size ATVs for use by children under 16 years old).

48 In addition to the aforementioned FHSA and FFA provisions, codified at 15 U.S.C. § 1264(d) and 15 U.S.C. § 1194(a), respectively, see e.g., 5 U.S.C. § 15c (codified provision of the Clayton Antitrust Act); 15 U.S.C. § 5712(a) (codified provision of the Telephone Disclosure and Dispute Resolution Act of 1992, providing for regulations concerning advertising for, operation of, and billing and collection procedures for, pay-per-call or "900 number" telephone services); 15 U.S.C. § 1603 (codified provision of the Telemarketing and Consumer Fraud and Abuse Prevention Act, providing for regulations defining and prohibiting deceptive, coercive, and invasive telemarketing acts or practices); and 15 U.S.C. § 6504 (codified provision of the Children's Online Privacy Protection Act, enabling parents to control what information is collected from their children online). These provisions authorize States or state attorneys general on behalf of their States and their citizens to bring civil suits to obtain violations of the relevant federal laws, in some cases, monetary damages as well as injunctive relief.

attorney general's action, or otherwise using such information in the other private civil actions.

Section 217 of the House Text would amend CPSC §24 (codified at 15 U.S.C. §2073) by renaming the section "Additional Enforcement of Product Safety Rules and of Section 15 Orders" and adding a new subsection expressly authorizing state attorneys general to bring an action in any federal district court where the defendant is found or transacts business to obtain injunctive relief to enforce consumer product safety rules and recall or corrective action orders. Certain conditions would apply, including notice to the CPSC and the U.S. Attorney General, as well as the defendant; the CPSC right of intervention in the action; and the suspension of a state attorney general's lawsuit while there is a pending federal civil or criminal action.

Whistleblower Protections

Section 21 of the Senate Text would add a new section 40 to the CPSA which would establish a remedy for an employee of a manufacturer, private labeler, distributor, retailer, or government agency, who believes that he/she has been subjected to adverse employment actions in retaliation for providing information to the employer, Federal Government, or a state attorney general relating to a violation of any laws, rules, or orders enforced by the CPSC; for cooperating with a proceeding concerning such violation; or for objecting to or refusing to participate in any activity or policy that the employee reasonably believed would be a violation of laws, rules, or orders enforced by the CPSC. The remedy would not be available to an employee who, of his/her own volition, intentionally caused a violation of any laws, rules, or orders enforced by the CPSC.

The employee would be able to file a complaint with the Secretary of Labor, who would then notify the person named as responsible for the retaliation and give such person an opportunity to respond to the allegations. Within 60 days of the complaint filing, the Secretary of Labor would conduct an investigation to determine whether there is reasonable cause to believe the complaint has merit and issue findings in writing, with a preliminary relief order where merited. Within 30 days of notification of the findings, the defendant could object and request a hearing; otherwise, the order becomes final and not subject to judicial review. Standards for burden of proof and evidence would be established. Within 120 days of the hearing conclusion, the Secretary of Labor would have to issue a final relief order or denial. If the Secretary determines that retaliation has occurred, the Secretary shall order the person committing such retaliation to take affirmative action to abate such actions; to reinstate the complainant to his or her former position with compensation, including back pay, and other terms, conditions, and privileges of his/her employment; and to provide compensatory damages. At the request of the complainant, the Secretary could also assess a sum of the aggregate amount of all costs and fees reasonably incurred by the complainant for bringing the complaint.

If the Secretary of Labor finds that the complaint was frivolous or in bad faith, the employer may be awarded attorneys' fees to be paid by the complainant. If the Secretary has not issued a final decision within 210 days of the complaint filing or within 90 days of a written determination, the complainant could bring an action in a federal district court with jurisdiction, without regard to amount in controversy. The burdens of proof established for the administrative hearing by this legislation would apply. The court would have jurisdiction to grant all relief necessary to make the employee whole, including injunctive relief and compensatory damages (including reinstatement with the same seniority status, back pay with interest, and special damages such as reasonable attorney's fees, expert witness fees, and litigation costs).

Any person adversely affected by a final administrative order could appeal to the federal appellate court having jurisdiction in the area where the violation allegedly occurred or in which the complainant resided when the violation allegedly occurred. Such appeal would not stay the relief order unless so ordered by the court. If such direct appeal were not made, the administrative order could not be judicially reviewed in other proceedings.

The Secretary of Labor could file a civil action in a federal district court for the District of Columbia or where the violation occurred to enforce an order against a person who has failed to comply. The court could grant all appropriate relief, including, but not limited to, injunctive relief and compensatory damages. Also, a person for whom an order of relief was granted may bring an action in a federal district court to require compliance, with regard to the amount in controversy or diversity of citizenship of the parties. The court may award court costs and fees as appropriate.

Notwithstanding other provisions of the new whistleblower section, a federal employee would be limited to remedies available under chapters 12 and 23 of title 5 of the U.S. Code, governing prohibited personnel practices and employee rights of action for reprisal for federal employees, for any violation of this section.The House Text does not contain whistleblower protections, which have been criticized by CPSC Acting Chair Nord.[49]

Enhanced Inspection, Public Notice, and Recall

Public Disclosure of Information and Database

CPSA §6 (codified at 15 U.S.C. §2055) provides for certain safeguards for the public disclosure of information on products that are identified as specific products of named manufacturers. Proprietary/trade secret information may not be disclosed and information protected from disclosure by the Freedom of Information Act may not be disclosed. The manufacturer must be notified and given the opportunity to review information to be disclosed with regard to confidentiality and accuracy within a minimum period of time prior to disclosure, unless the CPSC finds that public health and safety require a lesser period of notice and publishes this finding in the Federal Register. If the CPSC disagrees with the manufacturer and decides to disclose allegedly confidential or inaccurate information over the objections of the manufacturer, the manufacturer may sue in federal district court to enjoin disclosure. Certain types of information disclosure are exempt from these safeguards, including information regarding an imminently hazardous product, a violation of the CPSA, a rulemaking proceeding, an adjudicatory proceeding, or other proceeding under the CPSA. Consumer advocates and retailer critics of this provision assert that these safeguards have unnecessarily hindered the disclosure of safety and recall information, while industry advocates seek stronger protections with regard to substantiation and disclosure of information on product categories with problems that are not common to all manufacturers.[50]

---

[49] Congress DailyPM (February 20, 2008), available at [http://nationaljournal.com/pubs/ congressdaily/].

[50] See, e.g., GAO Report GAO/HEHS-97-147, *Consumer Product Safety Commission: Better Data Needed to Help Identify and Analyze Potential Hazards* 28-32 (1997), for a discussion of the disclosure rules and policies and the views of industry and consumer advocate groups.

Section 7 of the Senate Text would reduce disclosure protections in several ways:

- • A manufacturer/private labeler must respond within 15 days after the CPSC notifies it of the opportunity to mark as confidential information that could permit the public to identify it as the manufacturer/labeler of a product. Although the CPSA currently does not specify a time within which the manufacturer/labeler must respond, CPSC regulations concerning disclosures under the Freedom of Information Act (16 C.F.R. § 1015.18) require a response in five working days to a notice concerning information previously submitted to the CPSC by the manufacturer/labeler. These regulations also require that a response must accompany information submitted after a CPSC notice of the opportunity to request confidentiality (the submission may indicate a final confidentiality response within 10 working days of the new submission).
- The provision would reduce from 30 to 15 days before disclosure the time within which the CPSC must notify a manufacturer/labeler of an intended disclosure and provide the opportunity to comment on accuracy and from 10 to five days the time within which the CPSC must notify a manufacturer/labeler that it will still disclose information claimed to be inaccurate because of a CPSC determination that the disclosure is accurate and fair.
- The CPSC would not have to publish a finding in the Federal Register that public health and safety require a lesser period of notice to the manufacturer; it could publish this in any manner.
- Exemption of certain information from the disclosure protections would be expanded to include violations of any rule or provision of any law enforced by the CPSC, not just the CPSA and information for which the CPSC publishes a finding that public health and safety requires disclosure with a lesser period of notice and comment on accuracy than normally required.
- The CPSC would be authorized to file a request for expedited consideration in a civil action to enjoin the disclosure of allegedly inaccurate information; however, such expedited consideration would not be available in an action to enjoin disclosure on the grounds that it is protected or confidential under other laws.

Section 205 of the House Text would also make three of the changes enumerated above; it would *not*, however, establish a time limit for the manufacturer/labeler to respond to an offer to mark information as confidential prior to potential disclosure, nor would it permit the CPSC to request expedited consideration in an action to enjoin disclosure of allegedly inaccurate information.

Section 7 of the Senate Text would also establish a publicly available database of reported deaths, injuries, illness, and risk of such incidents. The safeguard restrictions would not apply to the database information received by the CPSC from consumers, government agencies, health care professionals (physicians, hospitals and coroners), child service providers, public safety entities (police and firefighters), and other non-governmental sources. Reports would be available on the website database within 15 days of receipt and any information determined to be inaccurate would be promptly removed. Manufacturers, labelers, and retailers would be given the opportunity to respond to information concerning their products, and such comments may be included on the database. Names and addresses of consumers may not be disclosed.

Section 206 of the House Text would require the CPSC to assess the effectiveness of the Injury Information Clearinghouse, the existing database maintained pursuant to the CPSA §5(a) to provide information to the public on incidents involving consumer products resulting in injury, illness, or death, and to transmit to Congress, within 180 days of the enactment of the act, a detailed improvement plan for maintaining a searchable Internet database to make information more accessible and useful to consumers with due regard to personal privacy protection. Such report shall include a detailed implementation schedule, recommendations for necessary legislation, and plans for a public awareness campaign.

CPSC Acting Chair Nord has criticized the mandatory database under the Senate Text (to be established without a study concerning feasibility, effectiveness, or other concerns) because it would permit publication of information and complaints received from consumers before the CPSC had the opportunity to vet the information for validity or accuracy and also because its implementation would require 25 percent of the CPSC budget.[51] However, proponents of the database point out that the National Highway Transportation Safety Administration already has such a database including automobile complaints reported by consumers that is publicly accessible through its website.

The CPSC currently maintains the National Electronic Injury Surveillance System (NEISS),[52] which is accessible to the public through the CPSC website,[53] the "NEISS injury data are gathered from the emergency departments of 100 hospitals selected as a probability sample of all 5,300+ U.S. hospitals with emergency departments. The system's foundation rests on emergency department surveillance data, but the system also has the flexibility to gather additional data at either the surveillance or the investigation level."[54] The database mandated by the Senate Text would be more comprehensive, drawing information from a variety of sources, including consumers. The U.S. Government Accountability Office (GAO) has concluded that the current data systems, including NEISS, upon which the CPSC depends in prioritizing its regulatory and enforcement activities, are inadequate.[55]

Substantial Product Hazard Reporting Requirement

Section 15 of the CPSA currently requires manufacturers, distributors, and retailers to inform the CPSC when they learn that a product distributed by them fails to comply with a consumer safety rule/standard, poses a substantial product hazard, or creates an unreasonable risk of injury or death. The language of the provision does not include notification of noncompliance with rules or standards promulgated by the CPSC under other acts in its jurisdiction. Section 12 of the Senate Text would amend CPSA §15 to require manufacturers, distributors, and retailers of any product or substance regulated by the CPSC, except for

---

[51] Congress DailyPM (for February 20, 2008, and for March 11, 2008), available at [http://nationaljournal.com/pubs/congressdaily/].

[52] CPSA §5(a) (1) requires the maintenance of an Injury Information Clearinghouse "to collect, investigate, analyze and disseminate injury data and information relating to the causes and prevention of death, injury and illness associated with consumer products...."

[53] See [http://www.cpsc.gov/library/neiss.html].

[54] From the CPSC website description of NEISS, available at [http://www.cpsc.gov/cpscpub/ pubs/3002.html].

[55] GAO Report GAO/HEHS-97-147, Consumer Product Safety Commission: Better Data Needed to Help Identify and Analyze Potential Hazards 13-19 (1997).

motor vehicle equipment, to notify the CPSC about products that do not comply with any rule/standard promulgated by the CPSC under any act. Section 208 of the House Text would similarly amend CPSA § 15, but expressly refers only to rules/standards promulgated by the CPSC under the FHSA, FFA, and PPPA, leaving out the RSA and other free-standing legislation requiring action by the CPSC with respect to single products (although the latter usually direct the CPSC to take action for which it already has authority under the main statutes under its jurisdiction).

Section 208 of the House Text would further provide that a notification to the CPSC that a product fails to comply with a rule promulgated under the FHSA, FFA, or PPPA may not be used as the basis for criminal prosecution under the FHSA except for offenses requiring a showing of intent to defraud or mislead. The House report clarifies that this applies when such notification would constitute the *sole* basis for criminal liability without a requirement of knowledge, willfulness, or intent. The CPSA criminal offenses require knowledge and willfulness and the FFA offenses generally require willfulness, although the FFA offense of failure to notify the CPSC of the export of a non-compliant product does not. Other acts administered and enforced by the CPSC do not provide for criminal penalties.

Enhanced Public Notice of Substantial Product Hazards

Section 210 of the House Text would authorize the CPSC to order a manufacturer, distributor, or retailer of a product presenting a substantial product hazard to give public notice of such hazard via its website, notice to third-party internet sellers of the product, announcements in languages other than English, and announcements on radio and television where the CPSC determines that a substantial number of consumers may not be reached by other notice. The Senate Text made no similar provision.

Enhanced Authority for Corrective Action Plans and Recalls

Section 13 of the Senate Text would amend CPSA § 15(d) (codified at 15 U.S.C. §2064(d)) concerning corrective action plans by:

- authorizing the CPSC to order corrective actions it determines are in the public interest and removes the ability of the manufacturer, distributor or retailer to choose which corrective action it may take;
- requiring affirmative CPSC approval in writing of any corrective action plan that a person *may* be required to submit as part of a CPSC order, replacing the current passive requirement that the plan be satisfactory to the CPSC;
- authorizing the CPSC to order an amendment of a corrective action plan if it finds that the approved plan is ineffective; and
- authorizing the CPSC to revoke its approval of an action plan if it finds that a person has failed to comply substantially with its obligations under the action plan and prohibiting such person from distributing the product after receiving notice of such revocation.

Section 209(b) of the House Text would amend CPSA § 15(d) (codified at 15 U.S.C. §2064(d)) concerning corrective action plans by:

- requiring a CPSC corrective action order to include a requirement that the person to whom the order applies must submit a plan for such action as promptly as practicable to the CPSC for affirmative approval in writing;
- authorizing the CPSC to order an amendment of a corrective action plan if it finds that the approved plan is ineffective or inappropriate and requiring it to consider whether a repair or replacement changes the intended functionality of the product; and
- authorizing the CPSC to revoke its approval of an action plan if it finds that a person has failed to comply substantially with its obligations under the action plan.

This provision would also add a new subsection 15(i) to the CPSA requiring the CPSC to establish guidelines for the information to be included in any recall/corrective action notice or imminent hazard notice that would aid consumers in identifying/understanding the specific product recalled, the nature of the associated hazard, and any available remedies.

Section 209(a) of the House Text would amend CPSA §15(c and f) (codified at 15 U.S.C. §2064(c and f)) to enhance recall authority by:

- expanding the scope of the hazard notification authority to include products against which an imminent hazard action has been filed in federal court as well as products determined by the CPSC to be a substantial hazard;
- authorizing the CPSC to order a manufacturer, distributor, or retailer to cease product distribution; notify other persons involved in transporting, storing, handling, or distributing the product to cease distribution; and to notify appropriate state and local health officials;
- requiring the CPSC to rescind any order concerning an allegedly imminently hazardous product if a federal district court determines that the product is not an imminently hazardous product; and
- clarifying that the requirement for a hearing prior to the issuance of an order to cease distribution and notify the public of a hazard does not apply to an order concerning product against which the CPSC has filed an imminent hazard action.

The Senate Text does not include a similar provision.

Requirements for Recall Notice Content

Section 42 of the Senate Text would amend CPSA §15 by adding a new subsection (i) that would establish the information to be contained in a recall notice of a substantial product hazard. This information would include the model or stock keeping unit (S KU) number, common product names, photograph, description of action being taken, the number of product units for which such action is being taken, identification of the hazard, identification of the manufacturers and distributors of the product, facility where the product was made, the dates the product was manufactured and sold, details of any deaths or injuries associated with the product, remedies available to the consumer, and any other information the CPSC deems necessary. The CPSC may require that a recall notice be distributed in languages other than English. Within one year of the date of enactment of this legislation, the CPSC shall make available to the public progress reports and incident updates concerning corrective action plans under CPSA § 15(d), injury and death statistics for products determined to be a substantial product hazard, and the number and type of consumer communications to the CPSC for each product for which the CPSC ordered repair, replacement, or refund under CPSA § 15(d).

The House Text does not include a similar provision.

Identification of Supply Chain

Section 14 of the Senate Text and §212(a) of the House Text both would amend CPSA §16 (codified at 15 U.S.C. §2065), regarding inspection and recordkeeping, to require importers, retailers or distributors of a consumer product or other product or substance regulated by the CPSC to identify the manufacturer upon the request of a CPSC officer or employee. They would similarly require manufacturers to identify each retailer or distributor whom the manufacturer directly supplied with a consumer product and each subcontractor involved in the manufacture of such product or from whom the manufacturer obtained a component of such product. Section 2 12(b) of the House Text would also amend CPSA §17 (codified at 15 U.S.C. §2066) regarding imported products, by requiring the CPSC to condition importation on the manufacturer's compliance with inspection and recordkeeping requirements and to provide agencies with information necessary for such agencies to assist the CPSC in conducting product surveillance to prevent unsafe imports, consistent with disclosure restrictions of CPSA §6. Currently, the CPSC has discretion to take these actions with regard to imports.

Financial Responsibility

Section 19 of the Senate Text would add a new section 39 to the CPSA, authorizing the CPSC to promulgate regulations requiring the posting of an escrow, proof of insurance, or other security from a person who has committed multiple significant violations of any statutes or rules enforced by the CPSC or from a manufacturer or distributor of any consumer product or other product regulated by any statute enforced by the CPSC. Such security must be sufficient to cover the costs of a recall or of holding and destroying the product if such action is required by the CPSC.

Section 201(b)(3) of the House Text provides for a study including an assessment of the feasibility of mandating bonds for serious hazards and repeat offenders and Commission inspection and certification of foreign third-party and proprietary testing facilities.

Annual Reporting Requirement

Section 223 of the House Text would amend the annual reporting requirement to include information on the number and summary of recall orders issued under CPSA §§12 and 15, a summary of voluntary actions taken by manufacturers, and an assessment of such orders and actions. The Senate Text does not include such a provision.

Inspection of Certified Proprietary Laboratories

Section 211 of the House Text would amend CPSA §2065 (15 U.S.C. §2065) to authorize CPSC officers and employees to inspect certified proprietary laboratories. The Senate Text has made no similar provision.

Safety of Imported Products

Study of Authority over Imported Products

Section 43 of the Senate Text would require GAO, within 180 days of the enactment of this legislation, to conduct a study of the authorities of the CPSA to assess their effectiveness in preventing the importation of unsafe consumer products; to develop a plan to improve the effectiveness of the CPSC in preventing such importation; and to submit a report of its findings to Congress, including recommendations related to inspection of foreign manufacturing plants by the CPSC and requiring foreign manufacturers to consent to the jurisdiction of U.S. courts with respect to enforcement actions by the CPSC.

Section 224 of the House Text would require the CPSC to study the effectiveness of the current authority to ensure the safety of imported products under CPSA § 17 in order to develop a specific strategy to increase the effectiveness of the CPSC's ability to prevent importation of unsafe products. The CPSC would be required to submit a report, including recommendations regarding any additional authority and legislation needed to implement this strategy, to Congress within nine months after the enactment of this legislation.

Cooperation with U.S. Customs and Border Protection

Section 36 of the Senate Text would require the CPSC within six months of enactment of this legislation to enter into an agreement with the Secretary of Homeland Security for the assignment by the CPSC of at least one full-time equivalent employee to the National Targeting Center of CBP. Such employee(s) shall be responsible for cooperating with CBP employees in identifying products intended for importation into the United States that pose a high risk to consumer safety before such products are actually imported. The CPSC may waive this requirement if it determines that such assignment would not improve the effectiveness of the CPSC in identifying such products before importation. The House Text contains no similar provisions.

Development of Methodology to Identify Unsafe Imports

Section 37 of the Senate Text would require the CPSC, within one year of enactment of this legislation, to develop a risk assessment methodology to identify consumer product shipments that are intended for import into the United States and are likely to include products that would be refused admission because, under CPSA §17 (15 U.S.C. §2066), they do not comply with an applicable consumer product safety rule; do not comply with labeling and certification requirements relating to applicable product safety standards; are an imminently hazardous product; or have a product defect which constitutes a substantial product hazard. This methodology would be required, as far as practicable, to use the International Trade Data System (ITDS) established under the Tariff Act of 1930 to evaluate and assess information about shipments of consumer products intended for import into the

United States before importation. Funds necessary to carry out this provision are authorized to be appropriated. The House Text contains no similar provisions.

Seizure and Destruction of Unsafe Imports

Section 38 of the Senate Text would require the CPSC, within six months of enactment of this legislation, to publish a list of product defects that constitute a substantial product hazard and, as appropriate, update such list and provide a copy to the Secretary of Homeland Security.

This provision would also amend several provisions to *require* rather than *permit* the CPSC to take certain actions with regard to imports. CPSA § 17(e), permitting CBP to destroy products refused importation into the United States in lieu of exportation, would be amended to require the Secretary of Homeland Security to. *ensure* the destruction of such products unless such products are exported within ninety days of the notice of admission refusal or such additional time as may be permitted by regulations. CPSA § 17(g), permitting the CPSC to condition importation on a manufacturer's compliance with inspection and recordkeeping requirements, would also be amended to require the CPSC to condition importation on compliance. CPSA §17(h)(2), permitting the CPSC to provide information necessary for other cooperating agencies to assist with the product surveillance program, would be amended to require the CPSC to do so. A new provision would be added clarifying that nothing in CPSA §17 should be construed to prevent the Secretary of Homeland Security from prohibiting importation or ordering the destruction or exportation of a product under any other laws. The House Text contains no similar provisions.

Database of Safety Standard Violators Available to U.S. Customs and Border Protection

Section 39 of the Senate Text would require the CPSC to establish and maintain a database containing information concerning manufacturing facilities and suppliers that have violated a consumer product safety rule, the nature and date of the violation, and information from such manufacturing facility or supplier refuting responsibility for a violation or showing remedial measures undertaken to correct a violation. The CPSC would be required to take reasonable steps to provide notice to a manufacturing facility or supplier of inclusion in the database and the reasons for inclusion and to establish a process by which they may provide responses for inclusion in the database.

The database would be available on a real-time basis to the CBP for use in determining whether an imported container has consumer products that violate a consumer product safety standard and whether action should be taken with respect to products in such container.

The CPSC and CBP would be prohibited from disclosing information in the database or providing access to the database to any person other than a law enforcement agency or U.S. intelligence agency where necessary to prevent a crime or to detect, prevent, or respond to a national security threat. This disclosure restriction would not apply to information collected, received, or maintained for purposes other than inclusion in the database. The database is exempt from the disclosure requirements of the Freedom of Information Act or Privacy Act.

The database information could not be the sole basis for the imposition of civil or criminal penalties on a person by the CPSC or by the CBP. Funds necessary to carry out this provision are authorized to be appropriated. The House Text contains no similar provisions.

Ban on Importation of Certain Toys

Section 44 of the Senate Text would amend CPSA §17 to prohibit the importation of toys, as classified under specified headings of the U.S. Harmonized Tariff Schedule, manufactured by a company which has a persistent pattern of manufacturing such toys with defects constituting substantial product hazards or which has manufactured toys that present a risk of injury to the public of such magnitude that the CPSC has determined that a permanent import ban with respect to such toys of such manufacturer is justified. The CPSC would be required to submit information identifying the manufacturer to the Secretary of Homeland Security. The CPSC would be required to submit an annual report to Congress identifying toys whose importation has been banned under this provision and their manufacturers.

The House Text contained no similar provision.

Children's Product Safety

Lead Content and Measurement.

Section 22 of the Senate Text would

- treat any children's product (defined as intended for a child *7 years old or younger*) that contains lead exceeding the standard as a banned hazardous substance under the FHSA within one year of enactment of this legislation;
- reduce the lead content standard for children's products in two phases over three years, with an initial standard permitting trace amounts of *0.03% (300ppm)* of product component weight or less (or a lower amount established by the CPSC), to be reduced to *0.01% (100ppm)* of product component weight three years after the date of enactment of this legislation or, if found not feasible by the CPSC, the lowest amount technologically feasible;
- permit the CPSC to establish even lower standards as are technologically feasible;
- permit certain limited exceptions for inaccessible components and electronic devices;
- require the CPSC to promulgate guidelines concerning what will be considered inaccessible components within two years of enactment of this legislation and require interim determinations to be in accordance with the guidelines in this section (not physically exposed due to a cover or case (not including paint, coating, electroplating) and will not become physically exposed in normal, reasonably foreseeable use and abuse);
- permits the CPSC to exempt lead crystal from the children's product ban if the Commission determines, after notice and a hearing, that the lead content will not result in absorption of lead into the human body nor have an adverse impact on public health and safety;
- lower the regulatory standard for lead in paint from *0.06% (600ppm)* of paint weight to *0.009% (90ppm)* of paint weight one year after the date of enactment of this legislation; and
- provide that any CPSC rule/standard inconsistent with ASTM F963 (adopted as a mandatory rule by §29 of the Senate Text) shall supersede it to the extent of the inconsistency.

Section 101 of the House Text would:

- treat any children's product (defined as intended for a child *12 years old or younger*) that contains lead exceeding the standard as a banned hazardous substance under the FHSA within 180 days of enactment of this legislation;
- decrease the permissible trace amount of lead content from the current level in three phases over four years, from *600ppm* by product part weight initially, to *300ppm* by product part weight two years after enactment, to *100ppm* by product part weight four years after enactment or, if found not feasible by the CPSC, the lowest amount feasible;
- permit the CPSC to establish even lower standards for any children's products as are technologically feasible and more protective;
- require a periodic review by the CPSC to consider strengthening the standard, if feasible;
- permit the CPSC to exempt certain products if it determines that the lead content in such products will not result in lead absorption by humans or an adverse effect on public health or safety;
- permit limited exceptions for inaccessible components and special requirements for electronic devices;
- lower the regulatory standard for lead in paint from *0.06% (600ppm)* of paint weight to *0.009% (90ppm)* of paint weight 180 days after the date of enactment of this legislation and reduce the paint standard for all children's products to $0.009mg/cm^2$, requiring the CPSC to further lower the latter standard if feasible within three years of the enactment of this legislation; and
- permit the CPSC to extend the effective dates for such standards for up to 180 days if necessary for implementation given specific factors and if such extension does not affect public health and safety.

The alternative measurement (units-of-mass-per-area) used for the paint standard for children's products in the House Text is intended to facilitate certain testing methods that can be conducted outside a laboratory such as at inspection sites at ports of entry or in stores, according to H.Rept. 110-501 at 30-31 (2007). In comparison, §23 of the Senate Text would require the CPSC to study the feasibility of establishing an alternative measurement standard for lead content (units of mass per area) similar to the standard used by the Department of Housing and Urban Development and the Environmental Protection Agency that is statistically comparable to the current parts-per-million standard.

Third Party Certification of Children's Products.

Section 10 of the Senate Text would:

- require compliance testing and certification of a children's product by an independent third-party laboratory beginning 60 days after the CPSC establishes an interim procedure;

- define children's products as consumer products designed or intended for use by a child *7 years old or younger* and establish criteria for determining whether a product is a children's product;
- require the CPSC, within one year of enactment of this legislation, to establish procedures and standards for product certification, accreditation and audits of third-party laboratories, and the labeling of certified children's products;
- permit the CPSC to extend the certification requirements for children's products to other products;
- establish interim certification procedures for children's products within thirty days of enactment of this legislation;
- permit the CPSC to certify, upon request, manufacturer/labelerowned laboratories that are protected from undue influence by the owner and also for provisional certification;
- establish interim guidelines for provisional certification of laboratories;
- provide for decertification of laboratories upon a finding of undue influence by a manufacturer or private labeler; and
- prohibit imports of children's products without certification by a third-party laboratory.

Section 102 of the House Text would:

- expand the requirement for product compliance certification CPSA §14 (codified at 15 U.S.C. §2063) to include products subject to standards under other acts enforced by the CPSC;
- require compliance testing and certification of a children's product by an independent third-party laboratory or a proprietary laboratory within one year after enactment of this legislation;
- define children's products as consumer products designed or intended primarily for children *12 years old or younger* and establish criteria for determining whether a product is a children's product;
- establish guidelines for the certification of proprietary laboratories ensuring that a laboratory is protected from undue influence by the manufacturer-owner; and
- provide for decertification if undue influence is found.

Tracking Labels for Children's Products

Section 11(b) of the Senate Text would require, effective one year after the date of enactment of this legislation, the manufacturer to place tracking labels, to the extent practicable, on children's products and, at the discretion of the CPSC, other consumer products and their packaging that would enable the ultimate purchaser to ascertain the manufacturer, production date, and production batch/run of the product. Section 103 of the House Text would similarly require such tracking labels, to the extent feasible, enabling identification of the location and date of production and any other information facilitating the specific source of the product.

Registration Forms and Standards for Durable Nursery Products

Section 34 of the Senate Text, the Danny Keysar Child Product Safety Notification Act, would require the CPSC to consult with consumer groups, child product manufacturers, and independent child product engineers to assess the effectiveness of voluntary consumer product safety standards for durable infant or toddler products (as defined in this section) and to promulgate safety rules that are substantially the same as or more stringent than such voluntary standards. The CPSC would be required to begin such rulemaking not later than one year after the enactment of this legislation and issue rules for at least two categories of durable nursery products every six months, in order of priority, until it has promulgated standards for all such product categories. It would be required to periodically review such rules to ensure the highest level of safety feasible.

The CPSC, not later than one year after the enactment of this legislation, would be required to promulgate regulations requiring manufacturers of durable infant or toddler products to provide consumers with postage-paid registration forms attached to the surface of such a product; to maintain, for at least six years after the product manufacturing date, a record of contact information for consumers registering with the manufacturer to improve recall effectiveness; and to place permanently the manufacturer information, model name and number, and production date on such products. Durable infant or toddler product would be defined as intended for use by children younger than *five years old*, including specified items such as furniture, carriers, strollers, gates, play centers, bath seats, etc. The provision would establish certain required elements for the registration forms, including a privacy statement that the information provided by the consumer will not be used for any purpose other than a recall or safety alert and an option to register via the Internet. Manufacturers would be prohibited from using or disseminating information collected via registration forms for any purposes other than a recall or safety alert and would be required to provide registration via the Internet. Nothing in the provision would require the manufacturer to collect, retain, or use any information unless provided by the consumer. The CPSC would be required to conduct a study and submit a report to Congress on the effectiveness of the registration form rules in facilitating recalls within four years of the enactment of this legislation.

If the CPSC determines that a recall notification technology can be used by a manufacturer of durable infant or toddler products that is as effective or more effective than registration forms in facilitating recalls, the Senate Text would require the CPSC to submit a report about this determination to the relevant congressional committees. A manufacturer who uses such technology in lieu of registration forms would be considered in compliance with certain regulatory requirements for registration forms. Within one year of the date of enactment of this legislation, the CPSC would be required to conduct a review of recall notification technology, assess the effectiveness of such technology in facilitating recalls of durable infant or toddler products, and submit a report to the relevant congressional committees. The CPSC would be required to promulgate regulations to carry out these provisions.

The Senate Text would also prohibit any commercial user, as defined in this provision, from manufacturing, selling, reselling, leasing, or providing for use any cribs that do not comply with the mandatory rule promulgated by the CPSC under this section. Commercial users would be persons who manufacture or sell cribs, sell or lease used/second-hand cribs, deal in used/second-hand cribs or otherwise hold themselves out to have knowledge or skill concerning cribs (such as child care facilities or homes). Such users would include, but not be limited to, hotels, motels, and day care centers.

Section 104 of the House Text, the Danny Keysar Child Product Safety Notification Act, would provide for requirements for the assessment and promulgation of consumer product safety standards for durable infant or toddler products (as defined in this section) that are substantially similar to those in the Senate Text. It would also require a similar timetable and periodic review of such consumer product safety rules. Additionally, it would provide substantially similar requirements for registration forms for durable infant or toddler products. However, the House Text does not include a provision concerning the assessment and use of recall notification technology. It also does not include a ban on the manufacture, sale, resale, lease, or provision for use of noncompliant cribs.

Labeling for Certain Toy and Game Advertising

Section 11(a) of the Senate Text would amend the FHSA required choking hazard labelling for certain toys and games to require similar cautionary statements in advertising that provides a direct means of purchase, including advertising on Internet websites, in catalogues, or other advertising materials. The distribution of a noncompliant advertisement would be a prohibited act under the CPSA.

Section 105 of the House Text would similarly require, effective 180 days after enactment of this legislation, that any advertisement that provides a direct means for purchasing a toy or game for which a choking hazard label is required, including advertising on the Internet, catalogues, or other materials, must include a similar cautionary statement on or immediately adjacent to such advertisement. The statement would have to be in the language primarily used in the advertisement, website, or catalogue in a clear and conspicuous manner. The distribution of a noncompliant advertisement would be a prohibited act under the CPSA. The CPSC would be required to promulgate regulations concerning size and placement of such cautionary statements within 180 of enactment of this legislation.

Study of Disparities in Injury/Death Rates of Minority Children

Section 24 of the Senate Text would require GAO, within 90 days of the date of enactment of this legislation, to initiate a study assessing racial/ethnic disparities in the risks and incidence of preventable injuries and deaths linked to the use of cribs, mattresses, swimming pools, toys, and other products intended for use by children. Minority populations in the study would include Black, Hispanic, American Indian, Alaskan Native, Native Hawaiian, and Asian/Pacific Islander children in the United States. GAO would be required to report its findings to the relevant congressional committees not later than one year after the date of enactment of this legislation, including recommendations for minimizing disparities and increased risks among minority children, for awareness and prevention campaigns targeting minority populations, and for educations initiatives to reduce current statistical disparities. The Senate text would authorize no additional appropriations for the study and report.

Section 106 of the House Text contains similar requirements for a report on the racial disparities in risks and incidence of preventable injuries and deaths among minority children. However, it would not include Native Hawaiians as a separate group (as opposed to being included in Asian/Pacific Islanders). Also, it would require GAO to consult with the CPSC as necessary. The House text also would not authorize any additional appropriations for the study and report.

Toy Safety Standard

Section 29 of the Senate Text would provide that the current version of ASTM International Standard F963-07, Consumer Safety Specification for Toy Safety, shall automatically be deemed a mandatory consumer product safety standard issued by the CPSC. ASTM International, the standard setting organization (with offices in Pennsylvania), would be required to notify the CPSC if it proposes revisions to this standard more than 60 days after enactment of this legislation. The proposed revision would be incorporated into the CPSC rule and the revised standard would then be considered a CPSC rule effective 30 days after the notification was received from ASTM International, unless the CPSC notifies ASTM International within 60 days of receiving such notification that it has determined that the proposed revision does not improve the safety of the products covered by the standard. The existing standard, without the proposed revision, would continue to be the CPSC rule.

Section 107 of the House Text would not automatically adopt the ASTM International toy standard; instead it would require the CPSC to study and assess the effectiveness of this standard. The CPSC would be required to report its findings to Congress within two years of the enactment of this legislation including the feasibility of requiring testing of all toys to such standards and whether promulgating mandatory rules similar or more stringent than the voluntary standard would benefit public health and safety. The study would include an assessment of the effectiveness of the voluntary standard with respect to intestinal blockage and perforation hazards posed by the ingestion of magnets. If the CPSC finds that there is substantial noncompliance causing an unreasonable risk of injury to children, the CPSC shall expedite rulemaking to adopt such voluntary standard as a mandatory standard.

Other Provisions Concerning Specific Product Standards

In General

The Senate Text contains several provisions concerning consumer safety product standards or studies concerning specific consumer products that are not included in the House Text, including

- child resistant portable gasoline containers (§28);
- all-terrain vehicles (ATVs) (§30);
- garage door openers (§31);
- carbon monoxide poisoning (§32);
- cigarette lighters (§33);
- ban on specified phthalates (chemical plasticizers used in toys and other children's products) and certain alternatives in certain children's products and child care articles
- equestrian helmets (§41); and
- study of the use of formaldehyde in textile and apparel (§45).

Although the House Text does not include such provisions because its sole focus is reform of the CPSC and its authority and enforcement powers, language is included in

H.Rept. 110-501 (2007) directing the CPSC to take action concerning single-product issues, including

- the promulgation by the CPSC of a final rule on ATV safety standards and bans with regard to children of different age groups;
- the promulgation by the CPSC of a final rule on cigarette lighters;
- a public awareness campaign concerning smoke alarms and smoke detection systems;
- a consideration of a safety standard concerning the warning labels for the lead content of ceramic food containers or serving ware;
- a consideration of CPSC authority to regulate pet toys that could be used by children and possible rules concerning the lead content and use of lead paint in such pet toys;
- a consideration of the tipping hazards of home appliances and furniture and possible safety standards; and
- a study of injuries and deaths related to toy guns and possible rules for marking of toys guns distinguishing them from actual firearms.[56]

The House Committee on Energy and Commerce noted in its report that it became aware of the potential dangers posed by phthalates and asbestos in toys late in the legislative process and would take up these issues in subsequent hearings and legislation.

Automatic Residential Garage Door Opener

The standard for automatic residential garage door openers/operators in the Senate Text apparently was the only single-product provision that became a source of contention on the floor regarding what standard to use for safety technology.[57] It resulted in a floor amendment to §31 of S. 2663, whose text was then substituted into H.R. 4040. S. 2663 as introduced would have required an external secondary entrapment protection device that uses "non-contact" technology, which would mean that the use of technology by which closing doors reverse course upon contact with a person or object would not satisfy the safety rule. The garage door operator itself is required to reverse course, but it is also required to have a back-up or secondary mechanism that functions independently of the door operator and with either non-contact (photosensor) or contact (door-edge touch sensor) technology to ensure that the door reverses course when someone or something is in the path of the closing door. Opponents of §31 in S. 2663 as introduced alleged that it favored one particular type of secondary entrapment protection technology used by one manufacturer to the exclusion of other equally safe secondary entrapment protection technology. Proponents asserted that since the primary entrapment protection device in the operator functioned by direct contact with someone or something in the path of the door, which did not always work effectively, it would be preferable to require a secondary protection device that did not require contact with an object providing 15 pounds of resistance or less.

---

[56] H.R. 5471, to require the Consumer Product Safety Commission to prescribe rules requiring distinctive markings on toy and look-alike firearms (cosponsored by Towns and Blackburn).

[57] 154 Cong. Rec. S1581-1584 (March 5, 2008) (remarks of Sen. DeMint, Pryor, and Durbin); Jim Snyder, "Safety Bill Gets Trapped under a Garage Door," *The Hill* (February 25, 2008).

The Senate adopted S.Amdt. 4141, a compromise that would exempt the manufacture of automatic residential garage door openers *without* a secondary, non- contact, entrapment protection device by a company that had produced such an opener before the date of enactment of this legislation if the Underwriters Laboratory, Inc. certified that the opener meets its Standards for Safety-UL 325 before the date of enactment of this legislation. It would also require the CPSC to review and, if necessary, revise its garage door opener standard, including the requirement for a secondary, non-contact, entrapment protection device, within one year of the date of enactment of this legislation, to ensure that the standard provides maximum protection for public health and safety. If such revised standard adopts the requirement for a secondary, non-contact, entrapment protection device, the exemption described above would not apply to garage door openers manufactured after the effective date of the revised standard.

In: Consumer Product Safety Issues
Editors: P. A. Varga, M. D. Pinter

ISBN: 978-1-60456-826-4

Chapter 10

PHTHALATES IN PLASTICS AND POSSIBLE HUMAN HEALTH EFFECTS*

Linda-Jo Schierow and Margaret Mikyung Lee

SUMMARY

Roughly a dozen chemicals known as phthalates are used to make the plastics found in thousands of consumer products, ranging from medical tubing to automotive dashboards to bath toys. These phthalates are not tightly held by the plastics and are released into the environment over time. Congress is concerned about possible human health effects from exposure to six of these chemicals: di-(2-ethylhexyl) phthalate (DEHP), dibutyl phthalate (DBP), benzyl butyl phthalate (BBP), diisononyl phthalate (DINP), diisodecyl phthalate (DIDP), and di-n-octyl phthalate (DnOP). DEHP, DBP, BBP, and (to less extent) DINP are known to be toxic to the reproductive systems of rodents. Recent experiments demonstrate that pre-natal exposure at a sufficient level to these same phthalates disrupts the normal action of hormones and can cause malformations of the reproductive organs of offspring (especially males).

Disruption of hormonal functions in humans is known to result in abnormal reproductive development. Many scientists believe that the phthalates toxic to rodents might be able to cause similar malformations in humans. However, human health effects of phthalate exposure have not been conclusively demonstrated. Very few studies have looked at possible effects in humans, but their results have been consistent with the results of rodent experiments. More research would be needed to test this hypothesis. Recent surveys have found almost universal exposure to phthalates. Individuals may be exposed to high enough levels of phthalates to cause reproductive abnormalities. Scientists at the National Toxicology Program have expressed "serious concern" about human male infants undergoing intensive medical procedures, and "concern" about development of human males less than a year old who are exposed to DEHP. In light of these concerns, the National Academy of Sciences is evaluating the risk of aggregate human exposure to multiple phthalates.

* This is an edited, reformatted and augmented version of CRS Report RL34572, dated July 29, 2008

Federal agencies have taken several actions, some as early as the mid 1980s, to evaluate and regulate phthalates, but no product to date has been banned outright. The agency responsible for regulating toys and most other child-care products is the Consumer Product Safety Commission (CPSC). In March 2008, the Senate approved an amendment to H.R. 4040, the Consumer Product Safety Commission Reform Act, that would restrict the use of six phthalates in toys and child-care products. The House version had no phthalate amendment. On July 29, 2008, the conferees announced approval of an amended version of the Senate provision.

The scientific basis for concerns about human health risks appears to be strong in the case of some phthalates (such as DEHP), adequate with respect to others (perhaps DINP), and weak for the remaining chemicals (for example, DIDP and DnOP). The strongest evidence with respect to developmental effects has been produced since about the year 2000. The Senate amendment would codify the voluntary agreements reached by CPSC with product manufacturers and reduce exposure to one particular phthalate. New formulations for toys and child-care products may pose greater or fewer risks than current formulations.

INTRODUCTION

"Phthalates"[1] refers to a group of chemical compounds that are heavily produced and widely used to make the plastics found in thousands of consumer products. The most common use of phthalates is to increase the flexibility of polyvinyl chloride (PVC) and polyvinylidene chloride (PVDC) polymers. Phthalates are released from those products over time, and are dispersed to the air, water, soil, and living things. Some (but not all) of these phthalates are known to cause reproductive damage in rodents. Recent interest by governmental bodies, including Congress, in the potential adverse human health effects that might be related to phthalate exposure has focused on six phthalates that are produced and used in very large quantities.[2] The six phthalates are di-(2-ethylhexyl) phthalate (DEHP), dibutyl phthalate (DBP), benzyl butyl phthalate (BBP), diisononyl phthalate (DINP), diisodecyl phthalate (DIDP), and di-n-octyl phthalate (DnOP). H.R. 4040, the Consumer Product Safety Commission Reform Act, as amended and approved by conferees, would prohibit the sale of children's toys and child care articles that contain more than 0.1% of DEHP, DBP, or BBP. A similar ban would apply for DINP, DIDP, and DnOP until the Consumer Product Safety Commission issues a rule either establishing or eliminating the ban on a permanent basis.

[1] The ph is silent.

[2] A high-production-volume chemical is defined by the U.S. Environmental Protection Agency as a chemical produced in amounts exceeding 1 million pounds annually. Hundreds of millions of pounds of phthalates are produced annually in the United States.

BACKGROUND

Health Effects

Compared to some other chemicals in commerce, phthalates are not extremely toxic. That is, they do not cause acute illness after a short period of low-level exposure. However, controlled experiments with rodents have demonstrated that some phthalates[3] at high doses damage reproduction and development.[4] Moreover, if administered at sufficient levels[5] and at the appropriate time to pregnant females, some phthalates can cause malformations of the reproductive organs of offspring, especially males.[6] In rats, exposure during gestation to some phthalates can cause testicular cancer in mature offspring.[7] The higher the phthalate exposure, the more frequent and severe are the effects on the reproductive system. Rat fetuses are most susceptible, but older rats can also be affected at somewhat higher levels of exposure.[8]

Disruption of hormonal functions in humans is known to result in abnormal reproductive development. Many scientists believe that the phthalates toxic to rats and mice might be able to cause similar malformations in humans, because the male hormones affected by phthalates are important to the normal development of the male reproductive tract in *all* species of mammals. However, human health effects of phthalate exposure have not been conclusively demonstrated. Very few studies have looked at possible effects in humans,[9] but their results have been consistent with the results of rodent experiments. A study published in 2005 provided the first evidence of subtle developmental effects, similar to those seen in animal studies, in human male infants exposed prenatally to breakdown products of phthalates.[10] More research would be needed to determine with certainty the effects of phthalates in

3 Di-(2-ethylhexyl) phthalate, dibutyl phthalate, and benzyl butyl phthalate.

4 A high dose is relative and varies depending on the chemical. For DEHP, for example, a high dose might be considered to be roughly more than 100 milligrams of phthalate (mg) per kilogram of the exposed animal's body weight (kg) per day.

5 For example, 14-23 milligrams of DEHP per kilogram of body weight per day.

6 Gray, Jr., L. Earl, Joseph Ostby, Johnathan Furr, et al. 2000. Perinatal exposure to the phthalates DEHP, BBP, and DINP, but not DEP, DMP, or DOTP, alters sexual differentiation of the male rat. Toxicological Sciences, v. 58, n. 2, p. 350-365.

Mylchreest, E., R.C. Cattley, and P. M. Foster. 1998. Male reproductive tract malformations in rats following gestational and lactational exposure to Di(n-butyl) phthalate: an antiandrogenic mechanism? Toxicological Sciences, v. 43, n. 1, p. 47-60.

Parks, Louise G., Joe S. Ostby, Christy R. Lambright, et al. 2000. The plasticizer diethylhexyl phthalate induces malformations by decreasing fetal testosterone synthesis during sexual differentiation in the male rat. Toxicological Sciences, v. 58, p. 339-349.

Wilson, Vickie S., Christy Lambright, Johnathan Furr, et al. 2004. Phthalate ester-induced gubernacular lesions are associated with reduced insl3 gene expression in the fetal rat testis. Toxicology Letters, v. 146, p. 207-215.

7 Fisher, Jane S., S. Macpherson, N. Marchetti, et al. 2003. Human 'testicular dysgenesis syndrome': a possible model using inutero exposure of the rat to dibutyl phthalate. Human Reproduction, v. 18, n. 7, p. 1383-1394.

8 Ibid.

U.S. Congress. House of Representatives. Committee on Energy and Commerce, Subcommittee on Commerce, trade, and consumer Protection. Hearing on Safety of Phthalates and Bisphenol-A in Everyday Consumer Products. June 10, 2008. Written testimony of Leon Earl Gray, Jr., Senior Reproductive Biologist and Toxicologist, U.S. EPA.

9 Main, Katharina M., Gerda K. Mortensen, Marko M. Kaleva, et al. 2006. Human breast milk contamination with phthalates and alterations of endogenous reproductive hormones in infants three months of age. Environmental Health Perspectives, v. 114, p. 270-276.

Swan, Shanna H., Katharina M. Main, Fan Liu, et al. 2005. Decrease in anogenital distance among male infants with prenatal phthalate exposure. Environmental Health Perspectives, v. 113, n. 8, p. 1056-1061.

10 Swan et al. 2005.

humans. Additional information about health effects is provided below in the section "The Six Phthalates."

Human Exposure

Results of the National Health and Nutrition Examination Survey (NHANES) indicated almost universal American exposure to low levels of the most common phthalates, usually multiple phthalates.[11] Women tend to have greater exposure than men, but children appear to be the group most exposed to DEHP, DBP, and BBP. Children also are more exposed to DnOP, but these exposures appear to be lower than those for DEHP, di-n-butyl phthalate, and BBP. Levels of a breakdown product of DINP were not detectable in children surveyed (but see section below on DINP) and DIDP was not measured in 2001-2002.[12] Studies of amniotic fluid have also documented exposure to multiple phthalates for human fetuses.[13] More generally, babies may be the most heavily exposed group. [14]

Phthalates are dispersed throughout the air, water, soil, and living things in the developed world. According to the Department of Health and Human Services (DHHS), food probably is the major source of exposure to some phthalates for the general population.[15,16] However, air also appears to be important.[17] For the human fetus, maternal exposure leads to prenatal exposure through the placenta. Personal care items, including baby lotion and powder, may be significant sources of exposure for infants.[18] For some individuals, certain medications also may be important sources.[19] Finally, medical devices may dominate exposure sources for critically ill patients. Individuals, such as newborns in intensive care, may be exposed to levels of DEHP much closer to, but still less than 1% of, levels that cause reproductive harm in rats.[20]

---

[11] DHHS. 2005. Third National Report on Human Exposure to Environmental Chemicals. p. 253. [http://www.cdc.gov/exposurereport/pdf/thirdreport.pdf].

[12] Ibid., p. 282.

[13] Silva, M.J., J.A. Reidy, A.R. Herbert, et al. 2004. Detection of phthalate metabolites in human amniotic fluid. *Bulletin of Environmental Contamination and Toxicology,* v. 72, p. 1226-123 1.

[14] Wormuth, Matthias, Martin Scheringer, Meret Vollenweider, et al. 2006. What are the sources of exposure to eight frequently used phthalic acid esters in Europeans? *Risk Analysis*, v. 26, n. 3, p. 803-824.

[15] DHHS, ibid., p. 253.

[16] Fromme, H., L. Gruber, M. Schlummer, et al. 2007. Intake of phthalates and di-(2- ethhylhexyl) adipate: results of the integrated exposure assessment survey based on duplicate diet samples and biomonitoring data. *Environment International*, v. 33, n. 8, p. 1012-1020.

[17] Adibi, Jennifer J., Robin M. Whyatt, Paige L. Williams, et al. 2008. Characterization of phthalate exposure among pregnant women assessed by repeat air and urine samples. *Environmental Health Perspectives*, v. 116, n. 4, p. 467-473.

[18] Sathyanarayana, S. C.J. Karr, P. Lozano, et al. 2008. Baby care products: possible sources of infant phthalate exposure. *Pediatrics*, v. 121, n. 2, p. e260-268.

[19] Hauser, Russ, Susan Duty, Linda Godfrey-Bailey, et al. 2004. Medications as a source of human exposure to phthalates. *Environmental Health Perspectives*, v. 112, n. 6, p. 75 1- 753.

[20] Weuve, Jennifer, Brisa N. Sanchez, Antonia M. Calafat, et al. 2006. Exposure to phthalates in neonatal intensive care unit infants: urinary concentrations of monoesters and oxidative metabolites. *Environmental Health perspectives*, v. 114, n. 9., p. 1424-1431.

Calafat, Antonia M., Larry L. Needham, Manori J. Silva, et al. 2004. Exposure to di-(2- ethylhexyl) phthalate among premature neonates in a neonatal intensive care unit. *Pediatrics*, v. 113, n. 5., p. e429-e434.

Phthalates do not bioaccumulate in the body or the environment; rather they break down rapidly. However, exposure to phthalates is continuous and substantial in the modern world.[21]

The Six Phthalates

The Senate bill would require regulation of six phthalates, following the example of the European Union and the state of California. They differ from one another in structure, uses, and toxicities. The extent to which they have been studied varies widely. Compared to the other commercially produced phthalates, these six, arguably, are more studied, more toxic, or more prevalent in consumer products and the environment. All six have been evaluated by the National Toxicology Program (NTP), an interagency program administered through the National Institute of Environmental Health Sciences/National Institutes of Health (NIEHS/NIH).[22]

Di-(2-ethylhexyl) Phthalate (DEHP)

DEHP is the most abundantly produced and the most studied phthalate.[23] It is used primarily to improve the flexibility of "vinyl" (that is, polyvinyl chloride (PVC) plastic). DEHP is found in medical devices, such as plastic tubing used for catheters and intravenous drug and fluid delivery, and many home and garden products. The NTP has expressed "serious concern[24] that certain intensive medical treatments of male infants may result in DEHP exposure levels that adversely affect development of the male reproductive tract."[25] In addition, the NTP expressed "concern for effects of DEHP exposure on development of the male reproductive tract for infants less than one year old," "some concern for effects of DEHP exposure on development of the reproductive tract of male children older than one

---

[21] Silva, Manori J., E. Samandar, J.L. Preau, Jr., et al. 2007. Quantification of 22 phthalate metabolites in human urine. *Journal of Chromatography. B, Analytical Technologies in the Biomedical and Life Sciences*, v. 860, n. 1, p. 106-1 12.

[22] Center for the Evaluation of Risks to Human Reproduction (CERHR), National Toxicology Program, DHHS. "CERHR Chemicals." [http://cerhr.niehs.nih.gov/chemicals/ index.html].

23 For example, some important scientific studies include Akingbemi, Benson T., Renshan Ge, Gary R. Klinefelter, et al., 2004, Phthalate-induced Leydig cell hyperplasia is associated with multiple endocrine disturbances, *Proceedings of the National Academy of Sciences*, v. 101, n. 3, p. 775-780; Gray, Jr., L. Earl, Joseph Ostby, Johnathan Furr, et al., 2000, Perinatal exposure to the phthalates DEHP, BBP, and DINP, but not DEP, DMP, or DOTP, alters sexual differentiation of the male rat, *Toxicological Sciences*, v. 58, n. 2, p. 350-365; Lin, H., Renshan S. Ge, Gary R. Chen, et al., 2008, Involvement of testicular growth factors in fetal Leydig cell aggregation after exposure to phthalate in utero, *Proceedings of the National Academy of Sciences of the United States of America,* v. 105, n. 290, p.72 18-7222; Parks, Louise G., Joe S. Ostby, Christy R. Lambright, et al., 2000, The plasticizer diethylhexyl phthalate induces malformations by decreasing fetal testosterone synthesis during sexual differentiation in the male rat, Toxicological Sciences, v. 58, p. 339- 349; Swan, Shanna H., Katharina M. Main, Fan Liu, et al., 2005, Decrease in anogenital distance among male infants with prenatal phthalate exposure, *Environmental Health Perspectives*, v. 113, n. 8, p. 1056-1061; and Wilson, Vickie S., Christy Lambright, Johnathan Furr, et al., 2004, Phthalate ester-induced gubernacular lesions are associated with reduced ins 13 gene expression in the fetal rat testis, *Toxicology Letters*, v. 146, p. 207- 215.

[24] The expression of "serious concern" is the greatest level of concern on a qualitative scale used by NTP. In order, from greatest to least, the levels of concern are: serious concern, concern, some concern, minimal concern, and negligible concern.

[25] Center for the Evaluation of Risks to Human Reproduction, National Toxicology Program, U.S. Department of Health and Human Services. *NTP-CERHR Monograph on the Potential Human Reproductive and Developmental Effects of Di(2-Ethylhexyl) Phthalate (DEHP).* NIH Publication No. 06-4476. November 2006. p. vii-viii. [http://cerhr.niehs.nih.gov/ chemicals/dehp/DEHP-Monograph.pdf].

year," and also some concern for developmental effects for the offspring of pregnant women.[26] After this DEHP monograph was issued, several studies began to explore associations between DEHP and other health effects, such as effects on thyroid hormone levels, asthma, and obesity.[27] Additional research is warranted in these areas and it is too soon to draw any conclusions regarding the potential role of DEHP in causing such problems.

Dibutyl phthalates (DBP)

There are two DBPs: di-n-butyl and di-isobutyl phthalate. The former is more studied, especially in Europe. DBPs are used in latex adhesives, nail polish, cosmetics, some inks and dyes, insecticides, and pharmaceutical coatings. The most recent NTP monograph on DBP (di-n-butyl phthalate) found "clear evidence of adverse effects" on the developing male reproductive tract in rodents.[28] Furthermore, the NTP concluded, "Based on recent data ... the NTP believes it is reasonable and prudent to conclude that the results reported in laboratory animals indicate a potential for similar or other adverse effects in humans."[29] As a result, NTP has "some concern for DBP causing adverse effects to human development, particularly development of the male reproductive system."[30] In the seven years since this NTP monograph, numerous studies have bolstered these findings.[31] Of particular note is the study by Lehmann et al. (2004).[32] It established the relationship between exposure to DBP and

---

[26] Ibid.

[27] For example, see Meeker, John D., Antonia M. Calafat, and Russ Hauser, 2007, Di(2- ethylhexyl) phthalate metabolites may alter thyroid hormone levels in men, *Environmental Health Perspectives*, v. 115, n. 7, p. 1029-1034; Stahlhut, R.W., E. van Wijngaarden, T.D. Dye, et al., 2007, Concentrations of urinary phthalate metabolites are associated with increased waist circumference and insulin resistance in adult U.S. males, *Environmental Health Perspectives*, v. 115, n. 6, p. 876-882; Jaakkola, Jouni J.K., and Trudy L. Knight, 2008, The role of exposure to phthalates from polyvinyl chloride products in the development of asthma and allergies: a systematic review and meta-analysis, *Environmental Health Perspectives*, v. 116, n. 7, p. 845-853.

[28] Ibid. p.2.

[29] Center for the Evaluation of Risks to Human Reproduction, National Toxicology Program, United States . Department of Health and Human Services. *NTP-CERHR Monograph on the Potential Human Reproductive and Developmental Effects of Di-n-Butyl Phthalate (DBP).* 2001. p. 4. [http://cerhr.niehs.nih.gov/chemicals/-phthalates/dbp/DBP_Monograph_Final.pdf].

[30] Ibid. p. 2.

[31] For example, see Barlow, N.J., B.S. McIntyre, and P.M. Foster, 2004, Male reproductive tract lesions at 6, 12, and 18 months of age following in utero exposure to di(n-butyl) phthalate, *Toxicologic Pathology*, v. 32, n. 1, p. 79-90; Fisher, Jane S., S. Macpherson, N. Marchetti, et al., 2003, Human 'testicular dysgenesis syndrome': a possible model using in-utero exposure of the rat to dibutyl phthalate, *Human Reproduction*, v. 18, n. 7, p. 1383- 1394; Mahood, I. Kim, Nina Hallmark, Chris McKinnell, et al., 2005, Abnormal Leydig cell aggregation in the fetal testis of rats exposed to di(n-butyl) phthalate and its possible role in testicular dysgenesis, *Endocrinology*, v. 146, p. 6 13-623; Thompson, Christopher J., Susan M. Ross, and Kevin W. Gaido, 2004, Di(n-butyl) phthalate impairs cholesterol transport and steroidogenesis in the fetal rat testis through a rapid and reversible mechanism, *Endocrinology*, v. 145, p. 1227-1237; Wilson, Vickie S., Christy Lambright, Johnathan Furr, et al., 2004, Phthalate ester-induced gubernacular lesions are associated with reduced ins 13 gene expression in the fetal rat testis, *Toxicology Letters*, v. 146, p. 207-2 15; and Zhang,Y., X. Jiang, and B. Chen, 2004, Reproductive and developmental toxicity in F1 Sprague-Dawley male rats exposed to di-n-butyl phthalate in utero and during lactation and determination of its NOAEL, *Reproductive Toxicology*, v. 18, n. 5, p. 669-676.

[32] Lehmann, Kim P., Suzanne Phillips, Madhabananda Sar, et al. 2004. Dose-dependent alterations in gene expression and testosterone synthesis in the fetal testes of male rats exposed to di-(n-butyl) phthalate. *Toxicological Sciences*, v. 81, p. 60-68.

effects on synthesis of testosterone in fetal male rats. A 2006 study found that di-isobutyl phthalate had testicular and developmental effects similar to di-n-butyl phthalate and DEHP.[33]

Benzyl Butyl Phthalate (BBP or sometimes BzBP)

BBP is used in vinyl flooring, automotive trim, food conveyor belts, and artificial leather. The latest NTP monograph on BBP was released in 2003, but was based on papers published before 2001. NTP determined that the evidence from animal studies was clear that adverse developmental effects could result from exposure to BBP.[34] However, effects were seen only at high levels of BBP exposure and estimated human exposure was much lower, although detailed exposure data were lacking. NTP concluded that it had minimal concern for fetal and infant developmental effects due to estimated BBP exposure.[35] Papers published after 2001 confirm the developmental toxicity of BBP.[36] However, the testimony of Leon Earl Gray, Jr., Senior Reproductive biologist and Toxicologist with the U.S. Environmental Protection Agency, indicated that DEHP, DBP (both forms), and BBP were equivalent in toxicity, based on four studies.[37]

Diisononyl phthalate (DINP)

This phthalate is used primarily to improve the flexibility of plastics in such products as gloves, drinking straws, garden hoses, and toys. It has been used to replace DEHP in toys and other applications. DINP is the most commonly used phthalate for toys, according to the Phthalate Esters Panel of the American Chemistry Council, a trade group representing chemical manufacturers.[38] The NTP monograph on DINP was published in 2003, but like the other monographs that have not been updated, it primarily considers research published

---

[33] Borch, J., M. Axelstad, A.M. Vinggaard, et al. 2006. Diisobutyl phthalate has comparable anti-androgenic effects to di-n-butyl phthalate in fetal rat testis. *Toxicology letters*, v. 163, n. 3, p. 183-190.

[34] Center for the Evaluation of Risks to Human Reproduction, National Toxicology Program, U.S. Department of Health and Human Services. *NTP-CERHR Monograph on the Potential Human Reproductive and Developmental Effects of Butyl Benzyl Phthalate (BBP).* NIH Publication No. 03-4487. March 2003. p. 2. [http://cerhr.niehs.nih.gov/chemicals/phthalates/ bb-phthalate/BBP_Monograph_Final.pdf].

[35] Ibid. p. 4.

[36] Wilson, Vickie S., Christy Lambright, Johnathan Furr, et al. 2004. Phthalate ester- induced gubernacular lesions are associated with reduced ins 13 gene expression in the fetal rat testis. *Toxicology Letters*, v. 146, p. 207-215.

Aso, S. H. Ehara, K. Miyata, et al. 2005. A two-generation reproductive toxicity study of butyl benzyl phthalate in rats. *Journal of Toxicological Sciences*, Dec., v. 30, p. 39-58.

[37] U.S. Congress. House of Representatives. Committee on Energy and Commerce, Subcommittee on Commerce, trade, and consumer Protection. Hearing on Safety of Phthalates and Bisphenol-A in Everyday Consumer Products. June 10, 2008. Written testimony of Leon Earl Gray, Jr., Senior Reproductive Biologist and Toxicologist, U.S. EPA.

The studies referenced by Dr. Gray were: Gray, Jr., L. Earl, Joseph Ostby, Johnathan Furr, et al., 2000, Perinatal exposure to the phthalates DEHP, BBP, and DINP, but not DEP, DMP, or DOTP, alters sexual differentiation of the male rat, *Toxicological Sciences*, v. 58, n. 2, p. 350-365; Hotchkiss, A.K., L.G. Parks-Saldutti, J.S. Ostby, et al. 2004, A mixture of the 'antiandrogens' linuron and butyl benzyl phthalate alters sexual differentiation of the male rat in a cumulative fashion, *Biology of Reproduction*, v. 71, n. 6, p. 1852-1861; Howdeshell, Kembra L., Vickie S. Wilson, Johnathan Furr, et al., 2008, A mixture of five phthalate esters inhibits fetal testicular testosterone production in the Sprague Dawley rat in a cumulative, dose additive manner, *Toxicological Sciences Advance Access* (reprint received from the author); and Tyl,R.W., C.B, Myers, M.C. Marr, et al., 2004, Reproductive toxicity evaluation of dietary butyl benzyl phthalate (BBP) in rats, *Reproductive Toxicology*, v. 18, n. 2, p. 241-264.

[38] American Chemistry Council, Phthalates Panel. Phthalates and children's toys. [http://www.phthalates.-org/yourhealth/childrens_toys.asp].

before 2001. The scientific evidence for developmental effects from DINP exposure of rodents is adequate "to conclude that DINP might adversely affect development of the human fetus if the levels of exposure are sufficiently high," according to NTP.[39] But the evidence for effects was not as strong as for DEHP and DBP, and in one study that compared the effects of exposure to various phthalates, DINP was found to be less potent than DEHP or DBP by an order of magnitude.[40] In his recent testimony before a House subcommittee, Dr. Leon Earl Gray rated the relative potency for producing developmental harm of various phthalates. He gave DINP a rating of 0.15 relative to DEHP, DBP (both forms), and BBP which he rated 1.0.[41] A number of studies of the potential effects on rodents of DINP exposure were published after 2001, but they appear consistent with the earlier work with respect to developmental toxicity.[42] In addition, there is some evidence of enhanced allergic responses due to DINP exposure.[43]

A review by the European Commission (EC) in 2006 concluded:

> "[i]n light of the divergent scientific views ... and the conclusions of the assessment of the risk for consumers under this Regulation, and taking into account the uncertainties in the evaluation of exposure to DINP from toys and childcare articles, precautionary considerations support the consideration at Community level of proportionate restrictions ... for the use of DINP in toys and childcare articles. Such measures should be reviewed after 3-4 years, in light of further scientific developments.[44]

Several studies have been conducted to estimate the level of exposure of children to DINP in toys due to mouthing. One study by the Consumer Product Safety Commission (CPSC) estimated DINP exposure through measurement of the time children spent in mouthing behavior and an analysis of DINP movement out of various toys.[45] NTP used that

---

[39] CERHR, National Toxicology Program, U.S. Department of Health and Human Services. *NTP-CERHR Monograph on the Potential Human Reproductive and Developmental Effects of Di-isononyl Phthalate (DINP).* NIH Publication No. 03-4484. March 2003. p. 2.

[40] "Less by an order of magnitude" is roughly equivalent to one-tenth. Gray, Jr., L. Earl, Joseph Ostby, Johnathan Furr, et al. 2000. Perinatal exposure to the phthalates DEHP, BBP, and DINP, but not DEP, DMP, or DOTP, alters sexual differentiation of the male rat. *Toxicological Sciences*, v. 58, n. 2, p. 350-365.

[41] U.S. Congress. House of Representatives. Committee on Energy and Commerce, Subcommittee on Commerce, trade, and consumer Protection. Hearing on Safety of Phthalates and Bisphenol-A in Everyday Consumer Products. June 10, 2008. Written testimony of Leon Earl Gray, Jr., Senior Reproductive Biologist and Toxicologist, U.S. EPA.

[42] For example, Borch, J., O. Ladefoged, U. Hass, et al., 2004, Steroidogenesis in fetal male rats is reduced by DEHP and DINP, but endocrine effects of DEHP are not modulated by DEHA in fetal, prepubertal and adult male rats, *Reproductive Toxicology*, v. 18, n. 1, p. 53-61; and Wenzel, A., C. Franz, E. Breous, et al., 2005, Modulation of iodide uptake by dialkyl phthalate plasticisers in FRTL-5 rat thyroid follicular cells, *Molecular and Cellular Endocrinology*, v. 244, n. 1-2, p.63-71.

[43] Lee, M.H., J. Park, S.W. Chung, et al. 2004. Enhancement of interleukin-4 production in activated CD4+ T cells by diphthalate plasticizers via increased NF-AT binding activity. *International Archives of Allergy and Immunology*, v. 134, n. 3, p. 213-222.

[44] European Commission. Commission Communication on the results of the risk evaluation and the risk reduction strategies for the substances: Dibutylphthalate; e,4-Dichloroaniline; Di-isodecyl phthalate; 1,2-Benzenedicarboxylic acid, di-c9-11-branched alkyl esters, c9- rich; Ethylenediaminetetraacetate; Methyl acetate; Monochloroacetic acid; n-Pentane; Tetrasodium ethylenediaminetetraacetate. Official Journal of the European Union. April 13, 2006. Available through Eur-Lex at [http://eur-lex.europa.eu/-LexUriServ/-LexUriServ. do?uri=OJ:C:2006:090:0004:0028:EN:PDF].

[45] Babich, M.A., S.B. Chen, M.A. Greene, et al. 2004. Risk assessment of oral exposure to diisononyl phthalate from children's products. *Regulatory Toxicology and Pharmacology*, v. 40, n. 2, p. 151-167.

data to conclude that its concern was "minimal" for developmental effects in children.[46] NTP also had minimal concern for DINP causing adverse effects to human reproduction or fetal development.[47] However, this conclusion was controversial.[48] A Japanese study also looked at DINP release from toys and and time spent mouthing and found considerably higher exposures than the CPSC.[49]

As mentioned above, it appeared from the NHANES that exposure to DINP was negligible in children surveyed by the Centers for Disease Control and Prevention (CDC) in 200 1-2002.[50] However, this result was based on measurement of the metabolite monoisononyl phthalate (MINP), the traditional measure of exposure to DINP. A recent study of how the human body processes DINP found that MINP is a minor metabolic product, while mono(carboxyisooctyl) phthalate (MCIOP), mono(oxoisononyl) phthalate (MOINP), and mono(hydroxyisononyl) phthalate (MHINP) are the major metabolites in DINP-dosed rats.[51] The authors of this study concluded that estimates of exposure to DINP might be underestimates if based on MINP levels.

Di Isodecyl Phthalate (DI DP)

DIDP is another plasticizer used primarily in electrical cords, leather for car interiors, and PVC flooring. As for the other phthalates addressed by the NTP, the expert panel found "no *direct* evidence that exposure of *people* to DIDP adversely affects reproduction or development" [emphasis added], but "studies with rats have shown that exposure to DIDP can cause adverse developmental effects."[52] NTP concluded that exposures to DIDP were probably not high enough to cause concern, and that scientists had minimal concern for developmental effects in fetuses and children.[53] NTP also found that DIDP would not adversely affect human reproduction.[54] Studies published since 2001 have not conflicted with the NTP conclusions. However, several recent studies have found endocrine-disrupting effects following rodent exposure to DIDP.[55]

---

[46] CERHR, p. 3.

[47] Ibid.

[48] Shea, K.M. 2003. Pediatric exposure and potential toxicity of phthalate plasticizers. *Pediatrics*, v. 111, n. 6, Pt 1, p. 1467-1474.

[49] Sugita, T., Y. Kawamura, M. Tanimura, et al. 2003. Estimation of daily oral exposure to phthalates derived from soft polyvinyl chloride baby toys. *Shokuhin eiseigaku zasshi. Journal of the Food Hygienic Society of Japan,* v. 44, n. 2, p. 96-102.

[50] DHHS, ibid., p. 282.

[51] Silva, M.J., J.A. Reidy, J.L. Preau, Jr., et al. 2006. Oxidative metabolites of diisononyl phthalate as biomarkers for human exposure assessment. *Environmental Health Perspectives* v. 114, n. 8, p. 1158-61.

[52] Center for the Evaluation of Risks to Human Reproduction, National Toxicology Program, U.S. Department of Health and Human Services. *NTP-CERHR Monograph on the Potential Human Reproductive and Developmental Effects of Di-isodecyl Phthalate (DIDP).* NIH Publication No. 03-4485. April 2003. p. 1.

[53] Ibid. p. 3.

[54] Ibid.

[55] Wenzel, A., C. Franz, E. Breous, et al. 2005. Modulation of iodide uptake by dialkyl phthalate plasticisers in FRTL-5 rat thyroid follicular cells. *Molecular and Cellular Endocrinology*, v. 244, n. 1-2, p.63-71.

Turan, N., R.H. Waring, and D.B. Ramsden. 2005. The effect of plasticisers on "sulphate supply" enzymes. Molecular and Cellular Endocrinology, v. 244, n. 1-2, p. 15-19.

Harris, R., N. Turan, C. Kirk, et al. 2007. Effects of endocrine disruptors on dehydroepiandrosterone sulfotransferase and enzymes involved in PAPS synthesis: genomic and nongenomic pathways. Environmental Health Perspectives, v. 115, Supp. 1, p. 5 1-54.

Di-N-Octyl Phthalate (DnOP)

DnOP is used primarily to improve the flexibility of plastics. DnOP is found in mixtures of phthalates that are used to make flooring, tarps, pool liners, bottle cap liners, conveyor belts, and garden hoses. There are few studies on which to evaluate the potential toxicity or exposure to DnOP. NTP found limited evidence that DnOP might cause developmental effects in highly exposed rodents.[56] No evidence of reproductive effects was found in the available studies.[57] NTP concluded that it had negligible concern for effects on adult reproductive systems, but it was unable to form an opinion on an appropriate level of concern with respect to developmental risks, due to the lack of available exposure data and lack of toxicity data for exposure levels that might have relevance for human exposure.[58] An online search by CRS of publications using Medline revealed no new studies after 2001 that might better inform a risk evaluation.

Federal Evaluation and Regulation of Phthalates[59]

Depending on use, phthalates are potentially regulated by various regulatory agencies, including the Environmental Protection Agency (EPA), the Occupational Safety and Health Administration (OSHA), the Food and Drug Administration (FDA), and the Consumer Product Safety Commission (CPSC). EPA regulates various phthalates released to the environment under most of its statutes. For example, DEHP is regulated as a hazardous air pollutant, a drinking water contaminant, a water pollutant, and a hazardous waste. OSHA regulates worker exposure to phthalates.

The current focus of congressional concern, however, is federal regulation of consumer products from which phthalates might be released. Federal agencies have taken several actions, some as early as the mid 1980s, to evaluate and regulate phthalates. For example, EPA has required manufacturers of phthalates to conduct certain tests to better inform federal regulators. These test orders were withdrawn in 1996, when EPA determined that it had received the necessary information (which focused at the time on carcinogenic potential). To date, however, no phthalate- containing product has been banned outright.

Food and Drug Administration (FDA)

FDA-regulated products that may contain phthalates include (1) medical devices; (2) food contact substances, such as plastic wrap;[60] and (3) cosmetics. FDA reported in June 2008 that, in tandem with its review of the safety of bisphenol A (BPA) in the products it

---

56 Center for the Evaluation of Risks to Human Reproduction, National Toxicology Program, U.S. Department of Health and Human Services. *NTP-CERHR Monograph on the Potential Human Reproductive and Developmental Effects of Di-n-octyl Phthalate (DnOP).* NIH Publication No. 03-4488. May 2003. p. 1.

57 Ibid.

58 Ibid. p. 3.

59 This section was written by Sarah A. Lister, Specialist in Public Health and Epidemiology, Domestic Social Policy Division.

60 Manufacturers maintain that phthalates are not used in food wrap in the United States. See the trade group website [http://www.phthalates.org/pdfs/Phthalates_mvf.pdf].

regulates,[61] it is also conducting a comprehensive inventory of regulated products that contain phthalates.[62]

FDA regulates a wide variety of medical devices in commerce. Many of these products are made of, or contain PVC. These include intravenous fluid bags and lines, tubing used for procedures such as cardiac bypass and dialysis, and indwelling medical devices, such as peripherally inserted central catheters, or "PICC lines." According to reports, phthalates in tubing can leach into the fluids they contain and pass into the body, and can leach directly from indwelling devices. In 2001, FDA completed a safety assessment of DEHP, which was the softener most commonly used in PVC-containing medical devices.[63] The assessment underpinned a public health notification in 2002 in which FDA identified a number of medical procedures that posed the highest risk of exposure to DEHP, and recommended the use of alternatives to DEHP-containing medical devices if these procedures were to be performed on high-risk individuals.64 Depending on the procedure, these individuals include infants, boys, pregnant or nursing women, and, for some procedures, healthy adults. FDA recommended, however, that needed medical procedures not be deferred solely because of concerns about DEHP exposure.

Phthalates are not added to foods directly, but are regulated by FDA as *food contact substances* or *indirect food additives*, where they are components of packaging that may leach into foods or beverages. FDA permits the use of a variety of phthalates for these purposes.[65] In recent congressional testimony, FDA reported that:

> [FDA] has recently established a Phthalate Task Group (PTG) to review all available use and toxicology information associated with phthalate exposure from food contact use and to better characterize any potential risk from these uses. The primary focus of the PTG will be to determine the most realistic exposure estimation and risk associated with phthalate use in food packaging. The PTG will review and address past studies on phthalates and any new information available. If our review indicates that existing data no longer supports the continued safe use of these materials in food contact material, FDA will take appropriate regulatory action to remove these materials from the marketplace.[66]

Under the authority of the Federal Food, Drug, and Cosmetic Act (FFDCA), FDA regulates the safety of cosmetics and personal care products, such as nail polish, shampoo and

---

[61] See CRS Report RS22869, *Bisphenol A (BPA) in Plastics and Possible Human Health Effects*, by Linda-Jo Schierow and Sarah A. Lister.

[62] Statement of Norris Alderson, Ph.D., FDA Associate Commissioner for Science, before the Subcommittee on Commerce, Trade and Consumer Protection, House Committee on Energy and Commerce, hearing on "Safety of Phthalates and Bisphenol A in Everyday Consumer Products," June 10 , 2008, 1 10th Cong., 2nd sess., Washington, DC, hereafter referred to as FDA testimony.

[63] FDA, "Safety Assessment of Di(2-ethylhexyl)phthalate (DEHP) Released from PVC Medical Devices," undated, at [http://www.fda.gov/cdrh/ost/dehp-pvc.pdf].

[64] FDA, "FDA Public Health Notification: PVC Devices Containing the Plasticizer DEHP," July 12, 2002, at [http://www.fda.gov/cdrh/safety/dehp.html].

[65] For example, a search for "phthalate" in FDA's regulations at [http://www.accessdata.fda. gov/scripts/-cdrh/cfdocs/cfcfr/cfrsearch.cfm] yields almost two dozen approved phthalates, including DEHP, in adhesives used for food packaging.

[66] FDA testimony. FDA's definition of safety with respect to food additives and food contact substances is that "there is a reasonable certainty in the minds of competent scientists that the substance is not harmful under the intended conditions of use." 21 CFR § 170.3(i).

lotions,[67] many of which contain phthalates. Cosmetic products and ingredients are not subject to FDA premarket approval, but FDA can prohibit the marketing of a cosmetic product if it is adulterated, meaning, among other things, that it contains "any poisonous or deleterious substance which may render it injurious to users" under labeled or customary conditions of use.[68] According to FDA, the principal phthalates used in cosmetic products are DBP, dimethylphthalate (DMP), and diethylphthalate (DEP). FDA reports that at the present time, it "does not have compelling evidence that phthalates, as used in cosmetics, pose a safety risk. ...," but that it is conducting a survey of phthalate levels in certain cosmetic products to more accurately assess infant exposure.[69] Under the authority of the Fair Packaging and Labeling Act (Public Law 89-755), FDA requires that retail cosmetic products carry ingredient labels, which would include phthalates. However, the listing of individual fragrance ingredients, which may contain phthalates, is not required.[70]

Consumer Product Safety Commission

The Consumer Product Safety Commission (CPSC) may regulate phthalates in consumer products, including toys and other children's products, under either the Federal Hazardous Substances Act (FHSA, 15 U.S.C. §§1261 et seq.) or the Consumer Product Safety Act (CPSA, 15 U.S.C. §§2051 et seq.), two of the statutes that it administers and enforces. Consumer products as defined in the CPSA does not include food, drugs, medical devices, and cosmetics, which fall under the jurisdiction of the FDA. The FDA has jurisdiction over food containers with regard to substances that may leach into food from the container. CPSC has jurisdiction with regard to other defects, such as shattering or choking hazards. It also has jurisdiction over toys, children's furniture such as cribs, car seats, pacifiers and teething rings, and other children's consumer products.

Although the CPSC may regulate phthalates under either the CPSA or the FHSA, if it chooses to promulgate safety standards pursuant to the CPSA for a product that may be regulated under the FHSA, it must promulgate a rule finding that it is in the public interest to regulate the risk of injury under the CPSA.[71] Since the CPSA and the FHSA differ with regard to the rule-making procedures, the precise nature of the acts prohibited with regard to products or substances that fail to comply with safety standards, and other issues, there are advantages and disadvantages to promulgating standards under the CPSA versus the FHSA. For example, injunctive enforcement authority for States attorneys general is expressly provided by the FHSA, but not by the CPSA; the apparent ambiguity of the CPSA on this point has led to clarifying provisions in both the Senate and House versions of H.R. 4040, the CPSC reform legislation now in conference. Therefore, if a phthalates standard were to be promulgated under the FHSA, States attorneys general would have the authority to enjoin

[67] "Cosmetic" is defined as " (1) articles intended to be rubbed, poured, sprinkled, or sprayed on, introduced into, or otherwise applied to the human body or any part thereof for cleansing, beautifying, promoting attractiveness, or altering the appearance, and (2) articles intended for use as a component of any such articles; except that such term shall not include soap." Federal Food, Drug, and Cosmetic Act (FFDCA) section 201(i); [21 U.S.C. 321(i)]. For more information, see FDA, "Cosmetics: FDA Authority and Policy," at [http://www. cfsan.fda.-gov/~dms/cos-pol.html].

[68] FFDCA section 601; 21 U.S.C. 361.

[69] FDA, "Phthalates and Cosmetic Products," February 7, 2008, at [http://www.cfsan.fda. gov/~dms/cos-phth.html].

[70] Ibid.

[71] CPSA §30(d) (15 U.S.C. §2079).

violations of the federal standard, whereas arguably they would not have similar authority to enforce a similar standard under the CPSA.

Additionally, critics allege problems with the ability of the CPSC to establish safety standards in a timely fashion. For example, some assert that the rulemaking procedures under the CPSA and other acts under the CPSC' s jurisdiction, such as the FHSA, are unnecessarily onerous, requiring protections beyond those required by the Administrative Procedures Act. Also, the CPSA requires the CPSC to rely on voluntary standards where an adequate standard exists with which the industry widely and substantially complies. The voluntary nature of some safety standards limits the action that can be taken by the CPSC for violations. Under both the CPSA and the FHSA, the CPSC may order a recall and/or other remedies for products that violate a safety standard under the pertinent act and may inspect factories where products are made.[72] Products violating a safety standard may be denied importation.

Although the European Union and some states have enacted safety standards regarding phthalates in children's products, as discussed below, the CPSC has not promulgated such standards. In 1983, the CPSC determined that DEHP in children's products might result in substantial exposure of children to an animal carcinogen.[73] The CPSC is not permitted to initiate rule-making relating to risks of cancer, birth defects or gene mutations unless it first establishes a Chronic Hazard Advisory Panel to study the issue and make recommendations. The panel appointed to study the risks of DEHP concluded that it could put children at risk of cancer from mouthing of products containing DEHP. Accordingly, the CPSC worked with the children's products industry to reach a voluntary agreement banning the use of DEHP from pacifiers, rattles, and teethers.[74] Although other children's products were not included in the agreement between CPSC and the industry, most manufacturers substituted other phthalates for DEHP in other children's products. DINP was the substitute. Despite studies conducted by the industry in the late 1990s linking DINP to liver toxicity and cancer in rodents, the CPSC concluded that the risk to children from mouthing children's products was minimal. However, the CPSC achieved a voluntary agreement with the industry banning DINP and dioctyl phthalate from pacifiers and bottle nipples. The CPSC appointed a Chronic Hazard Advisory Panel on DINP and conducted other studies in response to a petition to initiate rule-making regarding phthalates in children's products. The panel concluded in 2001 that DINP posed a minimal or nonexistent risk of cancer to humans. After further consideration of the panel report and other studies, in 2003 the CPSC denied the petition to establish a safety standard for PVC containing phthalates in children's products intended for children five years of age and younger.[75]

---

[72] For a general discussion of the statutes administered and enforced by the CPSC, see CRS Report RL34399, *Consumer Product Safety Commission Reform: Senate and House Versions of H.R. 4040*, by Margaret Mikyung Lee.

[73] See the discussion of scientific studies for various phthalates above and the discussion of CPSC actions in *Young Children and Plastic Toys* and *Phthalates in Plastic Toys*, 8 Consumer Product Safety Review 3-5 (2003).

[74] Although the CPSC has promulgated safety standards for pacifiers and rattles at 15 C.F.R. parts 1510 and 1511, these primarily concern choking/mechanical hazards, not hazards from phthalates or other chemical constituents.

[75] Denial letter from the CPSC to the National Environmental Trust, dated February 26, 2003, available at [http://www.cpsc.gov/LIBRARY/FOIA/FOIA03/petition/Ageunder.pdf] (last visited July 11, 2008).

Regulation in the European Union (EU) and the States

European Union

Proposed federal legislation and several state statutes concerning phthalates apparently were modeled on EU laws. Under Council Directive 76/769/EEC,[76] as amended by Council Directive 2005/84/EC,[77] the EU currently prohibits the use (at concentrations greater than 0.1% by mass of the plasticized material) of DEHP, DBP, and BBP in toys and child-care articles and of DINP, DIDP, and DnOP in toys and child-care articles that can be mouthed by children. It also prohibits the sale of toys and child-care articles containing phthalates at a concentration exceeding the permitted level. It defines "childcare article" as meaning "any product intended to facilitate sleep relaxation, hygiene, the feeding of children or sucking on the part of children."

Council Directive 2005/84/EC notes the existence of Commission Decision 1999/815/EC[78] that banned phthalates in toys and child-care articles as a renewable emergency measure in response to phthalate studies conducted by various Member States and assessed by the Scientific Committee on Toxicity, Ecotoxicity and the Environment.[79] The language of the Decision differs from that of the Directive in that it limits the ban on the six phthalates to products that can be mouthed by children under three years of age, while the Directive does not contain the age limit. This Decision apparently was last extended in 2004 until September 20, 2005, but Council Directive 2005/84/EC refers to the Decision as being renewed regularly. Regardless of whether the Decision continues to be renewed, EU Member States were required to take necessary measures to comply with the standards described in the Decision, therefore, those measures would continue in effect. It appears that the Decision may not have been renewed in anticipation of Council Directive 2005/84/EC and the Regulation (EC) No. 1907/2006,[80] which was recently enacted and will repeal Council Directive 76/769/EEC effective June 1, 2009, while retaining the same phthalate in children's product standard, effective June 1, 2009. The new Regulation also retains the Directive's requirement that the European Commission re-evaluate this standard by January 16, 2010, in light of new scientific information and amend the standard accordingly, if justified.

While the Directive is binding law on the Member States of the EU, requiring them to take necessary measures to bring their respective national laws into compliance with the Directive standards, the new Regulation is directly binding on the Member States. This means they are obligated to comply with the standard and enforce it without any implementing legislation or rule at the national level.

One group of phthalates, DEHP, DBP, and BBP, is banned without limitation because they are classified as reproductive toxicants that present an unacceptable risk given the

---

[76] 1976 O.J. (L 262) 201 (originally enacted July 27, 1976).

[77] 2005 O.J. (L 344) 40 (enacted December 14, 2005). This Directive amended the 1976 Directive to add the restrictions on phthalates in children's products.

[78] O.J. (L 315) 46 (issued December 7, 1999). This Decision in turn was preceded by Commission Recommendation 98/485/EC of July 1, 1998, O.J. (L 217) 35, inviting Member States to take measures to ensure a high level of child health protection with regard to childcare articles and toys intended to be placed in the mouth by children less than three years of age and made of PVC containing any of the six phthalates.

[79] Information received via email communications with the Delegation of the European Commission in Washington, D.C., July 11, 2008.

[80] Arts. 139, 141, Annex XVII.51 & 52, 2006 O.J. (L 396) 1 (enacted December 18, 2006) (popularly referred to as REACH, the acronym for Registration, Evaluation, Authorisation, and Restriction of Chemicals).

general safety requirements of the European Union. The other group, DINP, DIDP and DnOP, are banned only for products that can be placed in the mouth by children. Despite inconclusive scientific evidence of harm, these phthalates were banned under the precautionary principle of the European Union given the potential risk posed to children.[81]

States

California, Vermont and Washington have recently enacted legislation establishing safety standards for phthalate content in children's articles. These standards all appear to have been at least partly modeled on current or earlier versions of the EU regulations. Aside from the age limit specified for the phthalate standard or for subcategories of children's products, if any, the existing state statutes do not define children's products generally in terms of an age ceiling for "child" or "children." Hawaii's Senate has adopted a resolution requesting the Hawaii Department of Health to monitor research being conducted regarding the risks posed by phthalates and bisphenol-A in consumer products and to report recommendations and proposed legislation before the 2009 legislative session.[82] Oregon's legislature has adopted a joint memorial urging Congress to regulate phthalates at the federal level as a substance in cosmetics, personal care products, and children's toys.[83] Additionally, several other states have introduced legislation concerning phthalates.

Beginning January 1, 2009, the California statute[84] will prohibit the manufacture, sale, or distribution in commerce of any toy or child-care article that contains DEHP, DBP, or BBP in concentrations exceeding 0.1% and of any toy or child-care article, intended for use by children under three years of age that can be mouthed, that contains DINP, DIDP or DnOP in concentrations exceeding 0.1%. The statute requires manufacturers to use the least toxic alternative when replacing phthalates in such products and also prohibits them from replacing phthalates with certain carcinogens (including substances known, likely to be, or suggestive of being human carcinogens) or reproductive toxicants identified in accordance with federal or California laws. "Toys" are defined as "all products designed or intended by the manufacturer to be used by children when they play" and "child care article" is defined as "all products designed or intended by the manufacturer to facilitate sleep, relaxation, or the feeding of children, or to help children with sucking or teething."

This statute appears to be partly modeled on an earlier version of the EU Council Directive 76/769/EEC with regard to the threshold concentration level for the ban and the definition of child-care articles and toys. It does not include the amendment made by EU Directive 2005/84/EC (December 14, 2005) adding "hygiene" to the scope of child-care articles, which effectively included items such as lotion, powder, baby oil, etc. It also includes the three-year-old age limit regarding products that can be placed in the mouth from the EU Commission Decision 1999/815/EC (December 7, 1999). It added "teething" to the scope of child-care articles, for which the EU includes "sucking" but not teething.

---

[81] Information received via email communications with the Delegation of the European Commission in Washington, D.C., July 11, 2008.

[82] S.Res. 68, 24th Leg. (Haw. 2008).

[83] S. Jt. Memorial 8, 74th Leg. (Or. 2007).

[84] Cal. Health & Safety Code §§108935-108939 (current on LexisNexis).

The Vermont statute[85] appears to be modeled on the California statute, but with some differences. Beginning July 1, 2009, it will prohibit the manufacture, sale, or distribution in commerce of any toy or child-care article intended for use by a child under three years of age that contains DEHP, DBP, or BBP in concentrations exceeding 0.1% and of any toy or child-care article intended for use by a child under three years of age that can be placed in the mouth and that contains DINP, DIDP, or DnOP in concentrations exceeding 0.1%. The Vermont statute provides for an under- three-years-old age limit for the first group of phthalates, unlike the California statute. The Vermont law adopts the California statutory definition of "child care article" and "toy" and additionally defines "phthalate" as "any one of a group of chemicals used as plasticizers to provide flexibility and durability to plastics such as polyvinyl chloride (PVC)." Like the California statute, it requires manufacturers to use the least toxic alternative to phthalates and prohibits them from substituting carcinogens (including substances known, likely to be, or suggestive of being human carcinogens) or reproductive toxicants identified by the EPA under federal law, but not under state laws, apparently because Vermont does not have such environmental laws identifying carcinogens or reproductive toxicants.

The Vermont law provides that a violation of the phthalates law shall be deemed a violation of the Vermont Consumer Fraud Act and that the provisions of that act concerning the enforcement authority of the Vermont Attorney General and the rights of private parties shall apply to violations of the phthalates law. It further clarifies that nothing in the phthalates law regulates firearms, ammunition, shooting ranges, or hunting/fishing equipment.

The Washington provision banning phthalates in children's products is part of a broader statute[86] concerning chemicals in children's products generally. It covers lead and cadmium content and also provides, among other things, for the identification of chemicals of "high concern" to children and children's products by the Washington Department of Ecology, for notification to the Department by a manufacturer that its children's product contains a chemical of high concern, and for a product safety education campaign to promote awareness of unsafe children's products. With regard to phthalates, beginning July 1, 2009, it prohibits a manufacturer, wholesaler, or retailer from manufacturing, knowingly selling, offering for sale, or distributing for sale or for use in the state a children's product or product component containing phthalates, individually or in combination, at a concentration exceeding 0.1% by weight (a thousand parts per million). It defines phthalates as meaning the six phthalates discussed in this report. It adopts the California definition of "toy," but does not refer to "child-care articles." Instead, it defines "children's product," which includes the California definition of a child-care article, expanding it to include children's clothing, and also includes toys, children's cosmetics (for children under the age of twelve), children's jewelry (for children under the age of twelve), and car seats (it also includes a list of items not considered "children's products").

---

85 Act of May 24, 2008, Act 171, 2008 Vt. Adv. Legis. Serv. 171 (phthalates in products for young children).

86 Act of June 12, 2008, ch. 288, 2008 Wash. Adv. Legis. Serv. 288) (relating to the children's safe products act).

PROPOSED LEGISLATION

On July 28, 2008, the House and Senate conferees for H.R. 4040, the Consumer Product Safety Improvement Act of 2008, announced that an agreement on a final text had been reached resolving the differences between the Senate and House versions of the bill. As of the date of this report, it remains uncertain whether a conference agreement will be completed in time for a floor vote in both chambers before the August recess. The conference committee and Senator Pryor published summaries of some of the agreed provisions. Because a final text for the agreed provisions was not made public, however, the details of the agreed provisions generally remain unclear. The summary below is based on details published by the conference and by the media.

The conference agreement would permanently ban the three phthalates whose toxicity is not disputed and would temporarily ban three other phthalates pending a review by a Chronic Hazard Advisory Panel (CHAP). It would prohibit children's toys or child care articles that contain more than 0.1 percent DEHP, DBP, or BBP. The sale of children's toys or child care articles containing concentrations of more than 0.1 percent of DINP, DIDP, or DnOP would be prohibited on an interim basis until a review by a CHAP. After the CPSC receives the report from the CHAP, it would determine, by rule, whether to continue the interim ban.

In March 2008, the Senate approved an amendment to H.R. 4040, which would have restricted the use of six phthalates in certain toys and child-care products. The House-passed version of H.R. 4040 had no phthalate amendment; the House Committee on Energy and Commerce noted in its report that it became aware of the potential dangers posed by phthalates in toys late in the legislative process and intended to take up the issue later.[87]

Section 40 of the Senate version of H.R. 4040, also referred to as the Feinstein- Boxer Amendment for the two California senators who introduced this specific amendment to the bill, was modeled on the California statute, with some changes, but placed the provisions in the context of the Federal Hazardous Substances Act (FHSA) framework. It adopted the California statutory definition of "child care article" and did not adopt its definition of "toy," but instead defined "children's product" as "a toy or any other product designed or intended by the manufacturer for use by a child when the child plays," which effectively included the same products as the California statute.

The provision would have treated as a banned hazardous substance under the FHSA any children's product or child-care article which (1) contains in any part any combination of DEHP, DBP or BBP in concentrations exceeding 0.1% or (2) is intended for use by a child that can be placed in a child's mouth and either contains any combination of DINP, DIDP or DnOP in concentrations exceeding 0.1% or contains any combination of any of the six phthalates in concentrations exceeding 0.1%. Any prohibitions under FHSA §4 (15 U.S.C. § 1263) would have applied to such products, including the introduction, receipt or delivery into interstate commerce of such products; failure to permit inspections of any factory, warehouse, or other establishment where such products are manufactured, processed, packed, or held; etc. Like the California statute, the provision would have prohibited a manufacturer from replacing phthalates with certain carcinogens (including substances known, likely to be, or suggestive of being human carcinogens) or reproductive toxicants identified by the EPA.

[87] H.Rept. 110-501, at 47 (2007).

However, unlike the California statute, the provision would not have required manufacturers to use the least toxic alternative as a substitute for phthalates.

Section 40 would have provided that neither it nor FHSA §18(b)(1)(B) (15 U.S.C. § 1261 note), concerning preemption of non-identical State or local regulations of banned hazardous substances, would preempt State or local laws applying to a phthalate other than the six described in the bill; applying to a phthalate described in the bill that is not otherwise regulated by the bill; requiring a warning of risk, illness, or injury regarding any phthalate; or prohibiting the use of alternatives to phthalates not prohibited under this bill. Generally, non-identical State or local regulations are preempted unless an exemption is granted by the CPSC upon request by the State or locality.

CONCLUSION

When the House and Senate consider the conference report accompanying H.R. 4040, Members might be expected to consider two central questions about phthalates and the proposed amendment. First, what is the scientific basis for health concerns about exposure to these chemicals? Second, would the provision reduce the risks without generating greater risks? These issues are discussed briefly here, based on the scientific and legal information presented above.

Scientific Basis for Health Concerns

The scientific basis for concerns about risks to human health appears to be strong in the case of some phthalates, adequate with respect to others, and weak for the remaining chemicals. The strongest evidence with respect to developmental effects has been produced since about the year 2000. Many of these studies were not available to the NTP or to CPSC when they reviewed the phthalate literature in 2000 or 2001. At that time, regulators focused on carcinogenic effects, rather than effects on fetal development. This more recent animal evidence strongly supports a claim that DEHP, DBP, and DBB can harm reproduction and damage fetal and juvenile development in rats.

The structure-activity relationship (between the molecular structure of the phthalates and developmental damage) is well understood, such that scientific concern focuses now on DEHP, DBP, and DBB.[88] DINP also is a developmental toxicant, but is only about 15% as potent as the most potent phthalates. To the extent that it is still studied, it is generally studied together with other more potent toxic phthalates to evaluate additivity of effects. Scientific evidence regarding the other phthalates mentioned in the amendment to H.R. 4040 is lacking. Limited evidence indicates that DIDP, DINP, and DnOP might have effects on the immune system, but these phthalates appear to be much less toxic to developing rodents.

---

88 According to Dr. Paul Foster, Acting Toxicology Branch Chief of the National Toxicology Program (personal communication, July 9, 2008), the structure that is known to produce toxic effects is a benzene ring of 6 atoms

Data indicate that people are exposed to many phthalates, especially DEHP, DBP (di-n-butyl), and DBB. Children appear to be most heavily exposed. Data are insufficient to judge exposure for DINP, DIDP, and DnOP. Individuals such as newborn babies in the intensive care units of hospitals face multiple and continuous phthalate exposures.

Scientists are just beginning to explore the additive effects of exposure to multiple phthalates, as well as to phthalates in combination with certain pesticides.[89] To date, studies suggest there may be additive effects of multiple phthalate exposures.[90] The National Academy of Sciences is evaluating the risk of aggregate human exposure to multiple phthalates, and is expected to report before the end of 2008.

Would the Amendment Reduce the Risks Without Generating Greater Risks?

By eliminating the use of six phthalates in child-care items and toys, the proposed amendment would codify the voluntary agreements reached by CPSC with product manufacturers (to keep DEHP and DINP out of nipples, pacifiers, and teething toys, and DEHP out of toys that might be mouthed) and reduce exposure to DINP, the only phthalate currently used in the United States to produce toys. The effect of banning use of the other phthalates is less clear, because their child-related uses are not known. However, because the law would prohibit their use as substitutes for DINP in toys, toys would be eliminated as a source of exposure.

For each use, different chemicals might be used in lieu of the six phthalates. Acetates might be used in some applications, phthalates other than the six specified might be used in others. The risks of such chemicals may be known or unknown. Given the lack of legal authority to require testing for a chemical proposed for most uses, and the cost of testing, new formulations of products might pose either more or less risk than the current formulations.

---

with two linear side chains in the ortho position of between 4 and 6 carbon atoms each. Longer or shorter side chains are not developmentally toxic or are much less toxic.

89 Rider, C.V., J. Furr, V.S. Wilson, et al. 2008. A mixture of seven antiandrogens induces reproductive malformations in rats. International Journal of Andrology v. 31, n. 2, p. 249-262.

Fowler, Paul A., David R. Abramovich, Neva E. Haites, et al. 2007. Human fetal testis Leydig cell disruption by exposure to the pesticide dieldrin at low concentrations. Human Reproduction, v. 22, n. 11, p. 2919-2928.

90 Hotchkiss, A.K., L.G. Parks-Saldutti, J.S. Ostby, et al. 2004. A mixture of the "antiandrogens" linuron and butyl benzyl phthalate alters sexual differentiation of the male rat in a cumulative fashion. Biology of Reproduction, v. 71, p. 1852-1861.

Howdeshell, Kembra L., Vickie S. Wilson, Johnathan Furr, et al. April 14, 2008. A mixture of five phthalate esters inhibits fetal testicular testosterone production in the Sprague Dawley rat in a cumulative, dose additive manner. Toxicological Sciences Advance Access. Reprint received from the author.

Rider, Cynthia V., Johnathan Furr, Vickie S. Wilson, et al. 2008, A mixture of seven antiandrogens induces reproductive malformations in rats. International Journal of Andrology, v. 31, p. 249-262.

In: Consumer Product Safety Issues
Editors: P. A. Varga, M. D. Pintér

ISBN: 978-1-60456-826-4

Chapter 11

BISPHENOL A (BPA) IN PLASTICS AND POSSIBLE HUMAN HEALTH EFFECTS*

Linda-Jo Schierow and Sarah A. Lister

SUMMARY

Bisphenol A (BPA) is used to produce certain types of plastic. Containers made of these plastics may expose people to small amounts of BPA in food and water. Some animal experiments have found that fetal and infant development may be harmed by small amounts of BPA, but scientists disagree about the value of the animal studies for predicting harmful effects in people. At least one regulatory decision in the face of the scientific disagreement has led to a congressional inquiry into the extent to which the decision was based on good science. Legislation proposed in April 2008, S. 2928, would prohibit use of BPA in some products intended for use by children. Legislation proposed in June 2008, H.R. 6228, would prohibit the use of BPA in food and beverage containers regulated by the Food and Drug Administration (FDA).

INTRODUCTION

Bisphenol A (BPA)[1] is a synthetic chemical compound produced in the United States in large quantities, approximately 2.3 billion pounds annually.[2] The dominant use is in manufacturing certain forms of plastic: relatively hard, clear polycarbonate (PC), and epoxy resins that are used to line food cans. Under certain conditions, BPA may *migrate* (i.e., be released) from PC containers and plastic-lined cans into the food or liquids they contain.

---

* This is an edited, reformatted and augmented version of CRS Report RS22869, dated September 8, 2008

[1] Bisphenol A also is commonly known as carboxylic acid. It is the single molecule that is chained together (polymerized) to form polycarbonate.

[2] U.S. Department of Health and Human Services (HHS), National Toxicology "Draft NTP Brief on Bisphenol A," April 14, 2008, [http://cerhr.niehs.nih.gov/chemicals/bisphenol/BPADraftBrief VF_04_14_08.pdf].

The widespread use of BPA and the potential for human exposure, together with accumulating scientific evidence about possible BPA toxicity, led the National Toxicology Program (NTP) at the National Institutes of Health (NIH) to select BPA for a comprehensive review. NTP released a draft "brief" on BPA on April 14, 2008.[3] Its conclusions prompted some to call for federal restrictions on certain BPA uses, and sparked congressional and media interest in the past and current positions of the Food and Drug Administration (FDA). FDA regulates BPA and other chemicals used in food containers and maintains that current uses of BPA are safe. A final NTP monograph on BPA was released September 3, 2008.[4]

Health Effects

Exposure to large amounts of BPA is acutely toxic to humans and animals, but levels of BPA exposure from plastics are low. The possibility of human health effects from exposure to low doses of BPA is controversial, although animal evidence of possible harmful effects has been mounting for about 10 years. It is clear that BPA is capable of interfering with the action of estrogen, an important regulator of reproduction and development. (Interference with hormonal action is often referred to as *endocrine disruption*.) Therefore, many recent studies have focused on the potential effects of low levels of BPA exposure on fetal or newborn rats or mice. Some of the developmental effects seen among rodents exposed to low doses of BPA include changes in brains and behaviors; precancerous lesions in the prostate and mammary glands; altered prostate and urinary tract development; and early onset of puberty.[5]

These low-dose experiments are difficult to conduct, in part because BPA is ubiquitous in the environment. Thus, different studies have produced different results. Scientists employed by BPA manufacturers and some independent contractors argue that the hundreds of studies conducted so far have produced inconsistent results and are insufficient justification for more stringent BPA regulation. Other scientists maintain that well-designed and executed studies of sufficient statistical power on sensitive strains of laboratory rodents have clearly demonstrated the toxicity of low doses of BPA in mammals, and justify actions to reduce exposure for potentially vulnerable human populations.

Some researchers have proposed that BPA may interfere with hormones that regulate functions other than reproduction. A recently released study found that low-level exposure to BPA inhibits the release of adiponectin from human adipose (fat) tissue. Adiponectin increases insulin sensitivity and helps regulate glucose metabolism.[6] The researchers hypothesize that environmental BPA exposure may increase susceptibility to obesity and diabetes. The body of research in this area is less extensive than that into BPA's potential effects on reproductive hormones.

---

[3] Ibid.

[4] U.S. Department of Health and Human Services (HHS), National Toxicology Program, Center for the Evaluation of Risks to Human Reproduction, "NTP-CERHR Monograph on the Potential Human Reproductive and Developmental Effects of Bisphenol A," NIH Publication No. 08-5994, September 2008, 321pp, at [http://cerhr.niehs.nih.gov/chemicals/bisphenol/bisphenol.pdf].

[5] NTP Monograph on Bisphenol A, p. 7-8.

[6] Hugo, Eric R., Brandebourg, Terry D., Woo, Jessica G., et al., "Bisphenol A at Environmentally Relevant Doses Inhibits Adiponectin Release from Human Adipose Tissue Explants and Adipocytes," *Environmental Health Perspectives*, online August 14, 2008, at [http://www.ehponline.org/members/2008/1 1537/1 1537.pdf].

Human Exposure

Bisphenol A exposure in the general population comes primarily from consumption of food and beverages.[7] The latest national survey by the Centers for Disease Control and Prevention (CDC) found BPA in the urine of more than 92% of the people studied, which included children six years of age and older and adults.[8] Among these people, the highest average concentrations were found in children.[9] The NTP monograph estimates that the highest daily intakes of BPA occur in infants and children.[10] BPA has been found in human breast milk, but the NTP report estimates that infants who are formula-fed have higher daily BPA intake levels than those who are breast-fed,[11] because there is more BPA in infant formula than in breast milk, and because BPA may increase when PC baby bottles are used for formula feeding, especially if the bottles are heated. These BPA exposure levels in humans "are similar to levels of [BPA] associated with several 'low' dose laboratory animal findings of effects on the brain and behavior, prostate and mammary gland development, and early onset of puberty in females," according to the final NTP monograph.[12]

Current Federal BPA Regulation

Depending on its use, BPA is potentially regulated by various regulatory agencies, including the Consumer Product Safety Commission, the Environmental Protection Agency, and the Occupational Safety and Health Administration. BPA-containing PC polymers and epoxy resins used in food containers — such as baby bottles and infant formula cans, respectively — are regulated by FDA as *food contact substances*.[13] FDA conducts research into the possible endocrine disrupting effects of BPA. Agency regulations and guidance for industry include recommendations and guidelines for studies of potential reproductive, developmental, and neurological toxicity that may result from exposure to food contact substances.[14] These sources do not suggest that there is a systematic review process to study such effects that may result from exposure to previously approved products. Some consumer groups have sought for more than 10 years to have FDA declare uses of BPA-containing food contact substances unsafe, especially in packaging for infant formula.

---

7 NTP Monograph on Bisphenol A, p. 1.

8 Calafat, Antonia M., Xiaoyn Ye, Lee-Yang Wong, et al., 2008, "Exposure of the U.S. Population to Bisphenol A and 4-tertiary-Octylphenol: 2003-2004," Environmental Health Perspectives, v. 116, n. 1, p. 39-44.

9 Ibid.

10 Certain occupational groups are estimated to have the highest human exposure levels. NTP Monograph on Bisphenol A, p. 2.

11 Ibid., p. 3.

12 Ibid., p. 7-8.

13 Applicable FDA regulations are at 21 CFR §§ 177.1580, 175.300(b)(3)(viii), 177.1440, and 177.2280. See also FDA's Food Contact Substance Notification Program, at [http://www.cfsan.fda.gov/~dms/fcnrpt.html], and FDA's Consumer Update, "Safety and Food Packaging" at [http://www.fda.gov/consumer/updates/food-packaging08 1908.html].

14 See, in particular, FDA, "Toxicological Principles for the Safety Assessment of Food Ingredients," (the "Redbook"), updated July 2007, at [http://www.cfsan.fda.gov/~redbook/redtoca.html].

Events Surrounding the Current Controversy

In early 2007, NTP convened an expert panel to conduct a comprehensive review of the scientific literature on BPA. The panel met during 2007 and issued its report on November 26, 2007.[15] It concluded that animal studies were sufficient to elicit "some concern" about possible effects of BPA exposure on the neurological development of human fetuses and newborns, but "minimal concern" about effects on the early onset of puberty or development of mammary or prostate cancer. (The expression of "some concern" is midway in a qualitative scale used by NTP. In order, from greatest to least, the levels of concern are serious concern, concern, some concern, minimal concern, and negligible concern.) Some scientists disagreed with these conclusions.

NTP's own scientists reviewed the panel report, as well as numerous studies that were not considered by the panel, many that were completed or published in late 2007 and early 2008. NTP then issued its draft BPA "brief" on April 14, 2008, which largely agreed with the panel report, but expressed a higher level of concern with respect to early puberty and effects on the mammary and prostate glands. The draft report concluded, "... the possibility that [BPA] may alter human development cannot be dismissed."[16] Specifically, the NTP report concluded that there is "some concern" for neural and behavioral effects in fetuses, infants, and children at current levels of human exposure, and "some concern" in those same groups for effects on the prostate gland, mammary gland, and on earlier age of puberty in females. Public comment on the draft brief was invited through May 23, 2008.[17]

On June 11, 2008, the NTP Board of Scientific Counselors met to review the draft report and public comments. The Board voted to lower the level of concern for BPA's effects on the mammary gland and on the onset of puberty in females.[18] This vote is reflected in the final version of the NTP brief, which was included in the NTP monograph and issued September 3, 2008.[19] Thus, the official NTP view is that current levels of human exposure to BPA warrant "some concern" for possible effects on the brain, behavior, and prostate gland in fetuses, infants, and children; "minimal concern" for effects on the mammary gland and an earlier age for puberty in female fetuses, infants, and children, and for workers exposed occupationally; and "negligible concern" for all other current exposures and reproductive or developmental effects.

Canada published its risk assessment of BPA in April 2008, finding that "... the main source of exposure [to BPA] for newborns and infants is through the use of polycarbonate baby bottles when they are exposed to high temperatures and the migration of [BPA] from cans into infant formula. The scientists concluded in this assessment that bisphenol A exposure to newborns and infants is below levels that may pose a risk, however, the gap

[15] HHS, National Toxicology Program, "Expert Panel Report on the Reproductive and Developmental Toxicity of Bisphenol A," at [http://cerhr.niehs.nih.gov/chemicals/bisphenol/ BPAFinalEPVF1 12607 .pdf].

[16] Draft NTP Brief on Bisphenol A, p. 9.

[17] National Institute of Environmental Health Sciences, "Since You Asked - Bisphenol A," at [http://www.niehs.-nih.gov/news/media/questions/sya-bpa.cfm].

[18] National Institute of Environmental Health Sciences, "Actions on the Draft NTP Brief on Bisphenol A by the NTP Board of Scientific Counselors (BSC), June 11, 2008," at [http://ntp.niehs.nih.gov/files/-BSCactions-BPA_508.pdf].

[19] NTP Monograph on Bisphenol A, p. vii.

between exposure and effect is not large enough."[20] The Canadian government has said that although exposure levels are below those that could cause health effects, they are close to those levels, and the government wants to be prudent and reduce exposures further. It announced its intention to reduce BPA exposure in infants and newborns by (1) banning PC baby bottles, (2) developing stringent migration targets for BPA in infant formula cans, and (3) working with industry to develop alternative food packaging and a code of practice.

Also in April 2008, the American Chemistry Council, which represents chemical manufacturing companies, called on FDA to update its review of the safety of BPA in food contact applications, saying, "The extensive body of scientific study regarding [BPA] is well documented and well reviewed. Nevertheless, recent media reports have raised concerns about the safety and use of polycarbonate plastic and epoxy resins, unnecessarily confusing and frightening the public."[21]

Shortly thereafter, FDA formed an agency-wide task force to review current information regarding BPA in all FDA-regulated products. In June 2008, FDA asked its Science Board, the advisory board to the FDA Commissioner, to establish a subcommittee to review research on BPA and exposures from food containers, and deliver its findings to the Board's annual meeting in the fall.[22] Subsequently, Rosa DeLauro, Chairwoman of the House Appropriations Subcommittee on Agriculture (which funds FDA), wrote to FDA urging that the Science Board review BPA exposures from medical devices as well.[23] In August 2008, FDA published a draft risk assessment of BPA in food contact applications, saying, "FDA has concluded that an adequate margin of safety exists for BPA at current levels of exposure from food contact uses. At a later date, FDA will publish a separate document that provides a safety assessment of BPA exposure from other FDA-regulated products."[24] The BPA Subcommittee of the FDA Science Board is due to meet on September 16, 2008, to review the draft risk assessment, the NTP Brief, and other information regarding the safety of current food-contact uses of BPA.

Various states have enacted, or are considering, legislation to restrict use of BPA in products intended for use by infants and children. Also, concerns about the effects of BPA are affecting decisions in the marketplace, such as those by Wal-Mart, Playtex Infant Care, and Nalgene to stop allowing BPA in the bottles they produce or sell.[25]

---

[20] Health Canada, "Government of Canada Takes Action on Another Chemical of Concern: Bisphenol A," press release, April 18, 2008, at [http://www.hc-sc.gc.ca/ahc-asc/media/nr-cp/ 2008/2008_59_e.html].

[21] American Chemistry Council, "ACC Calls on FDA to Update Review of Bisphenol A," press release, April 17, 2008, at [http://www.americanchemistry.com/s_acc/index.asp].

[22] FDA, "FDA's Chief Scientist Asks Science Board Subcommittee to Review Research on Bisphenol-A," press release, June 6, 2008, at [http://www.fda.gov/opacom/hpnews.html].

[23] Congresswoman Rosa L. DeLauro, "DeLauro Presses for Expanded FDA Inquiry of BPA Health Risks," press release, June 16, 2008, at [http://www.house.gov/delauro/news.html].

[24] FDA, "Draft Assessment of Bisphenol A for Use in Food Contact Applications," August 14, 2008, p. 2, at [http://www.fda.gov/ohrms/dockets/ac/08/briefing/2008-0038b 1_0 1_00_ index.htm]. FDA's definition of safety in this context is that "there is a reasonable certainty in the minds of competent scientists that the substance is not harmful under the intended conditions of use." 21 CFR § 170.3(i).

[25] See, for example, Connecticut legislation at [http://www.cga.ct.gov/2008/FC/2008HB-0560 1- R000670-FC.htm], and "Companies Move to Curb Risk From Chemical BPA," *Associated Press*, April 21, 2008.

Congressional Activity

On January 17, 2008, John D. Dingell, chairman of the House Committee on Energy and Commerce, and Bart Stupak, chairman of the Subcommittee on Oversight and Investigations, announced an investigation into the use of BPA in products intended for use by infants and children, and FDA's determination of the safety of current uses of BPA in FDA-regulated products.[26]

On April 29, 2008, legislation (S. 2928) was introduced that would prohibit the use of BPA in some products intended for use by a child seven years old or younger. On June 10, 2008, legislation (H.R. 6228) was introduced that would amend FDA's authority such that food or beverage containers composed of BPA, or that could leach BPA into food or beverages, would be considered adulterated and could not be marketed. The latter bill was introduced during a hearing of the House Energy and Commerce Subcommittee on Commerce, Trade, and Consumer Protection on the safety of BPA and phthalates.[27]

CONCLUSION

There is scientific consensus that exposure to high levels of BPA can cause adverse reproductive effects in mammals. It is less clear that low-dose exposures are harmful. There is, however, growing concern about low-dose exposures among the public, and among many scientists, sharpened by the fact that BPA exposures within the general population are, without question, highest in infants. The scientific debate about the safety of BPA is likely to continue, and further reaction in the policy, regulatory, and commercial arenas is expected.

---

[26] Correspondence related to the investigation is at U.S. Congress, House of Representatives, Committee on Energy and Commerce web site, at [http://energycommerce.house.gov/ Investigations/Bisphenol.shtml].

[27] House Energy and Commerce Committee, Subcommittee on Commerce, Trade, and Consumer Protection, hearing on "Safety of Phthalates and Bisphenol-A in Everyday Consumer Products," June 10, 2008, 1 10th Cong., 2nd Sess., Washington, D.C. Phthalates are a class of chemicals that are used to soften plastics, and that are found in a variety of consumer products. For more information, see CRS Report RL34572, *Phthalates in Plastics and Possible Human Health Effects*, by Linda-Jo Schierow and Margaret Mikyung Lee.

In: Consumer Product Safety Issues
Editors: P. A. Varga, M. D. Pintér

ISBN: 978-1-60456-826-4

Chapter 12

CONSUMER PRODUCT SAFETY COMMISSION: CURRENT ISSUES*

Bruce Mulock

SUMMARY

The Consumer Product Safety Commission (CPSC), one of the agency's two current Commissioners stated recently, is at a crossroads. According to CPSC Commissioner Thomas Moore, either Congress will reverse the agency's steady decline or the CPSC will become so weakened as to be irrelevant.

Created in the early 1970s, the CPSC is a small agency by Washington standards (slightly more than 400 employees and a $62.5 million FY2007 appropriation) charged with a seemingly overwhelming responsibility: to protect American consumers against death or injury from unsafe products, including imported products. A series of high- profile product recalls — most notably of children's toys imported from China — has focused attention on the CPSC's resources, including its legal authority, for the task at hand. Legislation (H.R. 4040 and S. 2663) to strengthen the Commission is being considered.

The CPSC is but one part of growing congressional concern about the protection being afforded American consumers from unsafe imports. Food (pet food as well as that intended for human consumption), drugs, and products fall under the jurisdictions of other federal regulatory agencies. This report will be updated as events warrant.

ABOUT THE AGENCY

The Consumer Product Safety Commission is perhaps best understood in the context of the surge of the consumer protection movement of the 1960s. Beginning with Ralph Nader's efforts to dramatize the problems of auto safety, a general interest in product safety was stimulated. The Congress responded by passing a series of laws dealing with auto safety, toy

---

* This is an edited, reformatted and augmented version of CRS Report RS22821, dated June 17, 2008

safety, radiation safety, fabric flammability, and child resistant packaging for hazardous household substances. These laws — as well as several dating from the 1950s — were, in the opinion of some observers, neither comprehensive in scope with respect to the wide range of consumer products nor adequate in protection.

While industry was originally opposed to a federal regulatory approach, by the early 1970s, the National Association of Manufacturers agreed there was a need for Congress to pass a product safety law.[1] A key reason was because manufacturers were increasingly having to contend with a complex and confusing assortment of state laws, and the prospect of having to deal with a unified regulatory approach appeared attractive.

Although there are other historical factors, the CPSC owes its existence primarily to the work of the National Commission on Product Safety (NCPS). Legislation was introduced in 1966 and the following year P.L. 90-146 authorizing the NCPS was approved. In March 1968, President Lyndon Johnson appointed the seven-member, bipartisan Commission to "conduct a comprehensive study and investigation of the scope and adequacy of measures now employed to protect consumers against unreasonable risk of injuries which may be caused by hazardous household products."[2]

The agency's statutory purposes are to (1) protect the public against unreasonable risks of injury associated with consumer products; (2) assist consumers in evaluating the comparative safety of consumer products; (3) develop uniform safety standards for consumer products and minimize conflicting state and local regulations; and (4) promote research and investigation into the causes and prevention of product-related deaths, illnesses, and injuries.

In addition to the authority assigned to the CPSC by its enabling legislation, major consumer programs were transferred to the new agency from the Food and Drug Administration, the Health, Education and Welfare Department (now known as Health and Human Services), and the Commerce Department. Included were the Federal Hazardous Substances Act of 1960, as amended by the Toy Safety Acts of 1969 and 1984 and the Child Protection Amendments of 1966, the Poison Prevention Packaging Act of 1970, the Flammable Fabrics Act of 1953, and the Refrigerator Safety Act of 1956.

The CPSC was designed to be relatively small. Rather than relying on thousands of inspectors, the agency relies on the onus placed on firms by its enabling statute. Section 15(b), for example, requires manufacturers to report unsafe products to the Commission. This general approach appears to have worked reasonably well for domestically produced consumer products, because of the nation's regulatory regime. When it comes to imported products, however, the absence of key factors may make safety oversight more problematic. In China, for example, inspections cannot — with anything approaching confidence — be left to Chinese factory owners. China reportedly suffers from the absence of a rigorous regulatory system, endemic problems of corruption, and a lack of a free press.[3]

---

1 "Product Safety: Stricter Law in Congressional Mill," *Congressional Quarterly*, Dec. 18, 1971, p. 2629.

2 P.L. 90-146, Section 2(a), Nov. 20, 1967.

3 Edwin S. Steinfeld, "The Rogue That Plays by the Rules," The Washington Post, Sept. 2, 2007, p. B3.

Recent Recalls

Beginning in spring 2007, newspaper stories began appearing with increasing regularity about dangerous and sometimes deadly defects in imported products —principally products produced in China.[4] First came reports about toothpaste manufactured in China containing chemicals unsafe for human consumption. Then, awareness of unsafe products was brought home to millions of pet owners when it was discovered that some pet food manufactured in China contained melamine contamination, which was blamed for causing kidney failure in hundreds of animals. Subsequently, health and safety concerns were raised about seafood, tires, and chemical and pharmaceutical products.[5]

Throughout the summer of 2007, additional recalls of toys appeared. Some involved danger because of loose magnets, some because of beads containing a toxic chemical if ingested, and toy ovens with the potential for finger entrapment and burn hazards. A large number of recalls involved jewelry and painted toys containing unsafe levels of lead.

Thus, it is not surprising that there is a concern among many American consumers: If all these problems have been detected or uncovered in spite of various underfunded and understaffed inspection systems, have we been seeing just the tip of the iceberg?

CPSC Funding Reflects Increasing Congressional Concern

Throughout the summer and fall of 2007, appropriators took note of the growing concerns about the adequacy of the agency's funding. At oversight and appropriations hearings in both the House and Senate, witnesses from consumer organizations, industry, and the agency were among those who advocated providing CPSC with more money to get the job done. The CPSC's FY2007 budget was $62.27 million. The House, following the recommendation of the Appropriations Committee, proposed increasing the agency's appropriation to $66.83 million for FY2008. For its part, the Senate proposed $70 million. In the end, under the omnibus funding measure (H.R. 2764), CPSC received $80 million for FY2008, an increase of $17.7 million for FY2008, or 28%.

The appropriators directed CPSC to use its additional funding to increase staff, workspace, and information technology resources. In particular, appropriators instructed the agency to hire in the areas of hazard identification and reduction as well as compliance and field operations.

Consumer groups and others have repeatedly expressed concerns over the CPSC's staffing level. In 1977, three years after the Commission opened its doors, it had a staff of 900. The staffing level has declined sharply over the past three decades from a high of 978 in 1980. The budget for FY2007 culminated a two-year reduction of full-time positions (FTEs) from 471 to 420. The Commission's request for FY2008 anticipated a decrease of an additional 19 FTEs. All indications are that the CPSC will substantially increase its staffing level over the next few years. In the House, H.R. 4040, the Consumer Product Safety

4 For more information, see CRS Report RS22713, Health and Safety Concerns Over U.S. Imports of Chinese Products: An Overview, by Wayne M. Morrison.

5 For an overview of issues concerning U.S. food imports from China, see CRS Report RL34080, Food and Agricultural Imports from China, by Geoffrey S. Becker.

Commission Modernization Act, passed unanimously (407-0) in December 2007. That bill would authorize $80 million for FY2009, $90 million for FY2010, and $100 million for FY201 1. S. 2045, currently pending in the Senate (discussed below), also calls for substantially increased authorizations.

Import Safety Working Group

As noted earlier, the CPSC is but one agency involved in the problem of unsafe imports. Food (pet food as well as that intended for human consumption), drugs, and products (e.g., tires and goods associated with business and commercial application) fall under the jurisdictions of other federal regulatory agencies.

On July 18, 2007, President Bush issued an Executive Order (E.O. 13439) establishing the Interagency Working Group on Import Safety.[6] The working group comprises senior administration officials. It includes the Secretaries of Agriculture, Commerce, Health and Human Services, Transportation, and Homeland Security; the Administrator of the Environmental Protection Agency; and the Chairman of the CPSC.

HHS Secretary Michael Leavitt, who chairs the working group, presented an *Import Safety Action Plan* to President Bush on November 6, 2007. Noting the sheer magnitude of the import-export enterprise, the plan concludes that the United States cannot inspect its way to safety. Rather, the plan presented a number of short- and long-term recommendations designed to improve the safety of imports entering the country. The recommendations include creating a stronger certification process, encouraging various "best practices," increasing transparency, exchanging import data, increasing U.S. presence overseas, enhancing standards, and strengthening penalties.

The Working Group presented President Bush with a "Strategic Framework" to increase import safety that called for several immediate steps. One of these steps was a directive to federal agencies to accelerate their participation in an automated "single window" system for reporting imports electronically. The aim is to achieve better coordination and efficiency, thus facilitating information exchange among government agencies and between the government and the importing community in real time.

Legislative Proposals to Strengthen the CPSC

Although major consumer groups such as the Consumer Federation of America, Consumers Union, and U.S. Public Interest Research Group had been voicing concern about the CPSC's declining budgets (in real terms) and staffing for decades, and had tried for years to persuade Congress to amend and strengthen the agency's powers, until the recent uproar over dangerous imported products, they had made little headway.[7]

---

6 For information on the efforts of the working group, including fact sheets and its "Action Plan for Import Safety," see [http://www.importsafety.gov/].

7 The pattern of legislation being driven by a sudden increase in public awareness is well established. More than a century ago, for example, Upton Sinclair's book *The Jungle* (1906), which dealt with conditions in the U.S. meat packing industry and caused a public uproar, contributed to the passage of the Pure Food and Drug Act and the Meat Inspection Act of 1906.

Last year's spate of recalls of products that poisoned pets, endangered infants and children from lead-painted toys and small magnets, and posed potential health threats to millions who consumed imported fish and other foods has led to major CPSC reform bills in both houses of Congress.

In the House, H.R. 4040, "The Consumer Product Safety Modernization Act of 2007," was favorably reported by the House Energy and Commerce Committee on December 18, 2007, and passed in the House the following day by a vote of 407-0. Among the bill's major provisions are the following:

- Would require mandatory third-party testing of children's products, and requires pre-market testing of many children's products for lead and other hazards by certified laboratories;
- Would prohibit the CPSC from using agency rulemaking actions to preempt state common law claims, and increase the cap on civil penalties to $10 million (current level: $1.8 million);
- Would authorize state attorneys general to enforce CPSC standards;
- Would create new lead standards for children's products; when fully phased in, the bill would lower the standard for lead from 600 parts per million to 100 parts per million or trace amounts. It would mandate that the agency periodically review and revise this standard to require the lowest amount of lead science and technology make feasible to achieve;
- Would require mandatory safety standards for nursery products (e.g., cribs and high chairs); calls for CPSC to examine the current voluntary safety standards for toys, starting with dangerous magnets, and if found to be inadequate, mandatory standards must be adopted;
- Would require manufacturers to place distinguishing marks on products and packaging of children's products to aid in recalls. It would prohibit the sale and export of recalled products; prohibit the export of products that violate U.S. consumer product safety rules; and
- Would provide CPSC with $80 million for 2009, $90 million for FY20 10, and $100 million for FY20 11. It would also restore the agency to a full panel of five commissioners, and include $20 million to modernize the agency's testing lab.

In the Senate, Commerce, Science and Transportation Committee Chairman Daniel Inouye and the committee's Ranking Member, Ted Stevens, announced on February 15, 2008, they had joined with Senator Mark Pryor, sponsor of S. 2045, "The CPSC Reform Act of 2007," to announce a new compromise bill.[8] The new bill, S. 2663, does not include several of S. 2045's more controversial provisions, including giving whistleblowers financial incentives for reporting defects and hazards in products made by companies with whom they were employed and giving state attorneys general power to enforce product safety standards. Those two provisions had represented key differences from H.R. 4040, and had been opposed by the Bush Administration. Nevertheless, important differences remain between the House and Senate bills. S. 2663 provides larger authorizations for the agency, for a longer time period. Further, the Senate bill contains higher civil penalties for violations. And, unlike H.R. 4040, the Senate bill contains a number of provisions aimed at single product issues. These include specific provisions focusing on a garage door opener standard, an all-terrain vehicle safety standard, completion of the agency's cigarette lighter rulemaking, and a rule aimed at child resistant portable gasoline containers.

Key provisions contained in the compromise Senate bill include the following:

- Would authorize giving state attorneys general the right to seek injunctive relief for unsafe products, such as the ability to ask a court to remove a product from store shelves — authority not included in H.R. 4040. S. 2663, however, does not allow states to sue for damages as S. 2045 would have;
- Would grant whistleblowers various protections. Unlike S. 2045, however, S. 2663 does not include providing financial rewards for disclosing information about dangerous products;
- Would restore the CPSC to five members, as with H.R. 4040. Quorum requirements for the agency have complicated efforts to conduct business; it has been functioning with only two of the three commissioners since July 2006;
- Would provide CPSC with $88.5 million in FY2009 and increase funding 10% each year through 2015, i.e., $96.8 million for FY2010, $106.5 million for FY201 1, $117.1 million for FY2012, $128.8 million for FY1013, $141.7 million for FY2014, and $155.9 million for FY2015;
- Would ban lead in all children's products;
- Would require third-party safety certification of children's products;
- Would streamline the product safety rulemaking process to be more timely by eliminating a mandatory "Advanced Notice of Proposed Rulemaking" step; and
- Would allow CPSC to expedite the disclosure of industry-provided information in the interest of public health and safety.

Senate and House conferees were appointed on April 29 and May 14, respectively, and, having missed an earlier target date of the week following the Memorial Day recess, now hope to reach a compromise before the July 4 holiday.

Even as the conferees work to reconcile important differences between the two CPSC reform bills, another high-visibility issue has moved to the front burner for several regulatory agencies, including the CPSC: the safety of phthalates and bisphenol (BPA) in children's products. At House Energy and Commerce Committee hearings on June 10, 2008, scientists presented evidence on the toxicity of a class of industrial chemicals known as phthalates, used to make plastic toys and childcare products. The testimony, which was not conclusive with regard to possible harm, may help conferees decide whether to include a ban of these chemicals in the final version of the CPSC reform legislation.[9] While spokesmen for both CPSC and the Food and Drug Administration said the evidence of danger from BPA was insufficient to corroborate concerns, Canada's health agency announced in April that it planned to put BPA on its list of toxic substances and would ban plastic baby bottles made with the chemical.

In: Consumer Product Safety Issues
Editors: P. A. Varga, M. D. Pintér

ISBN: 978-1-60456-826-4

Chapter 13

HEALTH AND SAFETY CONCERNS OVER U.S. IMPORTS OF CHINESE PRODUCTS: AN OVERVIEW*

Wayne M. Morrison

SUMMARY

China is a major source of U.S. imports of consumer products (such as toys) and an increasingly important supplier of various food products. Reports of unsafe seafood, pet food, toys, tires, and other products imported from China over the past year or so have raised concern in the United States over the health, safety, and quality of imported Chinese products. This report provides an overview of this issue and implications for U.S.-China trade relations and will be updated as events warrant.

In 2007, China overtook Canada to became the largest source of U.S. imports (at $322 billion); about 17% of all U.S. imports now come from China. Over the past year or so, numerous recalls and warnings have been issued by U.S. firms over various products imported from China, due to health and safety concerns. This has led many U.S. policymakers to question the adequacy of China's regulatory environment in ensuring that its exports to the United States meet U.S. standards for health, safety, and quality; as well as the ability of U.S. government regulators, importers, and retailers to identify and take action against unsafe imports (from all countries) before they enter the U.S. market.

WARNINGS, RECALLS, AND DETENTIONS

The Food and Drug Administration (FDA) in March 2007 issued warnings and announced voluntary recalls on certain pet foods (and products use to manufacture pet food and animal feed) from China believed to have caused the sickness and deaths of numerous pets in the United States. In May 2007, the FDA issued warnings on certain toothpaste products (some of which were found to be counterfeit) found to originate in China that

* This is an edited, reformatted and augmented version of CRS Report RS22713, dated September 22, 2008

contained poisonous chemicals. In June 2007, the FDA announced import controls on all farm-raised catfish, basa, shrimp, dace (related to carp), and eel from China after antimicrobial agents, which are not approved in the United States for use in farm- raised aquatic animals, were found. Such shipments will be detained until they are proven to be free of contaminants. On January 25, 2008, the FDA posted on its website a notice by Baxter Healthcare Corporation that it had temporarily halted the manufacture of its multiple-dose vials of heparin (a blood thinner) for injection because of recent reports of serious adverse events (including an estimated8 1 deaths and hundreds of complications) associated with the use of this drug. On February 18, 2008, the *New York Times* reported that a Chinese firm that produces an active ingredient used to produce heparin was not certified by the Chinese government to make the drug and had not undergone FDA inspection; many have speculated that the Chinese plant is likely the source of the problem.[1] On September 12, 2008, the FDA issued a health information advisory on infant formula in response to reports of contaminated milk-based infant formula manufactured and sold in China.[2]

The National Highway Traffic Safety Administration (NHTSA) in June 2007 was informed by Foreign Tire Sales Inc., an importer of foreign tires, that it suspected that up to 450,000 tires (later reduced to 255,000 tires) made in China may have a major safety defect (i.e., missing or insufficient gum strip inside the tire). The company was ordered by the NHTSA to issue a recall. The Chinese government and the manufacturer have maintained that the tires in question meet or exceed U.S. standards.

The Consumer Product Safety Commission (CPSC) issued alerts and announced voluntary recalls by U.S. companies on numerous products made in China in 2007. From January-December 2007, over four-fifths of CPSC recall notices involved Chinese products.[3] Over this period, roughly 17.6 million toy units were recalled because of excessive lead levels. Recalls were also issued on 9.5 million Chinese-made toys (because of the danger of loose magnets), 4.2 million "Aqua Dots" toys (because beads contain a chemical that can turn toxic if ingested) and 1 million toy ovens (due to potential finger entrapment and burn hazzards).[4]

U.S. IMPORTS OF PRODUCTS OF CONCERN FROM CHINA[5]

Table 1 lists products imported from China in 2007 that have been the subject of recent U.S. health and safety concerns, such as toys, seafood, tires, animal foods, organic chemicals and pharmaceuticals, and toothpaste. It indicates that China was a major source of imports for many of these products. For example, China was the largest supplier of imported toys (89% of total), seafood products (15%), and tires (26%); the 2nd largest foreign supplier of animal food products (24%); the 6th largest supplier of toothpaste (1%); and the 9th largest source of

---

[1] New York Times, China Didn't Check Drug Supplier, Files Show, February 16, 2008.

[2] On September 22, 2008, *China Daily* reported that melamine-tainted formula had to date killed three children and sickened 53,000 others (13,000 of which had to be hospitalized).

[3] According to the CPSC, there were 61 toy recalls in 2007 compared with 40 in 2006.

[4] For a list of company recalls of Chinese products, see the CPSC website at [http://www.cpsc.gov/cpscpub/-prerel/prerel.html].

[5] For an overview of issues concerning U.S. food imports from China, see CRS Report RL34080, *Food and Agricultural Imports from China*, by Geoffrey S. Becker.

imported pharmaceuticals and organic chemicals (3%). The table also indicates that, despite health and safety concerns, U.S. imports of most of the products listed (with the exception of toothpaste) increased in 2007 over 2006 levels. For example, toy imports from China grew by 33.4%.

Table 1. U.S. Imports of Selected Products from China in 2007

Product Description	Imports from China ($ millions)	China's Rank as a Source of Imported Product	Imports from China as a % of Total U.S. Imports (%)	Percentage Change in Imports in 2007 over 2006 (%)
Dolls, toys, and games	19,460.5	1	89.4	33.4
Fish and other seafood products	2,054.3	1	14.9	4.8
Tires	2,436.4	1	26.0	28.5
Animal foods	163.0	2	24.0	19.3
Toothpaste	1.3	6	1.0	-59.6
Organic chemicals and pharmaceutical products	3,235.3	9	3.4	25.0
Total imports from China	321,507.8	1	16.5	11.7

Source: USITC DataWeb using various classifications systems and digit levels.

CHINA'S POOR REGULATORY SYSTEM

China is believed to have a rather weak health and safety regime for manufactured goods and agricultural products. Problems include weak consumer protection laws and poorly enforced regulations, lack of inspections and ineffective penalties for code violators, underfunded and understaffed regulatory agencies and poor interagency cooperation, the proliferation of fake goods and ingredients, the existence of numerous unlicensed producers, falsified export documents, extensive pollution,[6] intense competition that often induces firms to cut corners, the relative absence of consumer protection advocacy groups, failure by Chinese companies to effectively monitor the quality of their suppliers' products, restrictions on the media,[7] and extensive government corruption and lack of accountability, especially at the local government level.

[6] For example, many fish farmers in China reportedly feed various drugs to the fish to help keep them alive in polluted waters.

[7] China's media often reports on health and safety problems, but rarely criticizes the central government for such problems.

Chinese officials contend that most Chinese-made products are safe and note that U.S. recalls for health and safety reasons have involved a number of countries (as well as U.S. products). They also argue that some of the blame for recalled products belongs to U.S. importers or by U.S. designers. They further contend that some U.S. products imported into China have failed to meet Chinese standards.[8] However, they have acknowledged numerous product health and safety problems in China, as reflected in reports that have appeared in China's state-controlled media. For example, in June 2004, the Chinese *People's Daily* reported that fake baby formula had killed 50 to 60 infants in China. In June 2006, the *China Daily* reported that 11 people had died from a tainted injection used to treat gall bladders. In August 2006, *Xinhua News Agency* reported that a defective antibiotic drug killed seven people and sickened many others.

China has announced a number of initiatives to improve and strengthen food and drug safety supervision and standards, increase inspections, require safety certificates before some products can be sold, and to crack down on government corruption:

- In May 2007, the *Xinhua News Agency* reported that former director of China's State Food and Drug Administration had been sentenced to death for taking bribes (equivalent to $850,000) in return for approving untested and/or fake medicines (he was executed on July 10, 2007). On the same day, the *Xinhua News Agency* reported that the Chinese government had announced that it would, by the end of 2007, complete regulations for setting up a national food recall system would ban the sale of toys that failed to pass a national compulsory safety certification.
- On June 27, 2007, the *China Daily* reported that a nationwide inspection of the food production industry had found that a variety of dangerous industrial raw materials had been used in the production of flour, candy, pickles, biscuits, black fungus, melon seeds, bean curd, and seafood. As a result, the government reportedly closed 180 food factories found to be producing unsafe products and/or making fake commodities. It also reported that in 2006, the government had conducted 10.4 million inspections, uncovering problems in 360,000 food businesses, and had closed 152,000 unlicensed food businesses.
- On July 4, 2007, the *China Daily* reported that the government had finished making amendments to all food safety standards and had established an emergency response mechanism among several ministries to deal with major problems regarding food safety.
- On August 9, 2007, *China Daily* reported that the government had pledged to spend $1 billion by 2010 to improve drug and food safety.
- On August 15, 2007, a spokesperson from the Chinese embassy in Washington, DC, said that China would require that every food shipment be inspected for quality by the government by September 1, 2007.
- On August 20, 2007, the Chinese government announced that it had created a 19-member cabinet-level panel to oversee product quality and food safety (headed by

---

[8] Since June 2007, China has seized, rejected, or complained about U.S. shipments of orange pulp dried apricots, frozen chicken, pork products, pacemakers, soybeans, frozen potato slices, and sardines. In September 2007, China reportedly ordered increased inspections of imported U.S. products. Some analysts contend that these actions are retaliation for U.S. recalls of Chinese products.

Vice-Premier Wu Yi) and would start a four- month nationwide campaign to improve the quality of goods and food.

- On December 5, 2007, the government stated that during the first 10 months of the year, it had shut down 47,800 food factories without operating licenses.
- On January 15, 2008, China announced it had inspected over 3,000 export-oriented toy manufacturers and had revoked licences for 600 firms that failed to meet quality standards.

Despite these efforts, reports of tainted products persist. For example in January 2008, dozens of people in Japan reportedly became ill from eating dumplings imported from China that contained pesticide. In addition, hundreds of cancer patients in China were reportedly sickened by tainted drugs produced by a major Chinese drug company. On September 16, 2008, *China Daily* reported that tainted baby formula had killed 2 infants and sickened more than 1,200 others.

The United States and China reached a number of agreements in 2007 to address health and safety concerns:

- On September 11, 2007, the CPSC and its Chinese counterpart, the General Administration of Quality Supervision, Inspection and Quarantine (AQSIQ), signed a Joint Statement on enhancing consumer product safety. China pledged to implement a comprehensive plan to intensify efforts (such as increased inspections, efforts to educate Chinese manufacturers, bilateral technical personal exchanges and training, regular meetings to exchange information with U.S. officials, and the development of a product tracking system) to prevent exports of unsafe products to the United States, especially in regard to lead paint and toys.
- On September 12, 2007, the NHTSA signed a Memorandum of Cooperation with its Chinese counterpart on enhanced cooperation and communication on vehicles and automotive equipment safety.
- On December 11, 2007, the U.S. Health and Human Services (HHS) announced that it had signed two Memoranda of Agreement (MOA) with its Chinese counterparts; the first covering specific food and feed items that have been of concern to the United States, and the second covering drugs and medical devices. Both MOAs would require Chinese firms that export such products to the United States to register with the Chinese government and to obtain certification before they can export. Such firms would also be subject to annual inspections to ensure they meet U.S. standards. The MOAs also establish mechanisms for greater information sharing, increase access of production facilities by U.S. officials, and creates working group in order to boost cooperation. On March 13, 2008, the FDA announced that it planned to place eight FDA staffers in China. Some members of Congress have proposed placing a CPSC official at the U.S. embassy in Beijing.

Economic Implications

Many Members of Congress have called for tighter rules (such as increased inspections, certification requirements, and mandatory standards for toys), and increased funding for U.S. product safety agencies. On December 19, 2007, the House passed H.R. 4040 (Rush): *the Consumer Product Safety Modernization Act*. On March 6, 2008, the Senate passed its version of H.R. 4040 as a substitute amendment (S. 2263: *the CPSC Reform Act*). House and Senate Conferees reached a compromise agreement on H.R. 4040 on July 28, 2008, and the bill was signed into law (P.L. 110-314) on August 14.[9]

Concerns over the health, safety, and quality of Chinese products could have a number of important economic implications. Both the United States and China have accused each other of using health and safety concerns as an excuse to impose protectionist measures and some observers contend that this issue could lead to growing trade friction between the two sides. International concerns over the safety of Chinese exports may diminish the attractiveness of China as a destination for foreign investment in export-oriented manufacturing, as well as for foreign firms that contract with Chinese firms to make and export products under their labels (such as toys). Efforts by China to restore international confidence in the health and safety of its exports through increased inspections, certification requirements, mandatory testing, etc., could have a significant impact on the cost of doing business in China, which could slow the pace of Chinese exports and hurt employment in the export sector. Moreover, international concerns over the safety of Chinese products could prove to be a setback to the government's efforts to develop and promote internationally recognized Chinese brands (such as cars), which it views as important to the country's future economic development. Thus, it is very likely the Chinese government will take this issue very seriously. However, it is unclear how long it will take for the central government to effectively address the numerous challenges it faces (especially government corruption and counterfeiting) to ensure that its exports comply with the health and safety standards of the United States and other trading partners. Additionally, a sharp decrease in purchases by U.S. consumers of Chinese products could negatively impact U.S. firms that import and/or sell such products and may raise prices of some commodities as firms attempt to rectify various safety problems.

The current crisis in China over melamine-tainted milk (which can cause kidney stones) and the growing number of children who have reportedly have become ill (tens of thousands) have seriously challenged the government's assertions that most products made in China are safe and that an effective regulatory regime has been established. The government announced on September 22, 2008, that China's chief quality supervisor had stepped down from his post over the incident. Other local and provincial officials have reportedly been sacked for trying to cover up incident. At least 22 Chinese baby formula companies have been found to have tainted products. Press reports indicate that other milk products made in China may have been contaminated as well.

---

[9] See CRS Report RL34399, Consumer Product Safety Improvement Act of 2008: H.R. 4040, by Margaret Mikyung Lee, and CRS Report RS22821, Consumer Product Safety Commission: Current Issues, by Bruce Mulock.

In: Consumer Product Safety Issues
Editors: P. A. Varga, M. D. Pintér
ISBN: 978-1-60456-826-4

Chapter 14

DRUG SAFETY: PRELIMINARY FINDINGS SUGGEST RECENT FDA INITIATIVES HAVE POTENTIAL, BUT DO NOT FULLY ADDRESS WEAKNESSES IN ITS FOREIGN DRUG INSPECTION PROGRAM*

WHY GAO DID THIS STUDY

The Food and Drug Administration (FDA) is responsible for overseeing the safety and effectiveness of human drugs that are marketed in the United States, whether they are manufactured in foreign or domestic establishments. FDA inspects foreign establishments to ensure that they meet the same standards required of domestic establishments. Ongoing concerns regarding FDA's foreign drug inspection program recently were heightened when FDA learned that contaminated doses of a common blood thinner had been manufactured at a Chinese establishment that the agency had never inspected. FDA has announced initiatives to improve its foreign drug inspection program.

In November 2007, GAO testified on weaknesses in FDA's foreign drug inspection program (GAO-08-224T). This statement presents preliminary findings on how FDA's initiatives address the weaknesses GAO identified. GAO interviewed FDA officials and analyzed FDA's initiatives. GAO examined reports and proposals prepared by the agency, as well as its plans to improve databases it uses to manage its foreign drug inspection program.

WHAT GAO FOUND

Recent FDA initiatives—some of which have been implemented and others proposed—could strengthen FDA's foreign drug inspection program, but these initiatives do not fully address the weaknesses that GAO previously identified.

* This is an edited, reformatted and augmented version of GAO report GAO-08-701T, dated April 22, 2008.

- GAO testified in November 2007 that FDA's databases do not provide an accurate count of foreign establishments subject to inspection and do provide widely divergent counts. Through one recent initiative, FDA has taken steps to improve its database intended to include foreign establishments registered to market drugs in the United States. This initiative may reduce inaccuracies in FDA's count of foreign establishments. However, these steps will not prevent foreign establishments that do not manufacture drugs for the U.S. market from erroneously registering with FDA. Further, to reduce duplication in its import database, FDA has supported a proposal that would change the data it receives on products entering the United States. However, the implementation of this proposal is not certain and would require action from multiple federal agencies, in addition to FDA. Efforts to integrate these databases have the potential to provide FDA with a more accurate count of establishments subject to inspection, but it is too early to tell.
- GAO testified that gaps in information weaken FDA's processes for prioritizing the inspection of foreign establishments that pose the greatest risk to public health. While FDA recently expressed interest in obtaining useful information from foreign regulatory bodies that could help it prioritize foreign establishments for inspections, the agency has faced difficulties fully utilizing these arrangements in the past. For example, FDA had difficulties in determining whether the scope of other countries' inspection reports met its needs and these reports were not always readily available in English.
- GAO also testified that FDA inspected relatively few foreign establishments each year. FDA made progress in inspecting more foreign establishments in fiscal year 2007, but the agency still inspects far fewer of them than domestic establishments. FDA dedicated about $10 million to foreign drug inspections in fiscal year 2007 and plans to dedicate about $11 million to such inspections in fiscal year 2008.
- Finally, GAO testified that FDA faced certain logistical and staffing challenges unique to conducting foreign inspections. FDA is pursuing initiatives that could address some of the challenges that we identified as being unique to foreign inspections, such as volunteer inspection staff and lack of translators. FDA has proposed establishing a dedicated cadre of staff to conduct foreign inspections, but the timeframe associated with this initiative is unclear. FDA plans to open an office in China and is considering establishing offices in other countries, but the impact that this will have on the foreign drug inspection program is unknown.

Mr. Chairman and Members of the Subcommittee:

I am pleased to be here today as you consider the Food and Drug Administration's (FDA) plans to improve its program for inspecting foreign drug manufacturers whose products are marketed in the United States. America has become increasingly dependent on drugs and drug ingredients manufactured in foreign countries. Ten years ago, we reported that FDA needed to improve its foreign drug inspection program.[1] Among other things, we noted that FDA had

[1] GAO, Food and Drug Administration: Improvements Needed in the Foreign Drug Inspection Program, GAO/HEHS-98-21 (Washington, D.C.: Mar. 17, 1998).

serious problems managing its foreign inspection data. We were also critical of the number of inspections FDA conducted at foreign manufacturers. In November 2007, we testified on the preliminary findings of our current work in which we identified weaknesses similar to those we found in our previous report.[2] Our preliminary findings suggested that FDA had weaknesses in its databases, including conflicting information on the number of foreign establishments subject to inspection;[3] had information gaps that weakened its process for selecting foreign establishments for inspection; conducted infrequent inspections of these establishments; and faced logistical and staffing challenges unique to foreign inspections. Recent developments involving heparin sodium, a commonly used blood thinner, have further heightened concerns about the safety of drugs and drug ingredients and FDA's ability to inspect foreign manufacturers of these products. In January 2008, FDA began an investigation after receiving reports of serious adverse events in people receiving this drug. The agency later learned that an active pharmaceutical ingredient (API) found in heparin sodium contained a contaminant and had been manufactured at a Chinese establishment never inspected by FDA.[4]

Recently, FDA has begun or proposed initiatives to strengthen its foreign drug inspection program.[5] You asked us to assess whether FDA's initiatives will improve its management of this program. My testimony today will focus on these initiatives and how they address the weaknesses we previously identified.

To obtain information about FDA initiatives and how they address weaknesses in its program for inspecting foreign drug manufacturers, we interviewed officials from FDA, including from its Center for Drug Evaluation and Research (CDER) and Office of Regulatory Affairs (ORA), which each have responsibilities for managing the foreign inspection program. We examined reports and proposals prepared by the agency on related initiatives. We also examined FDA's plans to improve databases it uses to manage its foreign drug inspection program, including its Field Accomplishments and Compliance Tracking System (FACTS), Operational and Administrative System for Import Support (OASIS), and Drug Registration and Listing System (DRLS).[6] Our November 2007 testimony included the number of inspections from FACTS as of September 26, 2007. To provide information to update those preliminary findings, we obtained FACTS data that contained information on fiscal year 2007 inspections conducted or entered into this database since our previous analysis. We also obtained fiscal year 2007 data from OASIS to determine the types of drug

2 GAO, Drug Safety: Preliminary Findings Suggest Weaknesses in FDA's Program for Inspecting Foreign Drug Manufacturers, GAO-08-224T (Washington, D.C.: Nov. 1, 2007). We also recently testified about similar weaknesses that we identified in FDA's program for inspecting foreign medical device manufacturers. GAO, Medical Devices: Challenges for FDA in Conducting Manufacturer Inspections, GAO-08-428T (Washington, D.C.: Jan. 29, 2008).

3 FDA regulations define an establishment as a place of business under one management at one general physical location. 21 C.F.R. § 207.3(a)(7) (2007). Drug firms may have more than one establishment.

4 An API is any component that is intended to provide pharmacological activity or other direct effect in the diagnosis, cure, mitigation, treatment, or prevention of disease. FDA defines inactive ingredients as any component of a drug product other than the API, such as materials that improve the appearance, stability, and palatability of the product. According to FDA officials, the agency typically only inspects establishments manufacturing inactive ingredients on a for-cause basis.

5 See, for example, Food and Drug Administration, Revitalizing ORA: Protecting the Public Health Together In a Changing World (Rockville, Md.: Jan. 2008).

6 We also previously examined the reliability of DRLS. We found that DRLS was reliable, to the extent that it accurately reflects information provided by foreign drug manufacturing establishments that register with FDA. However, we determined that these data do not necessarily reflect all foreign establishments whose drugs are imported into the United States. We do not present new information from DRLS in this testimony.

products manufactured in China and offered for entry into the United States. We assessed the reliability of these databases by (1) reviewing existing information about the data and the databases that produced them, (2) interviewing agency officials knowledgeable about the data, and (3) performing electronic testing of required data elements. We found the data in the FACTS database reliable for our purposes. In addition, we found that while OASIS is likely to over-estimate the number of foreign establishments involved in the manufacture of those drugs because of uncorrected errors in the data, it provides sufficiently reliable information about the types of drugs offered for entry into the United States. The information we present represents the best information available and is what FDA relies on to manage its foreign drug inspection activities. Our ongoing work is focused on human drugs regulated by CDER and not on biologics,[7] medical devices, veterinary medicines, food, or other items or products for which FDA conducts inspections. However, we obtained information from the center responsible for medical devices, the Center for Devices and Radiological Health (CDRH), to learn about a recent change to one of its databases that addresses problems similar to those in DRLS. We shared the facts contained in this statement with FDA officials. They provided technical comments, which we incorporated as appropriate. We conducted the work for our November 2007 testimony from September 2007 through October 2007, and we conducted our work for this statement from March 2008 through April 2008. All of our work is being performed in accordance with generally accepted government auditing standards. Those standards require that we plan and perform the audit to obtain sufficient, appropriate evidence to provide a reasonable basis for our findings and conclusions based on our audit objectives. We believe that the evidence obtained provides a reasonable basis for our findings and conclusions based on our audit objectives.

In summary, recent FDA initiatives—some of which have been implemented and others proposed—could strengthen FDA's foreign drug inspection program, but these initiatives do not fully address the weaknesses that we previously identified. For example, we found that FDA's databases do not provide an accurate count of foreign establishments subject to inspection. FDA plans to implement electronic registration for foreign establishments. Implementing such a process may reduce inaccuracies in FDA's database of registered establishments. However, this will not prevent foreign establishments that do not manufacture drugs for the U.S. market from erroneously registering with FDA. For example, in some foreign markets, foreign drug manufacturers may register with FDA because registration may appear to convey an "approval" or endorsement by the agency. To reduce duplication in FDA's import database, FDA supported a proposal to create a unique governmentwide identifier for all establishments whose products are imported into the United States. However, the implementation of this identifier is not certain and would require action from multiple federal agencies in addition to FDA. Efforts to integrate these databases have the potential to provide FDA with a more accurate count of establishments subject to inspection, but it is too early to tell. FDA has also taken steps that could help it select foreign establishments for inspection by obtaining information from foreign regulatory bodies. However, the agency has not fully utilized arrangements with foreign regulatory bodies in the past that would allow it to obtain such information. FDA has also made progress in conducting more foreign

[7] Biologics are materials, such as vaccines, derived from living sources such as humans, animals, and microorganisms. Some biologics are regulated by CDER and inspections related to those products are included in our work.

inspections, but it still inspects relatively few establishments. FDA is pursuing initiatives that could address some of the challenges that we identified as unique to foreign inspections. For example, the agency has proposed establishing overseas offices, beginning in China, but the impact that these offices will have on the foreign drug inspection program is unknown. To date, it is unclear whether the agency's proposals will increase the frequency with which FDA inspects foreign establishments or the quality of information it uses to select establishments to inspect.

BACKGROUND

FDA is responsible for overseeing the safety and effectiveness of human drugs that are marketed in the United States, whether they are manufactured in foreign or domestic establishments.[8] Foreign establishments that market their drugs in the United States must register with FDA. As part of its efforts to ensure the safety and quality of imported drugs, FDA may inspect foreign establishments whose products are imported into the United States. Regular inspections of manufacturing establishments are an essential component of ensuring drug safety. Conducting testing of finished dosage form drug products cannot reliably determine drug quality. Therefore, FDA relies on inspections to determine an establishment's compliance with current good manufacturing practice regulations (GMP).[9] These inspections are a critical mechanism in FDA's process of assuring that the safety and quality of drugs are not jeopardized by poor manufacturing practices.

Requirements governing foreign and domestic inspections differ. Specifically, FDA is required to inspect every 2 years those domestic establishments that manufacture drugs marketed in the United States,[10] but there is no comparable requirement for inspecting foreign establishments. FDA does not have authority to require foreign establishments to allow the agency to inspect their facilities. However, FDA has the authority to conduct physical examinations of products offered for import, and if there is sufficient evidence of a violation, prevent their entry at the border.

Within FDA, CDER sets standards and evaluates the safety and effectiveness of prescription and over-the-counter drugs. Among other things, CDER requests that ORA inspect both foreign and domestic establishments to ensure that drugs are produced in conformance with federal statutes and regulations, including current GMPs. CDER requests that ORA conduct inspections of establishments that produce drugs in finished-dosage form as well as those that produce bulk drug substances,[11] including APIs used in finished drug products. These inspections are performed by investigators and, on occasion, laboratory

---

[8] FDA regulations define manufacturing to include the manufacture, preparation, propagation, compounding, or processing of a drug. 21 C.F.R. § 207.3(a)(8) (2007).

[9] GMPs provide a framework for a manufacturer to follow to produce safe, pure, and high- quality products. See 21 C.F.R. pts. 210, 211 (2007).

[10] 21 U.S.C. § 360(h).

[11] A bulk drug substance is any substance that is represented for use in a drug that, when used in the manufacturing, processing, or packaging of a drug, becomes an active ingredient or a finished drug product. 21 C.F.R. § 207.3(a)(4) (2007).

analysts.[12] ORA conducts two primary types of drug manufacturing establishment inspections:

- Preapproval inspections of domestic and foreign establishments are conducted before FDA will approve a new drug to be marketed in the United States.[13] These inspections occur following FDA's receipt of a new drug application (NDA) or an abbreviated new drug application (ANDA) and focus on the manufacture of a specific drug.[14] Preapproval inspections are designed to verify the accuracy and authenticity of the data contained in these applications to determine that the manufacturer is following commitments made in the application. FDA also determines that the manufacturer of the finished drug product, as well as each manufacturer of a bulk drug substance used in the finished product, manufactures, processes, packs, and labels the drug adequately to preserve its identity, strength, quality, and purity.
- Postapproval GMP surveillance inspections are conducted to ensure ongoing compliance with the laws and regulations pertaining to the manufacturing processes used by domestic and foreign establishments in the manufacture of drug products marketed in the United States and bulk drug substances used in the manufacture of those products. These inspections focus on a manufacturer's systemwide controls for ensuring that drug products are of high quality. Systems examined during these inspections include those related to materials, quality control, production, facilities and equipment, packaging and labeling, and laboratory controls. These systems may be involved in the manufacture of multiple drug products.[15]

FDA has established arrangements with regulatory bodies in other countries to facilitate the sharing of information about drug inspections. FDA has entered into arrangements related to GMP inspections with Canada, Japan, the European Union, and others. The scope of such arrangements can vary. Some arrangements may allow FDA to obtain reports of inspections conducted by other countries, for informational purposes. Other arrangements may involve more than the exchange of information. For example, FDA and another country may enter into an arrangement to work towards the mutual recognition of each other's inspection standards or the acceptance of one another's inspections, in lieu of their own.

CDER uses a risk-based process to select some foreign and domestic establishments for postapproval GMP surveillance inspections. The process uses a risk-based model to identify

---

[12] ORA investigators lead inspections. Investigators are responsible for performing or overseeing all aspects of an inspection. ORA laboratory analysts are chemists or microbiologists and have expertise in laboratory testing.

[13] When FDA receives an application for drug approval, officials review the inspection history of each establishment listed on the application. According to FDA officials, if an establishment listed on the application has received a satisfactory GMP inspection in the 2 previous years and the agency has no new concerns, FDA will consider this inspection sufficient and will not perform a preapproval inspection of this establishment.

[14] FDA must approve an NDA in order for a new drug to be marketed in the United States. FDA reviews scientific and clinical data contained in these applications as part of its process in considering them for approval to be marketed. Approval for a generic drug is sought through an ANDA.

[15] In addition, FDA conducts for-cause inspections when it receives information indicating problems in the manufacture of approved drug products, as well as when it follows up on manufacturers that were not in compliance with GMPs during previous inspections.

those establishments that, based on characteristics of the establishment and of the product being manufactured, have the greatest public health risk potential should they experience a manufacturing defect. For example, FDA considers the risk to public health from poor quality over-the-counter drugs to be lower than for prescription drugs. Consequently establishments manufacturing only over-the-counter drugs receive a lower score on this factor in the risk-based process than other manufacturers. Through this process, CDER annually prepares a prioritized list of domestic establishments and a separate, prioritized list of foreign establishments.

FDA uses multiple databases to manage its foreign drug inspection program.

- DRLS contains information on foreign and domestic drug establishments that have registered with FDA to market their drugs in the United States. These establishments must also list any drugs they market in the United States. These establishments provide information, such as company name and address and the drug products they manufacture for commercial distribution in the United States, on paper forms, which are entered into DRLS by FDA staff.
- OASIS contains information on drugs and other FDA-regulated products offered for entry into the United States, including information on the establishment that manufactured the drug. The information in OASIS is automatically generated from data managed by Customs and Border Protection (CBP). The data are originally entered by customs brokers based on the information available from the importer.[16] CBP specifies an algorithm by which customs brokers generate a manufacturer identification number from information about an establishment's name, address, and location.
- FACTS contains information on FDA's inspections of foreign and domestic drug establishments. FDA investigators and laboratory analysts enter information into FACTS following completion of an inspection.

According to DRLS, in fiscal year 2007, foreign countries that had the largest number of registered establishments were Canada, China, France, Germany, India, Italy, Japan, and the United Kingdom. These countries are also listed in OASIS as having the largest number of manufacturers offering drugs for entry into the United States. Specifically, according to OASIS, China had more establishments manufacturing drugs that were offered for entry into the United States than any other country. According to OASIS, in fiscal year 2007, a wide variety of prescription and over-the- counter drug products manufactured in China were offered for entry into the United States, including pain killers, antibiotics, blood thinners, and hormones.

In November 2007, we testified on preliminary findings that identified weaknesses in FDA's program for inspecting foreign establishments manufacturing drugs for the U.S. market. Specifically, we found that, as in 1998, FDA's effectiveness in managing the foreign drug inspection program continued to be hindered by weaknesses in its data on foreign establishments. FDA did not know how many foreign establishments were subject to inspection. FDA relied on databases that were designed for purposes other than managing the

---

[16] Customs brokers are private individuals, partnerships, associations, or corporations licensed, regulated, and empowered by CBP to assist in meeting federal requirements governing imports and exports.

foreign drug inspection program. Further, these databases contained inaccuracies that FDA could not easily reconcile. DRLS indicated there were about 3,000 foreign establishments registered with FDA in fiscal year 2007,[17] while OASIS indicated that about 6,800 foreign establishments actually offered drugs for entry in that year. FDA recognized these inconsistencies, but could not easily correct them partly because the databases could not exchange information. Any comparisons of the data must be performed manually, on a case-by-case basis.

Table 1. Number of FDA Inspections of Foreign Establishments Involved in the Manufacture of Drugs for the U.S. Market, by Country for the 10 Most Frequently Inspected Countries, Fiscal Year 2002 through Fiscal Year 2007

Country	Number of inspections							Number of unique establishments inspected	Number of establishments[a]
	FY2002	FY2003	FY2004	FY2005	FY2006	FY2007	Total		
India	11	19	38	33	34	64	199	152	410
Germany	24	15	35	25	19	25	143	95	199
Italy	17	30	26	21	18	28	140	98	150
Canada	29	12	17	23	23	20	124	88	288
United Kingdom	17	21	15	18	15	16	102	84	169
China	11	9	17	21	17	19	94	80	714
France	14	15	13	12	16	24	94	71	162
Japan	11	13	14	21	13	22	94	82	196
Switzerland	12	12	11	17	9	17	78	50	83
Ireland	11	5	11	14	3	14	58	43	61
All other countries	63	38	63	61	45	83	353	276	817
Total	220	189	260	266	212	332	1,479	1,119	3,249

Source: GAO analysis of FDA data.

a This count represents the number of establishments FDA used to plan its fiscal year 2007 prioritized surveillance inspections. In preparing this list, FDA draws on information from DRLS. It also obtains information from previous inspections to help it identify establishments that are subject to inspection but are not required to register—such as the manufacturer of an API whose product is not directly imported into the United States. However, as a result of the inaccuracies in DRLS, FDA recognizes that this list does not provide an accurate count of establishments subject to inspection.

We also testified that FDA inspected relatively few foreign establishments.[18] Data from FDA suggested that the agency may inspect about 8 percent of foreign establishments in a given year. At this rate, it would take FDA more than 13 years to inspect each foreign

[17] This count includes foreign establishments that were registered to manufacture human drugs, biologics, and veterinary drugs; FDA was unable to provide the number of registered establishments specifically manufacturing human drugs.

[18] We updated information presented in our November 2007 testimony because that data did not include complete counts of inspections conducted in fiscal year 2007.

establishment once, assuming that no additional establishments require inspection. However, FDA could not provide an exact number of foreign establishments that had never been inspected. From fiscal year 2002 through fiscal year 2007, FDA conducted 1,479 inspections of foreign establishments, and three quarters of these inspections were concentrated in 10 countries. (See table 1.) Because some establishments were inspected more than once during this time period, FDA actually inspected 1,119 unique establishments. For example, of the 94 inspections that FDA conducted of Chinese establishments, it inspected 80 unique establishments across this six year period. The lowest rate of inspections in these 10 countries was in China, for which FDA inspected 80 of its estimated 714 establishments, or fewer than 14 establishments per year, on average.

We testified that, while enforcing GMP compliance through surveillance inspections was FDA's most comprehensive program for monitoring the quality of marketed drugs, most of FDA's inspections of foreign manufacturers occurred when they were listed in an NDA or ANDA. The majority of these preapproval inspections were combined with a GMP surveillance inspection. Although FDA used a risk-based process to develop a prioritized list of foreign establishments for GMP surveillance inspections, few were completed in a given year—about 30 in fiscal year 2007. The usefulness of the process was weakened by the incomplete and possibly inaccurate information on those foreign establishments that FDA had not inspected recently, as well as those that had never been the subject of a GMP surveillance inspection.

We also testified that FDA's foreign inspection process involves unique circumstances that are not encountered domestically. For example, FDA relies on staff that inspect domestic establishments to volunteer for foreign inspections. Unlike domestic inspections, FDA does not arrive unannounced at a foreign establishment. It also lacks the flexibility to easily extend foreign inspections if problems are encountered. Finally, language barriers can make foreign inspections more difficult than domestic ones. FDA does not generally provide translators to its inspection teams. Instead, they may have to rely on an English-speaking representative of the foreign establishment being inspected, rather than an independent translator.

Recent Initiatives May Help FDA Select Foreign Establishments for Inspection, but Weaknesses in Its Foreign Drug Inspection Program Are Not Fully Addressed

FDA has initiated several recent changes to its foreign drug inspection program, but the changes do not fully address the weaknesses that we previously identified. FDA has initiatives underway to reduce the inaccuracies in its registration and import databases that make it difficult to determine the number of foreign establishments subject to inspection, although to date these databases still do not provide an accurate count of such establishments. FDA has taken steps that could help it select foreign establishments for inspection by obtaining information from foreign regulatory bodies. However, the agency has not fully utilized arrangements with foreign regulatory bodies in the past that would allow it to obtain such information. FDA has made progress in conducting more foreign inspections, but it still inspects relatively few establishments. FDA is also pursuing initiatives that could address some of the challenges that we identified as being unique to foreign inspections, but implementation details and timeframes associated with these initiatives are unclear.

FDA Initiatives Could Improve Its Data, but Will Not Ensure an Accurate Count of Foreign Establishments Subject to Inspection

FDA has initiatives underway to reduce inaccuracies in its databases, but actions taken thus far will not ensure that the agency has an accurate count of establishments subject to inspection. As we previously testified, DRLS does not provide FDA with an accurate count of foreign establishments manufacturing drugs for the U.S. market. For example, foreign establishments may register with FDA, whether or not they actually manufacture drugs for the U.S. market,[19] and the agency does not routinely verify the information provided by the establishment. Beginning in late 2008, CDER plans to implement an electronic registration and listing system that could improve the accuracy of information the agency maintains on registered establishments. The new system will allow drug manufacturing establishments to submit registration and listing information electronically, rather than submitting it on paper forms. FDA hopes that electronic registration will result in efficiencies allowing the agency to shift resources from data entry to assuring the quality of the databases. However, electronic registration alone will not prevent foreign establishments that do not manufacture drugs for the U.S. market from registering, thus still presenting the problem of an inaccurate count.

Recently, another FDA center implemented changes affecting the registration of medical device manufacturers, an activity for which we previously identified problems similar to those found in CDER.[20] In fiscal year 2008, CDRH implemented, in addition to electronic registration, an annual user fee of $1,706 per registration for certain medical device establishments[21] and an active re-registration process.[22] According to CDRH, as of early April 2008, about half of the previously registered establishments have reregistered using the new system. While CDRH officials expect that this number will increase,[23] they expect that the elimination of establishments that do not manufacture medical devices for the U.S. market—and thus should not be registered—will result in a smaller, more accurate database of medical device establishments. CDRH officials indicated that implementation of electronic registration and the annual user fee seems to have improved the data so CDRH can more accurately identify the type of establishment registered, the devices manufactured at an establishment, and whether or not an establishment should be registered. According to CDRH

19 FDA officials pointed out that some foreign establishments register, for example, because registration may erroneously appear to convey an "approval" or endorsement by FDA in foreign markets.

20 GAO, Medical Devices: Challenges for FDA in Conducting Manufacturer Inspections, GAO-08-428T (Washington, D.C.: Jan. 29, 2008).

21 21 U.S.C. §§ 379i(13); 379j(a)(3), (b), (h). The registration user fee is $1,706 in fiscal year 2008 and will increase by 8.5 percent per year, to $2,364 in fiscal year 2012. Fees are available for obligation only to the extent and in the amount provided in annual appropriations acts. FDA's authority to assess registration fees terminates on October 1, 2012.

22 CDRH indicated that the center will deactivate the registrations of those establishments that fail to complete the annual registration. Officials noted that, in the past, many establishments that had previously registered had not updated those registrations in several years.

23 According to CDRH, in April, the center will send letters to establishments that have registered in the past but have not completed their registration for fiscal year 2008 advising them that they must register using the new system and must pay the registration fee, if applicable, to be considered registered.

officials, the revenue from device registration user fees is applied to the process for the review of device applications,[24] including establishment inspections undertaken as part of the application review process. CDER does not currently have the authority to assess a user fee for registration of drug establishments, but officials indicated that such a fee could discourage registrations of foreign manufacturers that are not ready, are not actively importing, or have not been approved to market drug products in the United States. Officials also suggested that such fees could be used to supplement the resources available for conducting inspections.

FDA has proposed, but not yet implemented, the Foreign Vendor Registration Verification Program, which could help improve the accuracy of information FDA maintains on registered establishments. Through this program, FDA plans to contract with an external organization to conduct on-site verification of the registration data and product listing information of foreign establishments shipping drugs and other FDA-regulated products to the United States. As of April 2008, FDA had solicited proposals for this contract but was still developing the specifics of the program. For example, the agency had not yet established the criteria it would use to determine which establishments would be visited for verification purposes or determined how many establishments it would verify annually. FDA currently plans to award this contract in May 2008. Given the early stages of this process, it is too soon to determine whether this program will improve the accuracy of the data FDA maintains on foreign drug establishments.

In addition to changes to improve DRLS, FDA has supported a proposal that has the potential to address weaknesses in OASIS, but FDA does not control the implementation of this change. As we previously testified, OASIS contains an inaccurate count of foreign establishments manufacturing drugs imported to the United States as a result of unreliable identification numbers generated by customs brokers when the product is offered for entry.[25] FDA officials told us that these errors result in the creation of multiple records for a single establishment, which results in inflated counts of establishments offering drugs for entry into the U.S. market. FDA is pursuing the creation of a governmentwide unique establishment identifier, as part of the Shared Establishment Data Service (SEDS), to address these inaccuracies.[26] Rather than relying on the creation and entry of an identifier at the time of import, SEDS would provide a unique establishment identifier and a centralized service to provide commercially verified information about establishments. The standard identifier would be submitted as part of import entry data where required by FDA or other government agencies. SEDS could thus eliminate the problem of having multiple identifiers associated with an individual establishment. The implementation of SEDS is dependent on action from multiple federal agencies, including the integration of the concept into a CBP import and export system currently under development and scheduled for implementation in 2010. In addition, once implemented by CBP, participating federal agencies would be responsible for bearing the cost of integrating SEDS with their own operations and systems. FDA officials are not aware of a specific timeline for the implementation of SEDS. Developing an implementation plan for SEDS is a recommendation of Interagency Working Group Import Safety's *Action Plan for Import Safety: A Roadmap for Continual Improvement*.

---

24 21 U.S.C. § 379i(8).

25 The algorithm currently used by customs broker to assign the manufacturer identification number does not provide for a number that is reliably reproduced or inherently unique.

26 The SEDS concept was developed by a working group with representatives from FDA, the Environmental Protection Agency, and the departments of Agriculture, Commerce, Defense, and Homeland Security.

Finally, FDA is in the process of implementing additional initiatives to improve the integration of its current data systems, which could make it easier for the agency to establish an accurate count of foreign drug manufacturing establishments subject to inspection. The agency's Mission Accomplishments and Regulatory Compliance Services (MARCS) is intended to help FDA electronically integrate data from multiple systems.

It is specifically designed to give individual users a more complete picture of establishments. FDA officials estimate that MARCS, which is being implemented in stages, could be fully implemented by 2011 or 2012. However, FDA officials told us that implementation has been slow because the agency has been forced to shift resources away from MARCS and toward the maintenance of current systems that are still heavily used, such as FACTS and OASIS. Taken together, electronic registration, the Foreign Vendor Registration Verification Program, SEDS, and MARCS could provide the agency with more accurate information on the number of establishments subject to inspection. However, it is too early to tell.

FDA Initiatives to Obtain Information on Foreign Establishments May Have Limited Impact on Its Selection of Establishments to Inspect

FDA has taken steps to help it select establishments for inspection by obtaining information on foreign establishments from regulatory bodies in other countries, despite encountering difficulties in fully utilizing these arrangements in the past. FDA has recognized the importance of receiving information about foreign establishments from other countries and has taken steps to develop new, or strengthen existing, information-sharing arrangements to do so. For example, according to FDA, the agency is enhancing an arrangement to exchange information with the Swiss drug regulatory agency. FDA officials have highlighted such arrangements as a means of improving the agency's oversight of drugs manufactured in foreign countries. For example, they told us that in selecting establishments for GMP surveillance inspections, they sometimes use the results of an establishment inspection conducted by a foreign government to determine whether to inspect an establishment.[27] FDA told us that it received drug inspection information from foreign regulatory bodies six times in 2007.

FDA has previously encountered difficulties which prevented it from taking full advantage of information-sharing arrangements with other countries. Obtaining inspection reports from other countries and using this information has proved challenging. In order for FDA to determine the value of inspection reports from a particular country, it must consider whether the scope of that country's inspections is sufficient for FDA's needs. Evaluation of inspections conducted by foreign regulatory bodies can be complex and may include on-site review of regulatory systems and audit inspections. Further, to obtain results of inspections conducted by its foreign counterparts, FDA must specifically request them—they are not automatically provided. While FDA has provided certain foreign regulatory bodies access to its Compliance Status Information System—which provides information from the results of FDA's inspections—foreign regulatory bodies have not established similar systems to

---

[27] FDA officials told us that they do not use the results of an inspection conducted by a foreign regulatory body to make decisions about whether to approve a new drug.

provide FDA access to data about their inspections. FDA indicated that such systems are under development in some countries and FDA has been promised access when they are available. However, currently, FDA cannot routinely incorporate the results of inspections conducted by foreign regulatory authorities into its risk-based selection process.[28] FDA officials stated that, in the past, they encountered difficulties using inspection reports from other countries that were not readily available in English. Consequently, the existence of such information-sharing arrangements alone may not help FDA systematically address identified weaknesses in its foreign inspection program.

Arrangements that have the potential to allow FDA to formally accept the results of inspections conducted by other countries have been prohibitively challenging to implement. Although these arrangements allow countries to leverage their own inspection resources, according to FDA officials, assessing the equivalence of other countries' inspections and the relevance of the information available is difficult. They added that complete reliance on another country's inspection results is risky. The activities associated with establishing these agreements may be resource intensive, which may slow FDA's implementation of them. For example, FDA told us that a lack of funding for establishing such an arrangement with the European Union effectively stopped progress. Although FDA has completed preliminary work associated with this arrangement, the agency has concluded that it will be more beneficial to pursue other methods of cooperating with the European Union. The agency has no plans at this time to enter into other such arrangements.

FDA's current efforts to obtain more information from foreign regulatory bodies may help it better assess the risk of foreign establishments when prioritizing establishments for GMP surveillance inspections. However, most foreign inspections are conducted to examine an establishment referenced in an NDA or ANDA. The agency conducts relatively few foreign GMP surveillance inspections selected through its risk-based process. Therefore, these efforts may be of limited value to the foreign inspection program if the agency does not increase the number of such inspections.

FDA Has Increased Its Inspections of Foreign Establishments, but Still Inspects Relatively Few

FDA has made progress in conducting more foreign inspections, but it still inspects relatively few establishments. FDA conducted more foreign establishment inspections in fiscal year 2007 than it had in each of the 5 previous fiscal years. However, the agency still inspected less than 11 percent of the foreign establishments on the prioritized list that it used to plan its fiscal year 2007 GMP surveillance inspections.[29] The agency also still conducts far fewer inspections of foreign establishments than domestic establishments. Its budget calls for incremental increases in funding for foreign inspections. FDA officials told us that, for fiscal year 2008, the agency plans to conduct more GMP surveillance inspections based on its

[28] 28 In addition to challenges in obtaining inspection reports, FDA may also be limited by the type of information available. For example, FDA may not be able to obtain inspection reports on API manufacturing establishments because other regulatory bodies may only inspect finished-dosage manufacturers.

[29] As a result of the inaccuracies in its data, FDA recognizes that this list does not provide an accurate count of establishments subject to inspection.

prioritized list of foreign establishments. FDA officials estimated that the agency conducted about 30 such inspections in fiscal year 2007 and plans to conduct at least 50 in fiscal year 2008.

If FDA were to inspect foreign establishments biennially, as is required for domestic establishments, this would require FDA to dedicate substantially more funding than it has dedicated to such inspections in the past. In fiscal year 2007, FDA dedicated about $10 million to inspections of foreign establishments.[30] FDA estimates that, based on the time spent conducting inspections of foreign drug manufacturing establishments in fiscal year 2007, the average cost of such an inspection ranges from approximately $41,000 to $44,000.[31] Our analysis suggests that it could cost the agency $67 million to $71 million each year to biennially inspect each of the 3,249 foreign drug establishments on the list that FDA used to plan its fiscal year 2007 GMP surveillance inspections. Based on these same estimates, it would take the agency $15 million to $16 million each year to inspect the estimated 714 drug manufacturing establishments in China every 2 years. According to FDA budget documents, the agency estimates that it will dedicate a total of about $11 million in fiscal year 2008 and $13 million in fiscal year 2009 to all foreign inspections.

In its fiscal year 2009 budget, FDA proposed instituting a reinspection user fee.[32] Reinspections are conducted to verify that corrective actions the agency has required establishments to take in response to previously identified violations have been implemented. FDA's proposal to institute a reinspection user fee would allow it to charge establishments a fee when the agency determines a reinspection is warranted. However, as proposed, the reinspection user fee would be budget neutral, meaning that the other appropriated funds the agency receives would be offset by the amount of collected reinspection fees. As a result, this proposal would not provide the agency with an increase in funds that could be used to pay for additional foreign inspections.

FDA Initiatives May Address Some Challenges Unique to Foreign Inspections, but It Is Too Early to Determine Their Effectiveness

FDA has recently announced proposals to address some of the challenges unique to conducting foreign inspections, but specific implementation steps and associated time frames are unclear. We previously identified the lack of a dedicated staff devoted to conducting foreign inspections as a challenge for the agency. FDA noted in its report on the revitalization of ORA that it is exploring the creation of a cadre of investigators who would be dedicated to conducting foreign inspections.[33] However, the report does not provide any additional details

---

[30] According to FDA budget documents, the agency dedicated about $43 million to inspecting domestic drug manufacturers in fiscal year 2007.

[31] According to FDA, the cost of conducting foreign inspections varies, depending on whether the type of inspection was a preapproval or GMP surveillance inspection, by the time spent at an establishment, by the number of FDA staff conducting the inspection, and by the costs associated with traveling to the country in which the establishment is located.

[32] FDA also proposed a reinspection user fee in its fiscal year 2007 and fiscal year 2008 budgets, but these proposals were not enacted.

[33] See, for example, Food and Drug Administration, *Revitalizing ORA: Protecting the Public Health Together In a Changing World* (Rockville, Md.: Jan. 2008).

or timeframes about this proposal. In addition, FDA recently announced plans to establish a permanent foreign presence overseas, although little information about these plans is available. Through an initiative known as "Beyond our Borders," FDA intends that its foreign offices will improve cooperation and information exchange with foreign regulatory bodies, improve procedures for expanded inspections, allow it to inspect facilities quickly in an emergency, and facilitate work with private and government agencies to assure standards for quality. FDA's proposed foreign offices are intended to expand the agency's capacity for regulating, among other things, drugs, medical devices, and food. The extent to which the activities conducted by foreign offices are relevant to FDA's foreign drug inspection program is uncertain. Initially, FDA plans to establish a foreign office in China with three locations—Beijing, Shanghai, and Guangzhou—comprised of a total of eight FDA employees and five Chinese nationals. The Beijing office, which the agency expects will be partially staffed by the end of 2008, will be responsible for coordination between FDA and the Chinese regulatory agencies. FDA staff located in Shanghai and Guangzhou, who will be hired in 2009, will be focused on conducting inspections and working with Chinese inspectors to provide training as necessary. FDA has noted that the Chinese nationals will primarily provide support to FDA staff including translation and interpretation. The agency is also considering setting up offices in other locations, such as India, the Middle East, Latin America, and Europe, but no dates have been specified. While the establishment of both a foreign inspection cadre and offices overseas have the potential for improving FDA's oversight of foreign establishments and providing the agency with better data on foreign establishments, it is too early to tell whether these steps will be effective or will increase the number of foreign drug inspections.

Agreements with foreign governments, such as one recently reached with China's State Food and Drug Administration, may help the agency address certain logistical issues unique to conducting inspections of foreign establishments. We previously testified that one challenge faced by FDA involved the need for its staff to obtain a visa or letter of invitation to enter a foreign country to conduct an inspection. However, FDA officials told us that their agreement with China recently helped FDA expedite this process when it learned of the adverse events associated with a Chinese heparin manufacturer. According to these officials, the agreement with China greatly facilitated its inspection of this manufacturer by helping FDA send investigators much more quickly than was previously possible.

Concluding Observations

Americans depend on FDA to ensure the safety and effectiveness of the drugs they take. The recent incident involving heparin underscores the importance of FDA's initiatives and its steps to obtain more information about foreign drug establishments, conduct more inspections overseas, and improve its overall management of its foreign drug inspection program. FDA has identified actions that, if fully implemented, could address some, but not all, of the concerns we first identified 10 years ago and reiterated 5 months ago in our testimony before this subcommittee. Given the growth in foreign drug manufacturing for the U.S. market and the current large gaps in FDA's foreign drug inspections, FDA will need to devote considerable resources to this area if it is to increase the rate of inspections. However, FDA's plans currently call for incremental increases that will have little impact in the near future to reduce the interval between inspections for these establishments. In addition, many of FDA's initiatives will take several years to implement and require funding and certain interagency or intergovernmental agreements that are not yet in place. Taken together, FDA's plans represent a step forward in filling the large gaps in FDA's foreign drug inspection program, but do little to accomplish short-term change.

Mr. Chairman, this completes my prepared statement. I would be happy to respond to any questions you or the other Members of the subcommittee may have at this time.

INDEX

B

C

E

F

G

H

I

M

N

O

S

T

U